Lyons' Encyclopedia of Valves

Lyons' Encyclopedia of Valves

JERRY L. LYONS P.E.
Senior Project Engineer Chemetron
Corporation Fluid Controls Division
and
CARL L. ASKLAND, JR.
Project Engineer
Essex Cryogenics Industries, Inc.

Van Nostrand Reinhold Company
New York/Cincinnati/Toronto/London/Melbourne

Van Nostrand Reinhold Company Regional Offices:
New York   Cincinnati

Van Nostrand Reinhold Company International Offices:
London   Toronto   Melbourne

Copyright © 1975 by Van Nostrand Reinhold Company Inc.

Library of Congress Catalog Card Number: 75-29293
ISBN: 0-442-24961-6

All rights reserved. No part of this work covered by the copyright hereon may be reproduced or used in any form or by any means—graphic, electronic, or mechanical, including photocopying, recording, taping, or information storage and retrieval systems—without permission of the publisher.

Manufactured in the United States of America

Published by Van Nostrand Reinhold Company Inc.
135 West 50th Street, New York, N.Y. 10020

Published simultaneously in Canada by Van Nostrand Reinhold Ltd.

15 14 13 12 11 10 9 8 7 6 5

Library of Congress Cataloging in Publication Data

Lyons, Jerry L
   Lyons' Encyclopedia of valves.

   Includes bibliographical references and index.
   1. Valves–Handbooks, manuals, etc.   I. Askland, Carl L., joint author.   II. Title.   III. Title: Encyclopedia of valves.
TS277.L95   1975   621.8'4   75-29293
ISBN   0-442-24961-6

Preface

It is the hope of the authors that this book will be of use and interest to a variety of persons concerned with the design, manufacture, selection, installation and use of valves and related fluid control devices. To make the book usable to persons of widely differing backgrounds, we have tried to separate the detailed technical data from the more general material by dividing the book into two parts.

Part One is devoted to a general description of basic valve types and a glossary of terms used by valve designers and others in the field of fluid control. A large number of photographs is included in this section to show some typical examples of the valves and other devices described. There is a large variety of a given type of valve in the industry, so the choice of example becomes quite arbitrary. In most cases we attempted to choose a standard example rather than a "special." The decision of what terms to include in the glossary became very difficult at times. We tried to cover the field of fluid controls in a fairly broad sense. Many of the terms included are from the fields of physics, chemistry, chemical engineering, mechanical engineering, etc., but are ones that we felt are applicable in some degree to the control of fluids. Also, many of the terms have different meanings in other fields. We have given only the definitions applicable to fluid controls. We note here that many terms are not standard usage or even scientifically correct but have acquired meaning in practice. We have not attempted to change such definitions since that is the job of standards organizations. Also, it should be remarked that in many cases similar valves are given different names by different manufacturers. Terminology is also found to differ between various segments of the valve industry such as commercial, aerospace, etc. In all cases, we have attempted the difficult task of taking a middle ground. The authors would appreciate hearing of appropriate terms not included here and also of any errors which escaped our scrutiny.

In part two we present some material of use to the designer of fluid control devices. This includes a discussion of factors to consider in valve design, some calculational methods devised or modified by the authors for sizing valves, and a section on general flow equations for valves and piping. Other sections cover material properties and standards for fluid power. No pretense is made to a complete discussion of fluid mechanics or control theory. The material chosen is that which the authors and many of our associates have found useful in the design and selection of valves. In this section most of the data is given in terms of British Engineering Units, since this is, and will remain for a time, the system used in the United States. Some of this material is also given in SI units for the convenience of those currently using them, and also because the change to SI is in progress now, and will accelerate in the future. However, since there is still considerable discussion about metric standards, we have not included ISO standards, except in a few cases. ANSI standards for several quantities have been included.

The authors are deeply indebted to the following valve manufacturers, engineering societies and individuals for materials and/or advice and comments contributed to the preparation of this book:

Allis-Chalmers, Valve Division, Milwaukee, Wisconsin

Alpha Systems, Inc., Smithtown, Long Island, New York

American Society of Mechanical Engineers, New York, New York

American-Darling Valve Division of American Cast Iron Pipe Co., Birmingham, Alabama

Anderson, Greenwood & Co., Bellaire, Texas

Anvil Products, Inc., Allison Park, Pennsylvania

Aqua Matic Inc., Rockford, Illinois

Atkomatic Valve Company, Inc., Indianapolis, Indiana

Atlas Valve Co., Newark, New Jersey

Automatic Valve Corporation, Farmington, Michigan

Bell & Howell, Control Products Div., Bridgeport, Connecticut

Bingham-Willamette Company, Portland, Oregon

Bloomfield Valve Corporation, New Britain, Connecticut

Cameron Iron Works, Inc., Houston, Texas

Canada Valve Limited, Kitchener, Ontario, Canada

Carbone, A Division of Carbone-Lorraine Industries Corporation, Boonton, New Jersey

A. W. Cash Valve Manufacturing Corporation, Decatur, Illinois

Chemetron Corporation, Fluid Controls Division, St. Louis, Missouri

Clack Corporation, Windsor, Wisconsin

The Clarkson Company, Palo Alto, California

Cla-Val Co., Newport Beach, California

Continental Disc Corporation, Riverside, Missouri

Copes-Vulcan, Inc., Lake City, Pennsylvania

Crane Co., Chicago, Illinois

Crissair Inc., El Segundo, California

Cryolab Division of C. T. I., San Luis Obispo, California

Dally Engineering Valve Company, Pittsburgh, Pennsylvania

Design News (Cahners Publishing Company), Boston, Massachusetts

De Zurik, A Unit of General Signal, Sartell, Minnesota

Double A Products Company, Manchester, Michigan

Dresser Industrial Valve & Instrument Division, Alexandria, Louisiana

Eaton Corporation, Controls Division, Cleveland, Ohio

Essex Cryogenics Ind., Inc., St. Louis, Missouri

Fabri-Valve, A Dillingham Company, Portland, Oregon

The Fairbanks Company, Binghamton, New York

Fisher Controls Company, Coraopolis, Pennsylvania

Fittings Limited, Oshawa, Ontario, Canada

Flexible Valve Corporation, South Hackensack, New Jersey

FMC Corporation, Fluid Control Operation, Houston, Texas

Gadren Machine Co., Inc., West Collingswood Heights, New Jersey

J. D. Gould Co., Indianapolis, Indiana

Goyen Controls of America, Ltd., Ocean, New Jersey

G.P.E. Controls Division of Vapor Corporation, Morton Grove, Illinois

Hammond Valve Corporation, Hammond, Indiana

Peter Healey Brass Foundry, Inc., Evansville, Indiana

High Pressure Equipment Co., Erie, Pennsylvania

Hills-McCanna Company, Carpentersville, Illinois

Honeywell Inc., Process Controls Div., Washington, Pennsylvania

Inferno Manufacturing Co., Shreveport, Louisiana

ISA (Instrument Society of America), Pittsburgh, Pennsylvania

ITT Hammel-Dahl/Conoflow, Warwick, Rhode Island

Jackes-Evans Mfg. Company, Jackson, Mississippi

Jamesbury Corporation, Worcester, Massachusetts

Jenkins Bros., New York, New York

Johnston & Jennings Company, Chicago, Illinois

Jordan Valve Co. Div. Richards Ind., Inc., Cincinnati, Ohio

Kamyr Inc., Glens Falls, New York

Keystone Valve Company, Houston, Texas

Ladish Co., Cudahy, Wisconsin

Liquid Level Lectronics, Houston, Texas

Lunkenheimer Co., Cincinnati, Ohio

Machine Design (Penton Publishing Corp.), Cleveland, Ohio

Maid-o-Mist, Chicago, Illinois

Manatrol Div. Parker Hannifin, Elyria, Ohio

Manufacturers Standardization Society of the Valve and Fitting Industry, Arlington, Virginia

Maxon Corporation, Muncie, Indiana

McDonnell & Miller ITT, Chicago, Illinois

Mueller Company, Decatur, Illinois

Mueller Steam Specialty Div. SOS Consolidated, Brooklyn, New York

Multiplex Mfg. Co., "Crispin" Valve Division, Berwick, Pennsylvania

National Fluid Power Association, Thiensville, Wisconsin

Nibco Inc., Elkhart, Indiana

Orbit Valve Company, Little Rock, Arkansas

Pacific Valves Inc., Long Beach, California

Posey Iron Works, Lancaster, Pennsylvania

Posi Seal International, Inc., Westerly, Rhode Island

Precision Plumbing Products, Inc., Portland, Oregon

The Protectoseal Company, Bensenville, Illinois

Red Valve Company, Inc., Carnegie, Pennsylvania

Rego, Division of Golconda Corporation, Chicago, Illinois

Reliable Automatic Sprinkler Co., Inc., Mt. Vernon, New York

Republic Mfg. Co., Cleveland, Ohio

RKL Controls, Inc., Hainesport, New Jersey

Rockwell International, Flow Control Division, Pittsburgh, Pennsylvania

Ross Valve Mfg. Co., Inc., Troy, New York

Skinner Precision Industries, Inc., New Britain, Connecticut

Smith Valve Corporation, Worcester, Massachusetts

Society of Automotive Engineers, New York, New York

Sporlan Valve Company, St. Louis, Missouri

Stratoflo Products, Inc., Fort Wayne, Indiana

Teledyne Republic Mfg. Co., Cleveland, Ohio

Triangle Valve Co. Limited, Wigan, England

Universal Valve Company, Elizabeth, New Jersey

Valve Manufacturers Association, McLean, Virginia

Velan Engineering Companies, Montreal, Quebec, Canada

Vernay Laboratories, Inc., Yellow Springs, Ohio

Henry Vogt Machine Co., Louisville, Kentucky

Robert H. Wager Co., Inc., Chatham, New Jersey

Wardrup Valve Co., Mountain View, California

Waterman Hydraulics, Chicago, Illinois

Whitey Valve Co., Emeryville, California

W. E. Williams Valve Co., Long Island City, New York

W-K-M Valve Division, ACF Industries, Inc., Houston, Texas

Yarway Corporation, Blue Bell, Pennsylvania

Contents

Preface ... v

PART ONE

Valves and Valve Terminology
I. The Major Valve Types ... 3
II. Terminology of the Valve Industry ... 12

PART TWO

Valve Engineering and Design Data
I. Design Factors ... 87
II. Sizing A Valve ... 90
III. Valve Spring Design ... 96
IV. Some Useful Formulas ... 103
V. Properties of Some Valve Materials ... 109
VI. Fluid Power Symbols and Standards ... 119
    A. American Standard Graphical Symbols for Pipe Fittings, Valves and Piping ... 119
    B. USA Standard Graphic Symbols for Fluid Power Diagrams ... 127
    C. American National Standard Symbols for Marking Electrical Leads and Ports on Fluid Power Valves ... 149
    D. American National Standard Interfaces for 4-Way General Purpose Industrial Pneumatic Directional Control Valves ... 153
    E. USA Standard Dimensions for Mounting Surfaces of Subplate Type Hydraulic Fluid Power Valves ... 154
    F. Mounting Surfaces for Subplate Type Hydraulic Directional Control Valves for 315 Bar Hydraulic Service ... 170
    G. Method of Diagramming for Moving Parts Fluid Controls ... 172
VII. Test Methods for Valves ... 211
    A. Pressure Drop Tests ... 211
    B. Capacity Tests ... 226
VIII. Valve Noise Calculation ... 239

Appendix

A. Metric-English Conversion Factors ... 249
B. Pipe Flow Data ... 270
C. Liquid and Gas Flow Charts ... 280

Index ... 287

Lyons' Encyclopedia of Valves

PART ONE

VALVES AND VALVE TERMINOLOGY

The Major Valve Types

The number of valves used for the control of fluids today is enormous, with valves ranging from very simple shutoff devices to extremely complex servosystems. They may range in size from very tiny metering valves used in aerospace applications to industrial and pipeline valves measuring several feet in diameter and weighing hundreds of pounds. Valves control the flow of all types of fluids ranging from air and water to corrosive chemicals, slurries, liquid metals and radioactive materials. They may operate at pressures in the vacuum region to pressures of 100,000 psi or more, and temperatures from the cryogenic region to those of molten metals. They may have lifetimes ranging from only one cycle to many thousands of cycles without requiring repair or replacement. Valves may have very strict leakage requirements in space vehicle applications where even the tiniest leak may be disastrous, or they may have very generous leakage requirements in many industrial uses where loss is of minor importance or recovery is very easy. Valves may be actuated by a variety of means such as manual, electrical, pneumatic, etc. They may respond in a prescribed manner to signals from pressure and temperature transducers and other types of sensors, or they may simply open or close independently of actuation signal strength.

Nearly all of the valves in use today can be considered as modifications of a few basic types. Valves may be classified in different ways such as by size, function, material, type of fluid carried, pressure rating, actuating member, and others. We prefer to classify valves according to the nature of closure members employed. We feel that this is the most fundamental scheme. Also, it is simpler since nearly all valves will fall into one of approximately eight categories. In this chapter, we discuss some of the properties and uses of these basic types of valves.

BALL VALVE

Description. The ball valve is basically a ported sphere in a housing. Rotation of the sphere by 90° changes the position from open to closed. The ball may be of fixed or floating design and full or reduced port. Ball valves are available in a variety of sizes and with a wide choice of actuating mechanisms.

Usage. Ball valves are used in a wide range of applications including flow control, pressure control and shutoff. They can be designed for use of corrosive fluids, cryogenic liquids, very viscous fluids and slurries, in addition to normal liquids and gases. They can be used at high pressures and temperatures.

Advantages. Ball valves generally have a very low pressure drop and low leakage. They are small in size and low in weight compared to other types with similar ratings. They are rapid opening and are relatively insensitive to contamination.

Disadvantages. The seats of ball valves are subject to extrusion if the valve is used for throttling. Fluid trapped in the ball in the closed

4 VALVES AND VALVE TERMINOLOGY

PARTS IDENTIFICATION

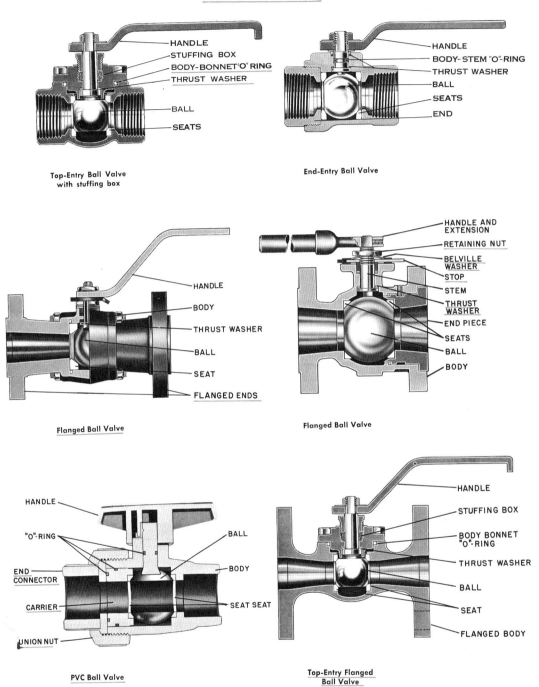

Fig. 1. Ball Valve. (*Courtesy of Lunkenheimer Co.*)

position may cause problems if the ball is not vented. Because of quick opening, they may cause water hammer in the system or cause surge pressures which may be undesirable.

BUTTERFLY VALVE

Description. The typical butterfly valve consists of a disc which can rotate about a shaft

THE MAJOR VALVE TYPES 5

in a housing. The disc closes against a ring seal to close off flow. Various actuating mechanisms such as the lever, cam, etc., are used to operate the valve.

Usage. Butterfly valves are generally used in low pressure system applications where leakage is relatively unimportant. They are normally used in large diameter lines.

Advantages. Butterfly valves have a very low pressure drop and are relatively lightweight. Also, the face to face dimension is usually quite small. The diameter of the valve can be

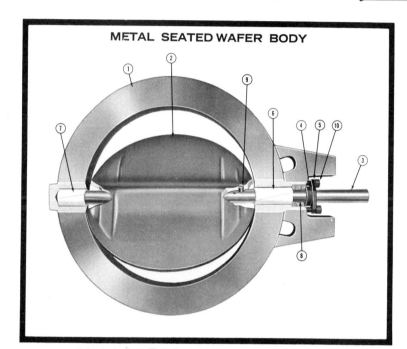

METAL SEATED WAFER BODY

1—Body
2—Disc
3—Shaft
4—Gland
5—Gland retainer
6—Upper bearing
7—Lower bearing
8—Packing
9—Pin
10—Gland screws

METAL SEATED BUTTERFLY VALVES
(a)

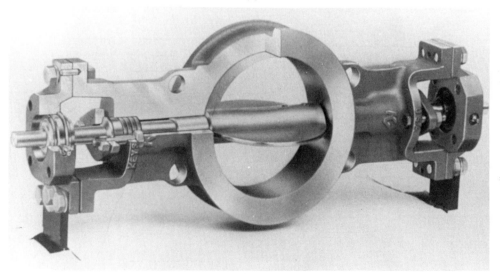

(b)

Fig. 2a,b. Butterfly Valve. (2a *Courtesy of Rockwell International*, 2b *Courtesy of Keystone Valve Co.*)

PARTS IDENTIFICATION

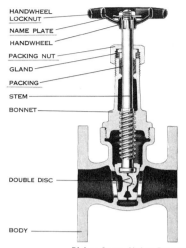

Rising Stem, Union Bonnet
Double Wedge Disc

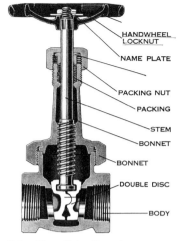

Rising Stem, Union Bonnet
Double Wedge Disc

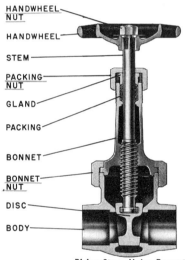

Rising Stem, Union Bonnet
Solid Wedge Disc

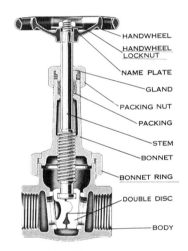

Rising Stem
Double Wedge Disc

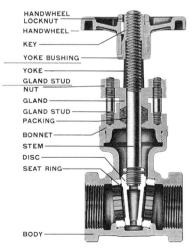

Outside Screw and Yoke
Solid Wedge Disc

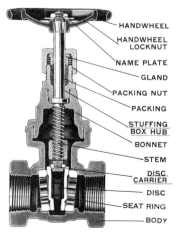

Non-Rising Stem
Single Wedge Disc
Renewable Seat Rings

Fig. 3. Gate Valve. (*Courtesy of Lunkenheimer Co.*)

of the same order as the diameter of the connecting pipes.

Disadvantages. The leakage in a butterfly is fairly high unless special seals are used. Seals are often damaged by high velocity flow. These valves usually require high actuation forces, and are generally limited to low pressure systems.

GATE VALVE

Description. Gate valves are characterized by a sliding disc or gate which is moved by the actuator perpendicular to the direction of flow. There are many variations in seat, stem and bonnet design in gate valves. They are available in a wide assortment of sizes and weights.

Usage. Gate valves are used primarily as stop valves, i.e., fully open or fully closed. They are not normally considered for throttling purposes. They are suited for high pressure and high temperature use with a wide variety of fluids. They are not usually used for slurries, viscous fluids, etc.

Advantages. Gate valves usually have a low pressure drop when fully open, provide a tight seal when fully closed, and are relatively free of contamination buildup.

Disadvantages. Gate valves are prone to vibration when in a partially open state and are also subject to seat and disc wear. Some large size gate valves are not recommended for steam service. Gate valves have slow response characteristics and require large actuating forces.

GLOBE VALVE

Description. The three major valves in the globe family are the globe, angle, and Y. They are all characterized by a closure member, usually in disc or plug form, which is moved by an actuator stem perpendicular to a ring shaped seat. The flow passes from the inlet port through the seat to the outlet port. The three types differ mainly in the orientation of the seat to the direction of flow through the valve.

Usage. Globe valves are used primarily for throttling purposes. They are widely used along with gate valves in power and process piping systems. The globe valve may be considered a general purpose flow control valve.

Advantages. The globe valve is generally faster to open or close than the gate valve. Their seating surfaces are less subject to wear and their high pressure drop makes them useful in controlling pressures.

Disadvantages. Their high pressure drop may be undesirable in many piping systems. In large sizes they require considerable power to operate, thus necessitating gearing, levers, etc. Globe valves are often heavier than other valves of the same flow rating.

PINCH VALVE

Description. Pinch valves are characterized by one or more flexible elements such as diaphragms, rubber tubes, etc., which can be moved together or against a stop to pinch off the flow.

Usage. Pinch valves are especially good in systems carrying slurries, gels, etc., since they have little tendency to build up contamination.

Advantages. Pinch valves are relatively low in cost, insensitive to contamination, have low pressure drops and can be tightly closed.

Disadvantages. The flexible members in pinch valves are subject to wear and hence periodic replacement. They are generally limited to low pressure, low temperature applications. Their use with corrosive materials is limited by the properties of the flexible element. Pinch valves generally require high actuation forces to close off.

POPPET VALVE

Description. The poppet valve is one in which the closure member moves parallel to the fluid flow and perpendicular to the sealing surface. The closure element is usually flat, conical or spherical on the sealing end. They may have many kinds of actuating elements including springs, screw systems, etc.

Usage. Poppet valves may be designed for almost any conceivable application. Their main

PARTS IDENTIFICATION

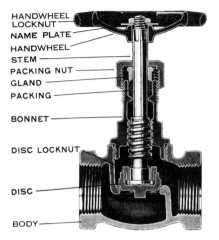

Screw-In Bonnet Globe

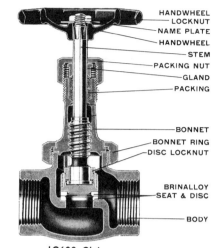

LQ600, Globe

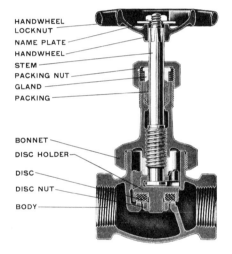

Non-Metallic Disc Globe

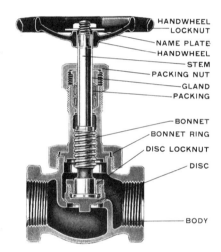

Union Bonnet Regrinding Globe

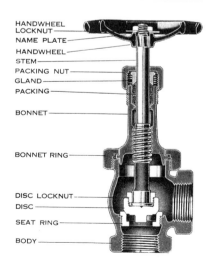

"Renewo" Angle

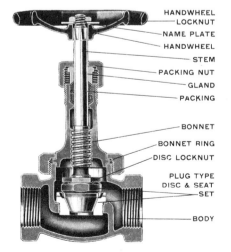

"Renewo" Globe, Plug Type

Fig. 4. Globe Valve. (*Courtesy of Lunkenheimer Co.*)

Fig. 5a,b,c. Pinch Valve. (5a *Courtesy of Red Valve Co.*, 5b,c *Courtesy of RKL Controls, Inc.*)

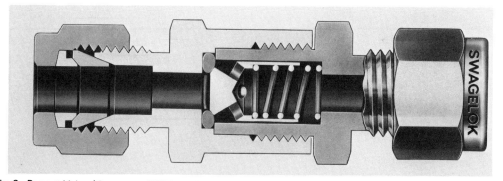

Fig. 6. Poppet Valve (*Courtesy of Whitey Valve Co.*) Copyright 1974 by Markad Service Company. All Rights Reserved.

uses are in pressure control, check, safety and relief functions.

Advantages. Poppet valves may provide large flow with very little actuator travel, excellent leakage control, and low pressure drop.

Disadvantages. Poppet valves are subject to pressure imbalances which may cause chattering in some applications. Some seating surfaces may be subject to contamination.

TAPER PLUG VALVE

Description. The taper plug valve is very similar to the ball valve, except that the closure member is a tapered plug instead of a ball.

Usage. Taper plug valves are useful in high temperature, low pressure applications. They perform many of the same functions as ball gate and globe valves.

10 VALVES AND VALVE TERMINOLOGY

PARTS IDENTIFICATION

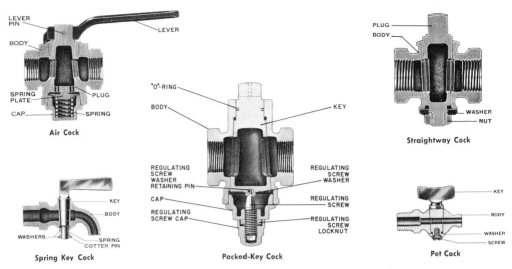

Fig. 7. Taper Plug Valve. (*Courtesy of Lunkenheimer Co.*)

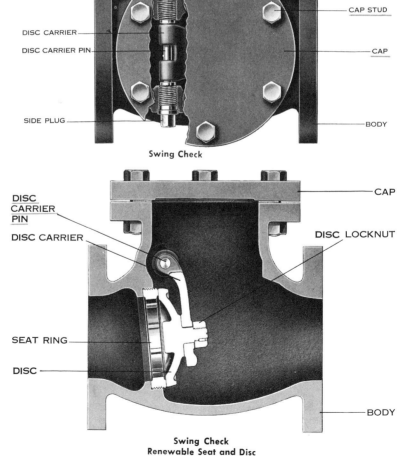

Swing Check
Renewable Seat and Disc

Fig. 8. Swing Valve. (*Courtesy of Lunkenheimer Co.*)

Advantages. Plug valves are normally small in size, requiring less headroom than most other valves. They are fairly low in cost and available in a wide range of materials. They provide a tight leakproof seal.

Disadvantages. Plug valves may be subject to binding and galling. They are not usually suited for steam service. Lubricated plug valves require periodic lubrication and the lubrication may react with the fluid being carried.

SWING VALVE

Description. Swing valves are similar to butterfly valves except that they are hinged on one edge rather than along a diameter. They may be actuated by flow, by torsion springs, by levers, etc.

Usage. Swing valves are primarily used as check valves to block flow in one direction.

Advantages. Swing valves have many of the same advantages as butterfly valves including low pressure drop, light weight, and relatively low cost.

Disadvantages. Swing valves may have high leakage, are subject to contamination buildup and introduce turbulence at low flow rates. The sealing surfaces may erode in high velocity flow.

11 Terminology of the Valve Industry

Ablation Wearing away of material by hot gases, plasmas, etc.

Abrasion The mechanical wearing away of one material by another material moving over it.

Absolute Filtration Rating The diameter of the largest particle that will pass through a filter under given conditions.

Absolute Pressure The total pressure measured from an absolute vacuum. It is the sum of gauge pressure and the prevailing atmospheric pressure measured on a barometer. The unit of pressure is the Pascal (Newton per square meter). Secondary units include dynes per square centimeter (metric), pounds per square inch (British), inches of water, millimeters of mercury.

Absolute Temperature Temperature measured from absolute zero. It is measured on the Kelvin (Metric) or Rankine (British) scales.

Absolute Viscosity The resistance of a fluid to relative motion of its parts. It is also called dynamic viscosity. Common units are Poise (metric), pounds per foot-second (lb/ft-sec) and Saybolt Universal Seconds.

Absolute Zero The zero point on the absolute or Kelvin temperature scale. Absolute zero is defined as the temperature at which a system undergoes a reversible isothermal process without transfer of heat. It is a limit point unattainable in practice. It is equivalent to $-273.15°C$ or $-459.69°F$.

Absorption The process of filtration by entrapping contaminants within the filtering media.

Acceleration The time rate of change of velocity. Units are meters/sec^2, ft/sec^2, etc.

Accumulator A device utilizing gravitational force, springs or the compressibility of fluids to store energy. It is used generally to supply peak demands in a system operating on an intermittent duty cycle. Common uses are: (1) Reserve pressure source, (2) Fluid (mass) source, (3) Dual-pressure circuits, (4) Leakage compensators, (5) Emergency power, (6) Pulsation or ripple dampener, (7) Transfer barrier, (8) Pressure-Volume compensator.

Accumulator, Hydropneumatic An accumulator in which the pressure on a stored liquid is supplied by compressed gas.

Accumulator, Hydropneumatic, Bladder A hydropneumatic accumulator in which a bladder is used to separate the liquid and gas.

Accumulator, Hydropneumatic, Diaphragm A hydropneumatic accumulator in which a flexible diaphragm is used to separate the stored liquid and compressed gas.

Accumulator, Hydropneumatic, Nonseparated A hydropneumatic accumulator in which a compressed gas operates directly on the stored liquid in the chamber.

Accumulator, Hydropneumatic, Piston A hydropneumatic accumulator in which a floating piston is used to separate the stored liquid from the compressed gas.

Accumulator, Hydropneumatic, Spring A hydropneumatic accumulator in which springs are used to apply force on the stored liquid.

Accumulator, Mechanical An accumulator in which a mechanical device is used to apply force on a stored fluid.

Accumulator, Mechanical, Spring A mechanical accumulator in which springs are used to apply force on a stored fluid.

Accumulator, Mechanical, Weighted A mechanical accumulator in which gravitational force is used to apply force on a stored fluid.

Accuracy An expression describing the ability of a measuring instrument to show the true value of a measured quantity. Conversely, it is the magnitude of the total error expected in the measurement. It is usually expressed as a percent of full scale reading of the measuring instrument. See also **Precision**.

Accuracy of Regulation The measure of the amount by which a controlled variable pressure changes from a predetermined value at the minimum flow, when the flow through the regulator is increased from the minimum controllable flow to maximum rated flow.

Acidity The concentration of H^+ ions in a solution. See also **pH**.

Acme Thread The type of screw thread with an included angle of 29° instead of the 60° of the U.S. thread. It is a cross between square and V threads. Acme threads are used, for example, on valve stems.

Acoustic Velocity The velocity of sound in a given medium. In air the speed is approximately 340 m/sec at standard temperature and pressure.

Action A term in the automatic control field which refers to that which is done to regulate a controlling element in an operation or process. This includes such things as on and off movements, derivative motion and acceleration.

Activation A term for a diaphragm actuator, pressure operated spring or pressure opposed diaphragm assembly for positioning an actuator stem in response to the operating pressure(s).

Actuator The part of a regulating valve that converts thermal, electrical, or fluid energy into mechanical energy to open or close the valve. See also **Airmotor**. (See Fig. 9)

Actuator, Electric See **Actuator**. (See Fig. 10)

Actuator, Pneumatic See **Airmotor**.

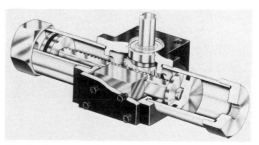

Fig. 9. Actuator. (*Courtesy of Chemetron Corporation, Fluid Controls Division.*)

Fig. 10. Electric Actuator. (*Courtesy of Chemetron Corporation, Fluid Controls Division.*)

Actuator Spring A spring used to provide resistance to the piston, diaphragm or other movement in a valve actuator. (See Fig. 30)

Actuator Stem A rod which extends from a diaphragm plate or piston, etc., permitting connection to external components.

Actuator Travel Characteristic The ratio of percent of rated travel to the actuator pressure. This may be an inherent or specified characteristic.

Adapter A seal support which is shaped to conform to the contour of the seal and its mating element.

Adapter, Female An adapter having a concave seal support.

Adapter, Male An adapter having a convex seal support.

Adapter, Pedestal A male adapter of T-shaped cross section that is used to support a U-shaped seal.

Additive One or more chemicals added to a fluid to change properties such as viscosity, freezing point, etc.

Adhesion The molecular attraction exerted between the surfaces of two bodies in contact.

Adiabatic A thermodynamic term meaning no flow of heat into or out of a system.

Adiabatic Flow Flow of a fluid under conditions where no heat enters or leaves the fluid. It is also called isentropic flow.

Adjustable Resistance A resistance which can be varied mechanically.

Adjusting Screw A screw used to regulate or control the compression setting of a valve spring. (See Fig. 62)

Adjusting Screw Cap A cover for the adjusting screw, used for the protection of the screw and to prevent accidental changes in setting. (See Fig. 63)

Adjusting Spring See **Actuator Spring**.

Adsorbent A filter medium that holds contaminants on its surface by molecular adhesions.

Aeration The process of mixing air, especially oxygen, into water or other liquids.

Aerodynamics The branch of physics dealing with the motion of air or other gases.

Aerometer An instrument used for determining the density of a gas.

Aerosol A suspension of very fine solid or liquid particles in a gas.

After-Cooler A device used to cool a compressed gas.

After-Effect The deflection remaining in a diaphragm after the load is removed.

After-Filter A filter following a compressed air dryer.

Air The naturally occurring mixture of nitrogen, oxygen, carbon dioxide and other gases.

Air Bleeder A device such as a needle valve, bleed plug, or capillary used to remove fluid from a high pressure point in a circuit.

Air Breather A device which permits air movement between the atmosphere and the component in which the device is installed.

Air Chamber A cavity filled with air used to equalize the flow of a liquid. See also **Accumulator**.

Air, Compressed Air which is under a pressure greater than the prevailing atmospheric pressure.

Air-Core Solenoid Valve A solenoid valve having a hollow core rather than a solid core. (See Fig. 11)

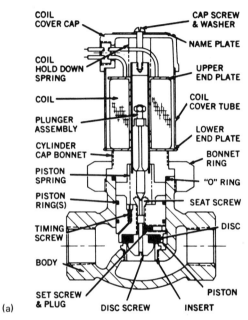

(a)

(b)

Fig. 11a,b. Air-Core Solenoid Valve. (*Courtesy of Atkomatic Valve Co., Inc.*)

Air, Dried Air with a moisture content lower than the maximum allowable for a specified application.

Air Driven Powered by compressed air.

Air, Free Air which is at prevailing atmospheric pressure and temperature and which is not contained, i.e., atmospheric air.

Air Motor A motor which is driven by compressed air. Also a synonym for diaphragm or piston actuator. See also **Actuator**. (See Fig. 12)

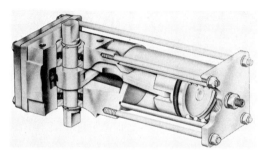

Fig. 12. Air Motor. (*Courtesy of Chemetron Corporation, Fluid Controls Division.*)

Air-Operated Controller A device that measures some variable quantity and connects the signal generated to a corresponding air pressure which can be used to operate a valve or other control device. Sometimes synonymous for positioner. (See Fig. 13)

Airpump See **Compressor**.

Air Range The operating pressure range or span.

Air Receiver A container in which a gas is stored under pressure to act as a source of pneumatic fluid power.

Air, Saturated Air at 100% relative humidity.

Air, Standard Air at a temperature of 68°F, pressure of 14.70 psia, and relative humidity of 36%. Some industries consider 60°F as the temperature of standard air. Specific gravities of gases are based on standard air which has a value of one (1). The density of standard air is 0.075 lb/ft^3.

Airtight Impermeable to air. It generally refers to a seal which seals tight enough to prevent leakage when tested with air.

Air-to-Close Action See **Normally Open**.

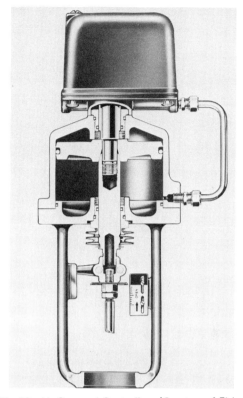

Fig. 13. Air-Operated Controller. (*Courtesy of Fisher Controls Company.*)

Air-to-Open Action See **Normally Closed**.

Alkalinity The concentration of OH$^-$ ions in a solution. See also **ph**.

Alloy A mixture of two or more metallic elements.

Ambient Temperature The temperature in the immediate vicinity of an object. It usually refers to room temperature.

American Standard Pipe Thread A type of screw thread used on pipe fittings and devices to give a positive seal. See also **National Pipe Thread**.

Amplifier A device which is used to increase volume rather than pressure. It is the opposite of a pressure intensifier.

Aneroid Barometer An instrument used for measuring atmospheric pressure. The element is usually an evacuated can or bellows to which an indicator is attached.

Angle Valve A type of globe valve in which the pipe openings are at right angles. (See Fig. 14)

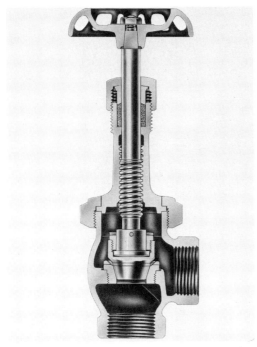

Fig. 14. Angle Valve. (*Courtesy of Jenkins Bros.*)

Aniline Point The lowest temperature at which a liquid can be completely blended with an equal amount of pure aniline.

Angular Acceleration The time rate of change of angular velocity.

Angular Momentum The momentum of a rotating system about an axis. For a mass particle:

$$\vec{L} = mvrx$$

where
- m = mass of the particle
- v = vector velocity of the particle
- r = vector from axis to particle

Anhydrous A material that is dry or without any absorbed water.

Annealed A metal that has been softened by exposure to high temperatures. This is usually done to make it easier to form or to relieve stresses.

Anodize A finish treatment of metals, especially aluminum, in which an oxide coat is formed on the surface electrochemically.

A.N.S.I. American National Standards Institute (formerly United States of America Standards Institute).

Anti-Extrusion Ring A ring used to bridge a clearance and thereby reduce extrusion of a seal. It is sometimes known as a backup ring.

Antifreeze A material or solution used to prevent freezing of another liquid by lowering its freezing point. A typical example is ethylene glycol when it is added to water.

Aqueous A term generally referring to a water based solution.

Archimedes' Principle A body wholly or partially submerged in a fluid is buoyed up with a force equal to the weight of the displaced fluid. This principle allows an easy determination of the density of a material, to tell if it is a pure material or an alloy, by weighing in air and again in a fluid such as water.

A.S.M.E. Boiler Code A set of specifications for the construction of boilers issued by the American Society of Mechanical Engineers.

Asperities Rough areas on a valve seat or other critical surface.

A.S.T.M. Standards Standard material specifications issued by the American Society for Testing Materials.

Atmospheric Pressure The pressure exerted by the atmosphere. Common units are psi, Pascals (Newtons/meter^2), Bars, and Dynes/cm^2. Atmospheric pressure decreases approximately exponentially with altitude.

Automatic Control Valve action not requiring any manual adjustment. Control is generally in response to an input signal such as pressure change, etc. See also **Automatic Regulator**.

Automatic Controller A self-actuating device or instrument which is used for regulating, measuring, positioning, etc. (See Fig. 15)

Fig. 15. Automatic Controller. (*Courtesy of Fisher Controls Company*).

Automatic Regulator Same as automatic controller. A self-operated regulating valve.

Automatic Reset The function of a control which automatically senses the magnitude and duration of a deviation from the control index and adjusts the valve to compensate.

Automatic Reset Response Response of a controller proportional to the deviation of the controlled variable from a mean or predetermined level.

Automatic Stop Check Valve A check valve which will stop, or check, at predetermined points as pressure increases. It is commonly used on multistage compressors or boilers. For other types of stop check valves, see Fig. 16.

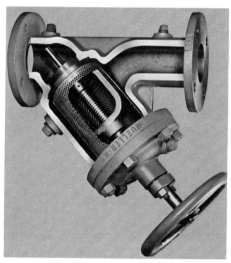

(a)

(b)

Fig. 16a,b. Automatic Stop Check Valve. (16a *Courtesy of Mueller Steam Specialty Division of SOS Consolidated,* 16b *Courtesy of Essex Cryogenics Industries, Inc.*)

Average Position Action An action for which there exists a predetermined relation between the controlled variable and the average position of a control element which moves between two fixed positions.

Axial Flow Flow of a fluid in a single direction such as streamline flow in a pipe. It is also nondivergent and nonrotational.

Back Connected Situation in which connections are made to the back or rear of a device.

Back Pressure The pressure on the upstream side of a valve seat.

Backup Ring See **Anti-Extrusion Ring**.

Baffle A series of plates used to change the direction of flow to prevent a straight through motion.

Baffle Plate An obstruction in a fluid system used to change the direction of flow.

Ball Check Valve A valve in which a spherical closure member is used to stop the flow of fluid in one direction, but which will allow flow in the opposite direction. (See Fig. 17)

Fig. 17. Ball Check Valve. (*Courtesy of Rockwell International.*)

Ball Valve A valve with a spherical closure member or gate. It is usually a quick opening valve and has a very tight shutoff characteristic. (See Fig. 18)

Fig. 18. Ball Valve. (*Courtesy of Chemetron Corporation, Fluid Controls Division.*)

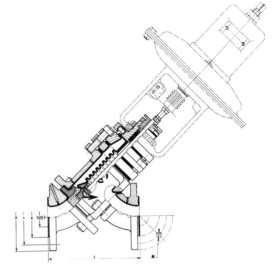

Fig. 19. Bellows Seal Valve. (*Courtesy of the Carbone Corp.*)

Band A raised area or collar used to reinforce the ends of some types of screwed fittings. See also **Proportional Band**.

Barometer A device used to measure prevailing atmospheric pressure. The two most common barometers are the aneroid, containing an evacuated chamber upon which air exerts pressure, and the mercury barometer.

Batch Process A process in which the flow of a controlled medium and the control agent is intermittent.

Batch Process Control A process in which work or materials remain stationary while being treated.

Baumé Scale A hydrometer scale for measuring the densities of liquids, particularly acid solutions.

Bearing Supports used to hold a revolving shaft in its proper position.

Bellows A flexible, thin-walled, circumferentially corrugated cylinder. It may have integral ends and can expand or contract axially under changing pressure.

Bellows Seal A seal for the valve stem in which the ends of the sealing material are fastened to bonnet and stem. The material expands and contracts with stem travel. (See Fig. 19)

Bellows Seal Valve A valve which uses a bellows in the neck for sealing instead of a packing. A packing may be used for a secondary seal. (See Fig. 19).

Bernoulli's Law A statement of the law of conservation of energy for a flowing fluid. For a frictionless fluid the energy due to pressure and velocity is constant at all points along a streamline.

$$E = \tfrac{1}{2}\rho v^2 + P$$

where

ρ = density
v = velocity
P = pressure

Bi-Directional A device designed for flow in both directions.

Bleed Off System A type of sensor with a fixed restriction and a valve in which supply pressure is fed to the fixed restriction.

Bleed On System A type of sensor with a fixed restriction and a valve in which supply pressure is fed to the valve.

Bleeder A valve for relieving pressure from a system.

Bleeder, Air A valve used to remove or release air pressure from a system. (See Fig. 20)

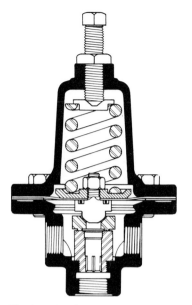

Fig. 20. Bleeder, Air. (*Courtesy of A. W. Cash Valve Mfg. Company.*)

Blind Flange A solid platelike fitting that is used to seal the end of a flanged end pipe. See also **Bottom Flange**.

Blow-Off System A piping system for blowing scale and other impurities out of boilers, tanks, etc.. It is also called a "blow-down" system.

Blow-Off Valve A valve specifically designed for use in blow-off lines. It is also called a "blow-down" valve.

Body Head See **Bonnet Assembly**.

Bolted Bonnet A bonnet which is connected to the neck flange with bolts. (See Fig. 3)

Bolted Gland A device which compresses the stuffing or packing in a stuffing box by means of tightening bolts. (See Fig. 3)

Bolted Stuffing Box See **Packing Box Assembly** and **Bolted Gland**.

Bonnet The part of a valve which connects the valve actuator to the body. It may also include the stem packing. (See Fig. 3)

Bonnet Assembly The combination of bonnet and valve actuator.

Bonnet Packing Material used around a stem and within the bonnet to prevent leakage. (See Fig. 3)

Bonnetless A term used for a pressure seal bonnet.

Borda Tube A tube about one diameter long projecting into a reservoir. It has a lower contraction coefficient than other configurations.

Bottom Bonnet See **Bottom Flange**.

Bottom Cap See **Bottom Flange**.

Bottom Cover See **Bottom Flange**.

Bottom Flange A plate covering an opening in the bottom of a valve body. It also denotes the flange for a bottom line connection in a three-way valve.

Bottom Plate See **Bottom Flange**.

Boundary Layer The layer of fluid at the wall which remains laminar when the flow becomes turbulent.

Boyle's Law The relation between pressure and volume of a fixed mass of gas for an isothermal process. The product of pressure and volume is constant if the temperature is held constant.

Brass To Iron A designation for a brass valve disc on an iron seat, or vice versa.

Brazing Ends The ends of a valve which are prepared for brazing.

Breathing Capacity The flow rate through an air breather.

Breechlock See **Pressure Seal Valve**.

British Thermal Unit The amount of heat necessary to raise the temperature of one pound of water from 63 to 64 °F.

Broken Stem A type of stem used on valves in the cryogenic and nuclear fields to prevent freezing or thermal expansion and seizing.

Bronze Trim Trim materials such as stem, disc, seat rings, etc., of a valve made of brass or bronze. It is also called *bronze mounted*.

Bubble Tight A valve seat which closes tight enough to prevent the leakage of visible gas bubbles.

Bulk Modulus The change in pressure per fractional unit change in volume of a fluid.

$$B = - \frac{dP}{dV}$$

where

$$V = \text{volume}$$
$$\frac{dP}{dV} = \text{rate of change of pressure with volume}$$

It is the reciprocal of the compressibility.

Buoyancy The tendency of an object to float in a fluid. It depends on the densities of the object and fluid. See also **Archimede's Principle**.

Burst Pressure The pressure which can be applied to a valve at room temperature for 30 seconds, or other specified temperature and time, without causing rupture.

Bushing A fitting which is used to reduce the size of an opening. See also **Sleeve**.

Butterfly Valve A valve in which a disc operates at right angles to the flow. The disc may close against a metal or resilient seal. (See Fig. 21)

Fig. 21. Butterfly Valve. (*Courtesy of De Zurik, A Unit of General Signal.*)

Butt Weld Ends Lips formed on the ends of the valve to butt against the connecting pipes. The lips on both valve and pipe are machined to form a groove to accommodate a backup ring for welding.

By-Pass A loop in a fluid system used for diverting flow around a part of the system.

By-Pass Valve A valve used to divert flow around a part of a system.

Cap A device used to close ports through which the flow usually passes.

Capacitance A change in the quantity of some variable contained per unit of change in some reference variable.

Capacitance, Electrical The ratio of charge stored on a conductor to its potential.

Capacitor An electrical device which is used to provide capacitance in a circuit. It can be used to reduce arcing, block direct current, compensate for inductance and act as a source of charge. It is used in a solenoid valve in the control circuit.

Capacity The maximum or minimum flows obtainable under prescribed conditions of operation such as pressure, temperature, etc..

Capillary A very thin tube, generally with an internal diameter in the range of a few thousandths of an inch.

Capillary Action The movement of a fluid through a capillary by means of the adhesion of the fluid molecules to the wall of the capillary.

Cascade Control An automatic control system in which several control devices feed into each other consecutively.

Cascade Control Loops A series of control devices or systems feeding one into another, each of which performs a control function. One or more loops can control the performance of other loops.

Cavitation A localized gaseous conditon in a liquid stream caused by sudden expansion or any other condition in which pressure falls below the critical pressure.

Chainwheel Operated Valve A valve which is operated by a chain driven wheel which opens and closes the seals. (See Fig. 22)

TERMINOLOGY OF THE VALVE INDUSTRY 21

Fig. 22. Chainwheel Operated Valve. (*Courtesy of De Zurik, A Unit of General Signal.*)

Channel Anything through which a fluid media flows.

Charles' Law The volume of a fixed mass of gas varies directly with absolute temperature if the pressure of the gas is held constant.

Check Valve A valve that will automatically stop flow in the back direction when the fluid in the line reverses. (See Fig. 23)

Chemical Affinity The natural attraction of one chemical species for another. A measure of their tendency to react.

Choke A flow restriction, generally a tube whose length is large compared to its cross sectional area.

Choked Flow Flow under conditions of critical pressure and temperature. The flow rate at this point cannot be increased by lowering the outlet pressure.

Circuit The directed route or path taken by a flow from one point to another.

Circuit Breaker A switch which is used to interrupt the flow in a circuit, usually acting automatically.

Circuit, Logic Control A circuit that gathers and processes information used to signal power controls.

Circuit, Pilot A secondary circuit used to control a primary circuit.

Circuit, Pressure Control A circuit used to adjust or regulate pressure in the system or any part of the system.

Circuit, Power Control A circuit that directs fluid power to working devices.

Circuit, Regenerative A circuit in which a pressurized media discharged from a component is returned to the system to reduce its input power requirements.

Circuit, Safety A circuit used to prevent accidental operation or to protect against overloads or other unsafe conditions.

Circuit, Sequence A circuit which establishes the order in which phases of a circuit occur.

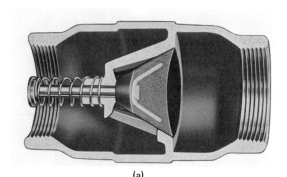
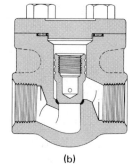

Fig. 23a,b. Check Valve. (23a *Courtesy of Stratoflo Products, Inc.*, 23b *Courtesy of Henry Vogt Machine Co.*)

Circuit, Servo A circuit which is controlled by automatic feedback in which the output of the system is measured and compared to the input signal and the difference between input and output controls the circuit.

Circuit, Speed Control A circuit in which components are arranged to regulate the speed of operation.

Circuit, Synchronizing A circuit in which multiple operations are required to occur at the same time or with predetermined time differentials.

Circuit, Unloading A circuit in which flow or pressure is relieved when delivery to the system is not required.

Clamp Gate Valve A gate valve in which the bonnet and body are held together by a U-bolt clamp. (See Fig. 24)

Clamp Type Bonnet See **Clamp Gate Valve**.

Clarifier A device used to remove solids from a hydraulic fluid.

Closed Circuit A complete circuit. One in which flow makes a complete trip and returns to the original starting point.

Closed Loop A group of automatic control units linked together with a process to form a chain. The effects of control action are constantly measured so that if the process deviates beyond desired limits, the control units will act to bring it back within those limits.

Closure A cap or plug used to close a fitting or port.

Closure Member A synonym for gate, disc, wedge, etc. It is the member used to close off a valve.

Cock A form of valve possessing a hole in a tapered plug which can be rotated to provide a flow path for fluid. It is also called a stopcock. (See Fig. 25)

(a)

(b)

Fig. 24. Clamp U-Gate Valve. (*Courtesy of the Lunkenheimer Co.*)

Fig. 25a,b. Cock. (25a *Courtesy of Peter Healey Brass Foundry, Inc.* 25b *Courtesy of the Mueller Co.*)

Coefficient Of Flow See **Flow Coefficient**.

Coefficient Of Linear Expansion The change in unit length, α, per degree rise in temperature.

$$\alpha = (1/L)\,dL/dT$$

where

L = length
$\dfrac{dL}{dT}$ = change in length with temperature

Cold Welding The welding or fusing of two pieces of metal due to mechanical force alone without external application of heat.

Collapse Pressure The difference between external and internal pressures that causes structural failure.

Composition Disc A valve seat made of a nonmetallic substance, generally an organic material.

Compressibility The fractional change in volume of a fluid per unit pressure change.

$$K = \frac{-1}{V}\frac{dV}{dP}$$

where

V = volume
dV = change in volume with pressure

It is the reciprocal of the bulk modulus.

Compressor A device for increasing the pressure of a gas.

Compressor, Multiple Stage A compressor with two or more stages in which the high pressure discharge from one stage is made to serve as the low pressure input to the next stage.

Compressor, Single Stage A compressor which has only one compression stage between inlet and outlet.

Computer A machine which performs calculations or compiles, selects or correlates data, usually very rapidly by means of a stored instruction and procedure. It generally refers to an electronic computer.

Condensate A liquid which has condensed from the vapor state. Condensate generally refers to water condensed from steam.

Conductor A device whose function is to contain and direct the flow of a fluid.

Conduit A tube, pipe, channel, etc., used for transporting a fluid.

Connector A device used to join a conductor to a component port or to one or more other conductors. See also **Coupling, Tee, Cross**.

Contaminant Material which contaminates or spoils the nature of another material for example, dirt or chemicals in a hydraulic fluid. Material that reduces the purity of another material.

Continuity Equation A statement of the law of conservation of mass. The rate of flow of mass into any fixed volume is equal to the rate of flow of mass out.

$$\nabla \cdot (\rho v) = \frac{-\partial \rho}{\partial t}$$

where

ρ = mass density
v = velocity of flow
t = time
∇ = del operator

Continuous Process Any process in which the flow of the controlled media is continuous or constant in time. The opposite of batch process.

Continuous-Process Control The process in which materials flow at a continuous rate while being processed.

Contraction Coefficient The ratio of the diameter of the vena contracta to the diameter of the orifice.

Control A device which regulates the operation of a component or system. See also **Controller**.

Control Agent The process, energy, or material for which the independent variable is a characteristic.

Control, Combination A combination of two or more basic controls in one control device.

Control, Cylinder A control in which the actuating device is a fluid cylinder.

Control, Electric A control which is actuated by an electric signal.

Control Flow The flow through valve control ports usually expressed in gallons per minute. Control flow is called no-load pressure flow when there is no load pressure drop, and load flow when there is a pressure drop.

24 VALVES AND VALVE TERMINOLOGY

Control, Hydraulic A control which is actuated by a liquid.

Control, Liquid Level A device used to control the level of a liquid in a container.

Control Loop A control composed of a number of devices each of which acts as an individual transfer system, joined together to form a network.

Control, Manual A control which is actuated by hand.

Control, Mechanical A control which is actuated by mechanical elements such as gears, cams, levers, etc..

Control Medium The type of energy used to vary the conditions of a process.

Control, Pneumatic A control which is actuated by gas pressure.

Control Point The value of a controlled variable to be maintained by a controller. The desired operating point.

Control, Pressure Compensated A control in

Control, Servo A control actuated by a feedback system which compares the output to a reference signal and makes corrections in the output to reduce the difference.

Control Signal The energy input to a device which causes it to make changes in output.

Control System The totality of components required for the control of some process.

Control System, Open Loop A control system in which the control action is independent of the output.

Control System, Closed Loop A control system in which the control action depends in some fashion on the output. See also **Feed Back Control System**.

Control Valve A valve which regulates the flow or pressure of a medium which affects some controlled process. Control valves are usually operated by remote signals from independent devices using control mechanisms powered electrically, pneumatically, electro-hydraulically, etc. (See Fig. 26)

Fig. 26. Control Valve. (*Courtesy of Yarway Corporation.*)

which a compensating device is operated by a pressure signal.

Control, Pump Controls applied to a variable delivery, positive displacement pump to adjust its volumetric output or direction of flow.

Control Response The output of a controller that is the result of a change in a controlled variable.

Controlled Medium A process or material in which some variable is controlled.

Controlled Variable A quantity which is controlled by a controlling device or system.

Controller An instrument actuated by a control signal (pneumatic, electric, etc.) for controlling process variables.

Controller Lag The delay in correcting a deviation in a controlled variable due to delay in transmission of a control signal.

Controller Response The action obtained from a controller in response to a change in the controlled variable.

Controlling Means A control device which makes an action to correct a deviation in a controlled variable from the desired value.

Convection The transfer of heat through the associated transfer of mass. It may be natural convection in which mass is caused to move by means of density gradients due to temperature differences, or forced convection in which mass is moved by mechanical means.

Convection, Forced Transfer of mass by mechanical means such as pumps, centrifugal blowers, etc.

Cooler A device used to extract heat from a fluid.

Corrective Action A change in flow of a control agent initiated by an automatic controller.

Corrosion Deterioration of materials (usually metals) due to chemical and/or electrical action.

Coupler A type of valve for the use of quick removal of an end item such as a drill motor, etc.. See also **Quick Disconnect Coupling**.

Coupling A straight connector for gas or liquid lines.

Coupling, Disconnect A coupling which can rapidly connect or disconnect fluid lines, usually without need of tools. See also **Coupler**.

Cross A hydraulic or pneumatic fitting with four openings at right angles.

Critical Point The pressure and temperature conditions for critical or sonic flow.

Critical Pressure The ratio of downstream to upstream pressure at the point of sonic flow.

Cryogenics The science dealing with the properties of matter at temperatures near absolute zero.

Cryogenic Valve A valve used to control the flow of cryogenic liquid. (See Fig. 27)

Cushion A device used to provide controlled resistance to motion.

Cushion, Cylinder A cushion built into a cylinder to restrict the flow at the outlet and thereby arrest the motion of the piston rod.

Cushion, Die A cushion installed with a die on a press to provide a controlled resistance against the work. The return motion of the cushion may be used to eject the work.

Cushion, Hydraulic A cushion in which the resistance is provided by a hydraulic cylinder.

Cushion, Hydropneumatic A cushion in which the resistance is provided by a hydraulic cylinder in which air pressure acts on the hydraulic fluid in the reservoir to return the cushion to its normal position.

Cushion, Pneumatic A cushion in which the resistance is provided by an air cylinder.

C_v See **Flow Coefficient**.

Cycle A series of events repeating in a fixed time interval.

Cycle, Automatic A cycle of operation which repeats indefinitely until stopped by outside means.

Cycle, Manual A cycle which is started manually and controlled through all phases.

Cycle, Semi-Automatic A cycle which is started with a given signal, proceeds through a predetermined sequence of events and stops with all elements in their initial position.

Cylinder A device used to convert fluid power into linear mechanical motion or force. It usually consists of movable elements, such as piston and rod, operating within a cylindrical bore or cavity.

Cylinder, Adjustable Stroke A cylinder which is equipped with adjustable stops at one or both ends to limit the amount of piston travel.

Cylinder, Cushioned A cylinder which has a piston-assembly deceleration device at one or both ends of the stroke.

Cylinder, Double Acting A cylinder in which the fluid can be applied to the movable element in either direction.

Cylinder, Double-End-Rod A cylinder which has a rod extending from each end.

Cylinder, Piston Type A cylinder in which the internal element has one or more diameters and the seal applied is of the expanding type.

26 VALVES AND VALVE TERMINOLOGY

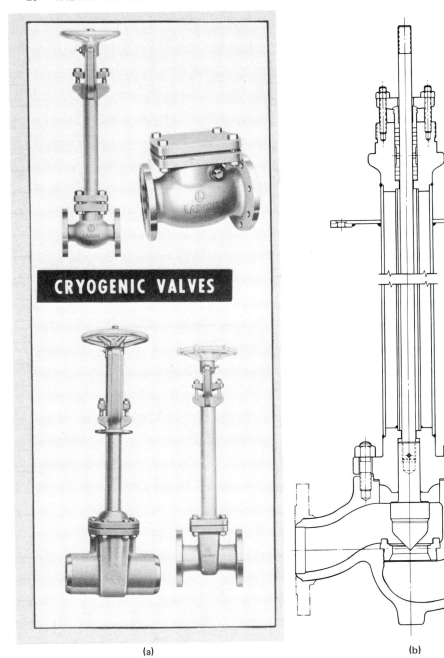

Fig. 27a,b. Cryogenic Valve. (27a *Courtesy of the Ladish Co.*, 27b *Courtesy of ITT Hamel-Dahl-Conoflow.*)

Cylinder, Plunger Type A cylinder in which the internal element has a single diameter and upon which the seal applied is a contracting type.

Cylinder Port The port through which the valve action is made common with either the inlet or the outlet port in multi-action valves.

Cylinder, Retractable Stroke A cylinder in which the position of stop can be temporarily changed to permit retraction of the piston assembly.

Cylinder, Rotating A cylinder in which a relative rotation of the cylinder housing and the piston and piston rod, plunger, or ram is required.

Cylinder, Single Acting A cylinder in which the fluid can be applied to the movable element in only one direction.

Cylinder Tandem Two or more cylinders having interconnected piston assemblies.

Cylinder, Telescoping A cylinder having multiple nested tubular segments which can provide a long stroke from a short retracted envelope.

Dampen To check or reduce vibration in a system.

Damper A device used to dampen or reduce the amplitude of a pressure or shock wave.

Darcy's Equation The equation relating pressure drop to flow friction through a pipe. There are several equivalent forms of the equation of which the most basic is:

$$P = \frac{\rho f L v^2}{144 D^2 g}$$

where

ρ = density(lb/ft³)
f = friction factor
v = velocity of flow(ft/sec)
D = I.D. of pipe(in.)
L = length of pipe(ft)
g = acceleration of gravity(32.2 ft/sec²)

The Darcy Equation is valid for laminar or turbulent flow of any type liquid through a pipe. See also **Hagen-Poiseuille Law** and **Friction Factor**.

Dashpot A fluid mechanical device used for damping oscillations. It consists of a cylinder filled with oil or other fluid and a piston connected to the part to be damped. The viscosity of the fluid causes a friction force on the piston which damps the vibrations.

Data Plate A plate carrying identification of valve type, pressure rating, etc.

Dead Band A specific range of values in which an input signal can be altered without causing a change in the output signal. See also **Dead Zone**.

Dead-End Shutoff A condition of no flow through a valve. It may be tested by determining the change in a controlled variable per unit time when discharging into a specified volume.

Dead-Tight Nonleaking.

Dead Time Any definite delay between two related events. It is also called process lag.

Dead Zone The range of a variable in which an instrument cannot detect change or initiate corrective measures.

Decay Rate The decrease of pressure with time.

Demand Side The part of a process being controlled.

Density Mass per unit volume. Common units are kilograms per cubic meter (SI metric), grams per cubic centimeter (CGS metric), pounds per cubic foot (British).

Derivative Action A control operation in which the speed with which a correction is made is proportional to the deviation from established limits.

Derivative Controller Action An action for which there is a predetermined relation between the derivative function of a controlled variable and the position of the final control element.

Derivative Response Time rate response.

Desired Value The value of the control variable which is to be maintained.

Deviation The difference at any specified time between the control and set points.

Dewar Flask A double-walled flask used for storing cryogenic liquids. The space between walls is evacuated to reduce heat conduction and usually aluminized to reduce thermal radiation absorption.

Dewpoint The temperature of a gas or liquid at which condensation or evaporation occurs.

Diagram, Pressure–Time A graph of pressure plotted against time.

Diaphragm A flexible material used to separate the control medium from the controlled medium and which actuates the valve stem. (See Fig. 30)

Diaphragm Actuator A valve operator in which pressure exerted on a diaphragm is used to position the valve stem. (See Fig. 28)

Diaphragm Assembly See **Diaphragm Actuator**.

Diaphragm Button See **Diaphragm Plate**.

Diaphragm Case The housing which contains the diaphragm and establishes one or more pressure chambers. (See Fig. 28)

28   VALVES AND VALVE TERMINOLOGY

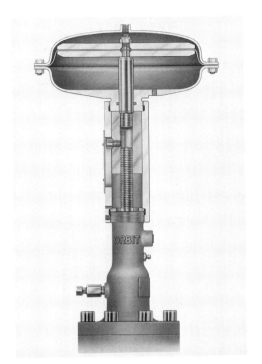

Fig. 28. Diaphragm Actuator. *(Courtesy of Orbit Valve Company.)*

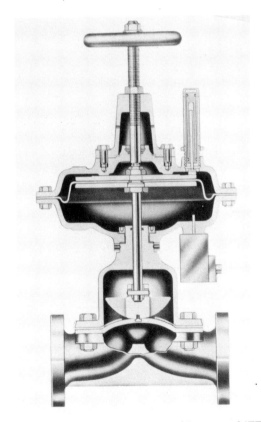

Fig. 29. Diaphragm Control Valve. *(Courtesy of ITT Hamel-Dahl-Conoflow.)*

Diaphragm Control Valve A control valve that is actuated by a diaphragm or which has a spring-diaphragm actuator. (See Fig. 29)

Diaphragm Cover See **Diaphragm Case.**

Diaphragm Disc See **Diaphragm Plate.**

Diaphragm Head See **Diaphragm Plate.**

Diaphragm Housing See **Diaphragm Case.**

Diaphragm, Metal A diaphragm made of metal such as stainless steel or bronze.

Diaphragm Motor A diaphragm actuator comprised of case, diaphragm, plate, spring, stem extension, yoke, spring seat, spring adjustor, travel indicator and scale, and hand wheel operator. (See Fig. 30)

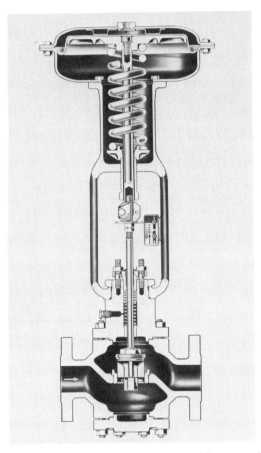

Fig. 30. Diaphragm Operated Valve. *(Courtesy of Fisher Controls Company.)*

Diaphragm Motor Operator See **Diaphragm Actuator.**

Diaphragm Motor Valve See **Diaphragm Control Valve.**

Diaphragm Operated Valve A valve which is operated by a diaphragm, generally a diaphragm control valve. (See Fig. 30)

Diaphragm Plate A plate used with a diaphragm for support purposes and to transmit force to the actuator stem. (See Fig. 28)

Direct-Acting Instrument An instrument in which the air pressure supplied to a controlled device increases as the quantity being measured by the instrument increases.

Direct-Acting Valve A normally open valve which requires fluid pressure to close it. (See Fig. 31)

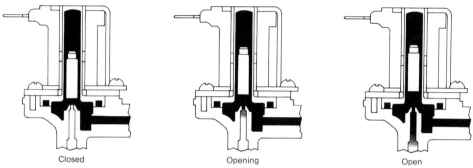

Closed   Opening   Open
Fig. 31. Direct-Acting Valve. (*Courtesy of Eaton Corporation.*)

Diaphragm Pressure Span The difference between high and low values of a diaphragm pressure range.

Diaphragm Rod See **Actuator Stem**.

Diaphragm, Slack A diaphragm made of synthetic material having a fabric layer. Convolutions are molded into the material or sufficient slack is provided so that convolutions are formed by pressure loading.

Diaphragm Stress Stress occurring in the plane of a membrane or other surface under deflection. It is the most important stress in the case of thin-walled pressure vessels.

Diaphragm Top See **Diaphragm Actuator**.

Differential The difference between two values of a measured quantity, generally the difference between the highest and lowest values of the quantity. Also called the delta of the quantity.

Differential Gap Action The action of a controller in which the output pressure remains a maximum (generally 20 psi) or a minimum (generally 0 psi) until the controlled measurement crosses a band or gap causing the output pressure to reverse. The measured variable must then span the gap in the opposite direction before the output signal is restored to its original condition.

Direct-Acting Controller A controller in which an increase in the controlled variable causes an increase in output pressure.

Direct Actuator A diaphragm actuator in which the actuator stem extends as diaphragm pressure increases.

Disc The part of the valve used to actually close off the flow of a fluid. See also **Closure Member**.

Discrete Units Individual or distinct units as opposed to continuous or irrational units.

Distance–Velocity Lag The lag or delay in the change of a measured value in a variable medium caused by the distance between the measuring points and the controlling points.

Dither A low amplitude, high frequency periodic electrical signal, sometimes superimposed on a servovalve input to improve system resolution.

Double Disc A two-piece disc used as the element in a gate valve. Upon contact with the seating faces in the valve the wedges located between the disc faces force them against the body seats to shut off the flow. (See Fig. 3)

Double Ported Valve A valve which has two ports to overcome line pressure imbalances. (See Fig. 31A)

Double Wedge Two wedges used as the seal elements in a valve. Turning the valve stem downward spreads the split wedges and each seals independently. (See Fig. 32)

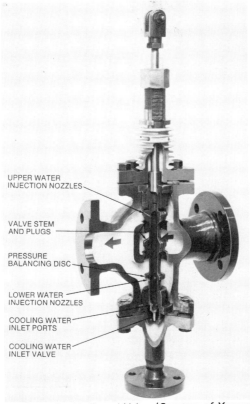

Fig. 31A. Double Ported Valve. (*Courtesy of Yarway Corporation.*)

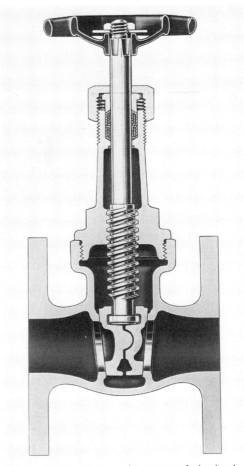

Fig. 32. Double Wedge. (*Courtesy of the Lunkenheimer Co.*)

Drainage Fitting A type of fitting used to drain fluid from pipes.

Dribble Range The difference between cracking pressure and the pressure at which liquid leakage begins. See also **Simmer**.

Drift A sustained change in value of a variable. It is also called wander.

Droop The amount by which a controlled variable such as pressure, temperature, etc., deviates from a set value at the minimum controllable flow when the flow through a regulator is gradually increased from minimum controllable flow to the rated maximum.

Drop Tight A valve that will not pass fluid droplets when it has been closed.

Dry Bulb Temperature The temperature of air measured on a thermometer for which the bulb is kept dry. This is one of the temperatures used to determine relative humidity. See also **Wet Bulb Temperature** and **Relative Humidity**.

Duplex Design Actuator An actuator which has an adjustable operating force and which operates on a differential air signal.

Durometer Hardness The hardness value of an elastomer as measured on a durometer scale.

Dwell The part of a stroke or cycle in which the pressure or feed is stopped.

Dynamic Imbalance The net or effective force produced on a valve plug at any prescribed open position by the fluid forces which act on it.

Dynamic Packing Packing used to provide a seal around a moving element such as a valve stem.

Eccentric Valve A valve which needs no resilient liner or lubricant to obtain a tight metal to metal seal. (See Fig. 33)

TERMINOLOGY OF THE VALVE INDUSTRY 31

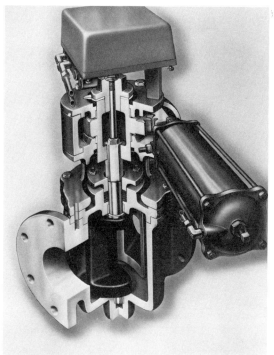

Fig. 33. Eccentric Valve. (*Courtesy of De Zurik, A Unit of General Signal*.)

Effluent The fluid leaving a filter or other device.

Elbow A fitting used to make a change in direction of a pipeline. It is also called an Ell.

Electric Motor Actuator A valve operator in which an electric gear motor is the major component.

Electro-Pneumatic Actuator A valve operator having an electrical control system that transduces the electric signal of a controller into a pneumatic input to the diaphragm housing.

Emulsion A stabilized mixture of two immiscible components such as water and oil.

End Connection The method of connecting components of a pipe system, e.g., flanged, threaded, butt welded, socket welded, etc..

End Load Rating The maximum axial force which can be applied to a filter element or other device without deformation or seal failure.

End Point Control Quality control obtained through continuous automatic analysis. Any deviation from control standards is corrected automatically.

Enthalpy The sum of internal energy and pressure energy of a fluid. It is a conserved quantity in a throttling process.

Entrained Air A mixture of air bubbles in a liquid having a tendency to separate from the liquid.

Entropy The ratio of the quantity of heat flowing in or out of a body to its absolute temperature in units of Joules per degree Kelvin.

Equivalent Length The length in pipe diameters of straight pipe which gives the same pressure drop as the valve under given flow conditions.

Erosion The wearing away of a valve seat due to high velocity flow. See also **Wire Drawing**.

Equal Percentage Flow Characteristics See **Percentage Flow Characteristic**.

Equal Percentage Plug A valve plug shaped to allow flow of a medium in direct proportion to the amount of plug lift. (See Fig. 34)

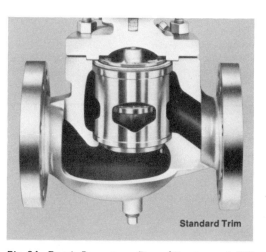

Fig. 34. Equal Percentage Plug. (*Courtesy of ITT Hamel-Dahl-Conoflow.*)

Equilibrium A balance condition. A condition of maximum or minimum potential energy.

Error The difference between the actual and desired values of a controlled variable.

Excitation The supplying of an input signal to a device which causes an output. To "turn on."

Exhaust Port The port connected to the downstream side of a fluid system.

32 VALVES AND VALVE TERMINOLOGY

Exhaust Valve The valve of a system (usually an engine) that provides an outlet for exhaust gases.

Expansion Joint A pressure-tight device that allows expansion or contraction of a piping system due to temperature changes.

Expansion Valve A valve, usually a diaphragm type, which is actuated by an expanding gas such as freon. This type of valve is widely used in refrigeration and air conditioning systems. (See Fig. 35)

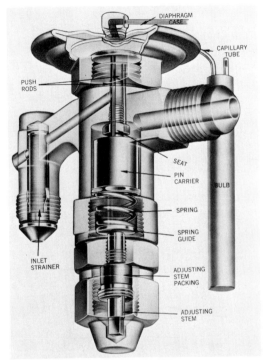

(a)

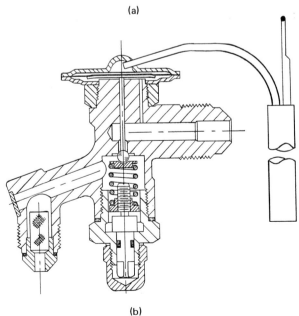

(b)

Fig. 35a,b. Expansion Valve. (35a *Courtesy of Sporlan Valve Co.*, 35b *Courtesy of Jackes-Evans Mfg. Co.*)

Extra Heavy A piping material, generally cast iron, which is suitable for pressures up to 250 psi.

Extrusion The distortion under pressure of a seal into the clearance region between metal parts.

Face-To-Face-Dimension The dimension of a valve or fitting from the face of the inlet port to the face of the outlet port.

Facing The finish of contact surface of flanged-end piping materials.

Fail Safe Valve A valve that either fails in the open position or closes to prevent a costly or dangerous situation within a system. (See Fig. 36)

Feed The part of a cycle in which work is done on the workpiece.

Feedback The process of bringing back information about the condition under control to compare it with a target value.

Feedback Controller A mechanism which measures the value of a controlled variable, compares it to a command or set value, and manipulates a controlled system in order to maintain a desired relationship between controlled variable and command.

Feedback Control System A control system which tends to maintain a prescribed relationship between two or more system variables by comparing functions of these variables and using their difference as the means of control.

Feedback Signal The signal which is returned to the input of a system and compared to a reference signal to establish an actuated signal which returns the controlled variable to the desired value.

Female Thread An internal thread in pipe fittings, valves, etc., for making screwed connections.

Filler Ring A ring which is used to fill the recess of a U or V-seal.

Filter A device used to separate contaminates from a fluid flowing through it.

Filter Element A device of porous materials such as paper, charcoal, etc., which performs the process of filtration. It is usually contained

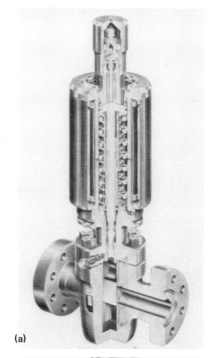

(a)

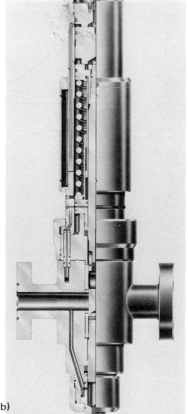

(b)

Fig. 36a,b. Fail Safe Valve. (*Courtesy of Cameron Iron Works, Inc.*)

34 VALVES AND VALVE TERMINOLOGY

in a can of some type and is generally replaceable.

Filter Media The porous materials which are used to perform the process of filtration.

Filter Media, Depth Porous materials which trap contaminants within a tortuous path or labyrinth.

Filter Media, Effective Area The total effective or functional surface of a porous media.

Filter Media, Surface Porous materials that trap contaminants primarily on the top surface.

Final Control Element The element in a control system that actually varies the control agent.

Fire Hydrant Valve A valve which, when closed, drains at an underground level to prevent freezing. (See Fig. 37)

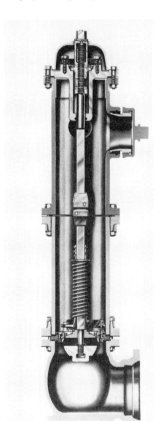

Fig. 37. Fire Hydrant Valve. (*Courtesy of American-Darling Valve.*)

Fitting The closure or connector for fluid lines and passages.

Fitting, Compression A fitting which grips and seals by means of a manually adjustable deformation.

Fitting, Flange A fitting which utilizes a radially extending collar for connection and sealing.

Fitting, Flared A fitting which grips and seals by means of a preformed flare at the end of the tube.

Fitting, Flareless A fitting which grips and seals by means other than a flare.

Fitting, Hose A fitting on the end of a hose used to connect it to another device.

Fitting, Welded A fitting which is attached by welding.

Flange A rim on the end of a valve, pipe or fitting for bolting onto another pipe element.

Flange Bonnet A valve bonnet which has a flange through which bolts connect it to a mating flange on the valve body. (See Fig. 4)

Flange Ends Flanges on a valve or fitting for joining to other fittings or pipes. The flange ends may be plain faced, raised face, small male and female, large male and female, small tongue and groove, large tongue and groove, and ring joint. (See Fig. 1)

Flap Valve A valve in which a hinged disc or flap provides the closing action. This valve is usually a nonreturn type and used for low pressure. (See Fig. 38)

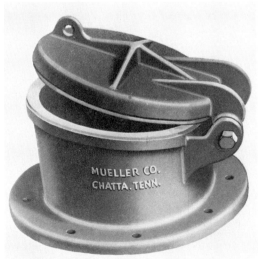

Fig. 38. Flap Valve. (*Courtesy of Mueller Co.*)

Flared Ends The ends of a pipe or other element which are shaped so as to have increasing diameter towards the end.

Flash Point The temperature at which a fluid first gives off enough vapor to ignite with a spark or flame.

Flat Faced Bonnet Joint The point of connection between a valve body and bonnet when the bonnet is seated on a flat surface instead of being recessed into the body.

Flat Full Face Gasket A flat gasket which covers the entire surface of the components being joined.

Flexible Wedge Disc A valve disc which has a solid core, but which is flexible on the outside.

Float Level Controller A control which is operated by a bulb floating on the surface of a liquid in a tank. The rising and falling of the liquid level causes the bulb to rise or fall thus opening or closing a valve or actuating some other device.

Float Valve A valve which opens or closes due to the rising or falling of a floating bulb on the surface of a liquid in a tank. (See Fig. 39)

Floating Action An action in which a predetermined relation exists between the value of a controlled variable and the rate of motion of a final control element.

Floating Average Position Action An action in which there is a predetermined relation between a controlled variable and the rate of change of the time average position of a final control element which is periodically moved between two fixed positions.

Floating Control A type of control action which varies the flow in relation to a value of a variable without a definite relationship between them other than direction of change.

Floating Controller Action An action in which there is a predetermined relation between values of a control variable and the rate of motion of a final control element which may or may not have neutral zones.

Floating Rate As applied to proportional plus floating controller action, it is the number of times per unit time that the effect or proportional position action is reproduced by proportional speed floating action.

Floating Speed The rate of movement of a final control element corresponding to a specified deviation. It is usually expressed in percent of full range of movement per unit of time.

Floating Time The reciprocal of floating rate. It applies to proportional plus floating controller action.

Flow Characteristic The relation between the flow through a valve and the percent of valve stem travel required for a given flow.

Flow Coefficient The number of U.S. gallons of water per minute at 60° F that will flow through a valve with a pressure drop of one pound per square inch under stated conditions of pressure and percent rated travel. It is also referred to as the C_v of a valve.

Flow Fatigue The ability of a device to resist structural failure due to flexing caused by differential pressure.

Flow, Laminar A condition of flow in which fluid moves in parallel layers. It occurs in situations where the Reynolds number is less than approximately 2000.

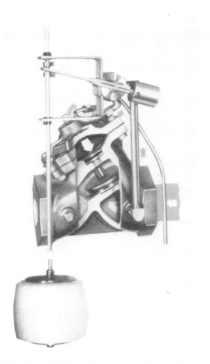

Fig. 39. Float Valve. (*Courtesy of Aqua Matic Inc.*)

Flow, Metered Flow at a controlled rate such as flow through a control valve.

Flow Rate The mass, weight or volume of a fluid flowing through a conductor per unit of time.

Flow, Steady State A flow condition in which variables such as pressure, temperature and velocity have constant values at any point in the fluid.

Flow, Streamline See **Flow, Laminar**.

Flow, Turbulent A flow condition in which the fluid moves in a random manner. It generally occurs when the Reynolds number is greater than approximately 4000.

Flow, Unsteady A flow condition in which quantities such as pressure, velocity and temperature change with time at some fixed point in the fluid.

Flowmeter A device which is used to indicate either flow rate, total flow or a combination of both.

Fluid The state of matter that is not solid and is able to flow and change shape. The term fluid includes both the liquid state and the gas or vapor state.

Fluid Conditioner A device which is used to control the physical characteristics of a fluid.

Fluid Coupling, Flexible A flexible hose or coiled tubing used for carrying fluid where flexibility is required.

Fluid Friction The friction in a fluid due to the viscosity of the fluid.

Fluid, Halogenated A fluid containing any of the halogen elements, fluorine, chlorine, bromine, astatine and iodine, or compounds of these elements.

Fluid Power Power transmitted and controlled through use of a pressurized fluid.

Fluid Power System A system which controls and transmits power by means of a pressurized fluid within an enclosed circuit.

Fluid Stability, Chemical The resistance of a fluid to chemical change.

Fluid Stability, Hydrolytic The resistance of a fluid to permanent change in any of its properties due to chemical reaction with water.

Fluid Stability, Oxidation The resistance of a fluid to permanent change due to chemical reaction with oxygen.

Fluid Stability, Thermal The resistance of a fluid to permanent changes caused by heating.

Fluidic A term pertaining to devices, systems and assemblies utilizing fluid components.

Fluidic Amplification, Flow The ratio of the change of flow in a specified load impedance connected to a control device to the change in flow applied to the controls of the device.

Fluidic Amplification, Power The ratio of the change of power in a specified load impedance connected to a control device to the change in power applied to the controls of the device.

Fluidic Amplification, Pressure The ratio of the change of pressure drop across a specific load impedance to the change in pressure drop applied across the controls of a device.

Fluidic Amplifier A device which enables one or more fluid dynamic signals to control a source of power and thus deliver an enlarged reproduction of the signal.

Fluidic Amplifier, Closed A fluidic amplifier which does not have a vent port.

Fluidic Amplifier, Impact Modulator A fluidic amplifier in which the impact plane position of two opposing streams is controlled to change the output.

Fluidic Amplifier, Open A fluidic amplifier which has vent ports.

Fluidic Amplifier, Stream Deflection A fluidic amplifier usually analog, which utilizes one or more control streams to deflect a power stream, thus changing the output.

Fluidic Amplifier, Turbulence A fluidic amplifier in which a power jet is at a pressure such that it is in the transition region of laminar stability and can be made turbulent by a secondary jet.

Fluidic Amplifier, Vortex A fluidic amplifier in which the angular velocity of a vortex is controlled to change the output.

Fluidic Amplifier, Wall Attachment A digital fluidic amplifier in which the control of the attachment of a stream to a wall changes the output.

Fluidic Device, Active The class of fluidic devices that require a power supply separate from the controls.

Fluidic Device, Passive The class of fluidic devices that operates on signal power alone.

Fluidics The engineering science devoted to the use of fluid dynamic phenomena to sense, control, process information, and/or actuate.

Fretting A type of corrosion occurring in valves due to small oscillatory tangential motion between two materials.

Friction The resistance to motion of an object under the action of an external force.

Friction Factor A factor appearing in the Darcy equation to describe the effect of friction of the fluid particles and consequent loss of pressure. For laminar flow it is a function of the Reynolds number, while for turbulent flow it depends on the Reynolds number and the nature of the pipe wall.

Front Connected The condition wherein connections are made to normally exposed surfaces of components.

Froude Number The ratio of the force of inertia to the force of gravity.

$$N_F = \frac{V^2}{Lg}$$

where

V = velocity
g = acceleration of gravity
L = characteristic length

Full Range The complete or entire range of values which a variable may assume.

Full Scale End Points The maximum positive and negative values of some controlled variable.

Galvanize A process of protecting steel from corrosion by applying a zinc coating to it.

Gasket A material used for sealing a joint in a piping system. It usually is a flat piece of elastomer, cork, asbestos compound or similar material and is used between mating flanges or similar surfaces. It provides a static permanent seal.

Gate Valve A valve in which a sliding disc or gate is moved by an actuator perpendicular to the direction of flow. They are normally used in the fully opened or fully closed position and not for throttling purposes. (See Fig. 40)

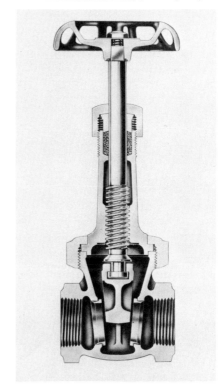

Fig. 40. Gate Valve. (*Courtesy of Jenkins Bros.*)

Gauge An instrument used for measuring some physical property such as pressure, temperature, etc.

Gauge, Bellows A gauge in which a bellows is the sensing element. A difference of pressure between inside and outside causes the bellows to expand or contract in an axial direction.

Gauge, Bourdon Tube A pressure gauge in which a bourdon tube is the sensing element. The curved bourdon tube straightens under increasing pressure and moves an indicator needle.

Gauge, Diaphragm A gauge in which a diaphragm is the sensing element. The amount of deflection of the diaphragm is proportional to the pressure being measured.

Gauge, Fluid Level A gauge used to indicate fluid level.

Gauge Isolator Valve Gauge isolator valves protect pressure gauges from damaging pressure surges, hydraulic shock and mechanical vibrations. Fluid is completely isolated from the gauge until the knob is pressed. By pressing the knob, fluids are connected directly to the gauge port giving instant and accurate readings on gauge. As the knob is released, the spring loaded valve closes automatically connecting the gauge post to drain and completely blocking the pressure port. (See Fig. 41)

Fig. 41. Gauge Isolator Valve. (*Courtesy of Manatrol Division, Parker Hannifin Corp.*)

Gauge, Piston A pressure gauge in which a piston operating against a spring is the sensing element.

Gauge, Pressure An instrument for measuring the pressure in a system. It generally measures pressure above atmospheric pressure. A typical pressure gauge consists of a bourdon tube connected to an indicator and a scale in a case.

Gauge Pressure The pressure measured above atmospheric pressure.

Gauge, Vacuum A gauge for measuring pressures below atmospheric pressure.

Gibbs Function Enthalpy minus the product of temperature and entropy.

$$G = H - TS$$

It is conserved in an isothermal, isobaric process.

Gland The cavity of a stuffing box into which the packing is stuffed.

Gland Follower The closure for a stuffing box.

Globe Valve A family of valves characterized by a closure member which travels in a line perpendicular to the valve seat. They are used primarily for throttling purposes and general flow control. (See Fig. 42)

Fig. 42. Globe Valve. (*Courtesy of Jenkins Bros.*)

Ground Wire A device to ground the immediate internal portion of a valve in contact with a fluid media to reduce explosion caused by static electricity.

Guide Bushing A ring used to guide a valve stem or actuator.

Hagen-Poiseuille Law The form of Darcy's equation for laminar flow.

$$P = .000668 \frac{\mu L v}{d^2}$$

where

P = pressure drop (psi)
μ = dynamic viscosity (centipoise)
v = velocity (ft/sec)
d = diameter (in.)

In this case the friction factor in Darcy's Law is the ratio:

$$f = \frac{64}{R}$$

where

R = the Reynolds number (less than 2000 for laminar flow).

Hammer Blow Handwheel A handwheel that provides additional operating torque for operating a valve in situations where a plain handwheel is insufficient but where gearing is unnecessary. (See Fig. 43)

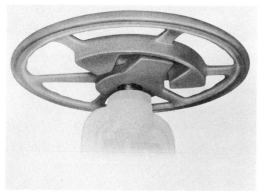

Fig. 43. Hammer Blow Handwheel. *(Courtesy of Jenkins Bros.)*

Hanger A device for supporting pipes to provide strain relief.

Head The height of a column of liquid above a specified point expressed in units such as feet of water, inches of mercury, etc. It is a measure of pressure exerted by the column of liquid.

Head, Friction The head required to overcome the friction in a fluid system.

Head, Static The height of a column of fluid above a fixed reference point.

Head, Static Discharge The static head measured from the centerline of the pump to the free discharge surface.

Head, Static Suction The static head from the surface of a supply source to the centerline of the pump.

Head, Total Static The static head measured from the surface of a supply source to the free discharge surface.

Head, Velocity The equivalent head through which a liquid would have to fall to attain a specified velocity.

Header The length of pipe or container to which are joined two or more pipe lines used to carry fluid from a common source to the various points of use.

Heat Exchanger A device that transfers heat from one fluid to another through a conducting wall.

Helmholtz Function The internal energy minus the product of temperature and entropy.

$$A = U - TS$$

It equals work done on a system in an isothermal process and is conserved for an isothermal, isochoric process.

Horsepower The unit of power in the British system of units. One horsepower equals 550 ft lb/sec.

Hose A flexible line used to conduct a fluid. It is usually made of an elastomeric or plastic material.

Hose, Wire Braided A hose which consists of a flexible material, such as rubber, reinforced with woven wire braid.

Hose End Valve A valve which has fittings for connection to a hose. It is generally intended for water service. (See Fig. 44)

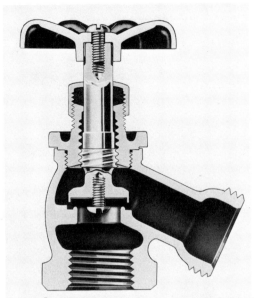

Fig. 44. Hose End Valve. (*Courtesy of Nibco Inc.*)

Hub End A type of end connection used on valves, fittings and pipe for water supply or sewage lines, which is connected by leading or caulking.

Humidity The amount of water vapor in the air at a given temperature. See also **Relative Humidity**.

Hydraulic Pertaining to use of liquids in circuits to provide force and/or movement.

Hydraulic Actuator A device used to convert hydraulic energy into mechanical motion.

Hydraulic Amplifier A fluid valving device such as a sliding spool, nozzle flapper or jet pipe with receivers, which acts as a power amplifier.

Hydraulic Controller A controller which is operated by means of a liquid such as water, oil, etc.

Hydraulic Cylinder Operator A valve which is operated by a cylinder from the power furnished by a liquid under pressure.

Hydraulic Gauge A gauge that is designed and constructed specifically for use at extremely high pressures where water or other liquid is used as the pressure medium.

Hydraulic Horsepower Power expressed in terms of the flow rate and pressure differential.

$$H.P. = .000583 \, QP$$

where

$H.P.$ = Hydraulic horsepower
Q = flow rate (gpm)
P = pressure differential (psi)

Hydraulic Motor Actuator A device which translates the rotational motion of a hydraulic motor into linear motion.

Hydraulic Piston Operation A piston that is hydraulically operated.

Hydraulic Radius The cross sectional area of the flowing fluid divided by the wetted perimeter of the container. It serves as an effective diameter for resistance calculations, etc., for flow in an open channel.

Hydraulics The branch of physics that deals with the mechanical properties of liquids such as water, oil, etc.

Hydrodynamics The branch of physics that deals with the motion of liquids under the action of forces.

Hydrokinetics The branch of physics that deals with the motions of fluids without regard to the forces causing the motions.

Hydrometer A device used to measure the specific gravity of a fluid. It is generally constructed to be partially immersed in the fluid and the depth to which it is immersed is proportional to the density of the fluid.

Hydropneumatic A term referring to the combination of pneumatic and hydraulic power in a system or device.

Hydrostatics The branch of physics that deals with the properties of liquids at rest.

Hygroscopic A substance which has a strong affinity for water.

Hypergolic Denoting a propellant or fuel that ignites spontaneously on contact with an oxidizer.

Hysteresis In a cyclic process, hysteresis is the failure to follow the same path in the forward direction as in the backward direction.

Ideal Gas An imaginary gas consisting of point masses having no internal forces. It is approximated by a real gas at high temperature and low pressure.

Increaser A fitting with a large opening at one end that is used to increase the size of a pipe opening.

Indicating Instrument A measuring instrument that is read by observing the position of a pointer on a scale.

Indicator Disc See **Travel Indicator**. (See Fig. 13)

Industrial Instrument Any device which measures and controls the values of some process variable.

Influent The fluid flowing into a valve or other device.

Inherent Diaphragm Pressure Range The low and high limits of the pressure applied to a diaphragm to produce greater travel of a plug when the valve body is under atmospheric pressure.

Inherent Flow Characteristic The flow characteristic when a constant pressure drop is maintained across the valve.

Inherent Rangeability The ratio of maximum to minimum flow within which the deviation of a specified flow characteristic does not exceed some specified limit.

Inhibitor A substance which serves to prevent chemical reaction such as oxidation, corrosion, etc.

Inlet Port The port connected to the upstream side of a fluid system. Fluid enters the valve or other device through the inlet port and exits through the outlet port.

Inner Valve See **Valve Plug**.

Inner Valve Seat See **Valve Seat Ring**. (See Fig. 3)

Input An incoming signal, pressure, flow, etc., of a control system or device.

Input Current The electrical current, usually measured in milliamps, supplied to a control valve or device which commands control flow.

Inside Screw, Nonrising Stem The type of stem, usually found in gate valves, in which the disc rises on the threaded part of the stem instead of the stem rising through a threaded portion of the bonnet. (See Fig. 3)

Inside Screw, Rising Stem The type of stem often found on gate and globe valves in which the stem rises as the handwheel is turned. Inside screw means that the threads are inside the bonnet. (See Fig. 3)

Installed Diaphragm Pressure Range The low and high values of pressure applied to the diaphragm to produce a specified travel with specified conditions within the valve body.

Installed Flow Characteristic The flow characteristic when the pressure drop across the valve varies as required by flow and other conditions in a system in which the valve is installed, as opposed to the inherent flow characteristic.

Instrument Any device that has the ability to measure, record, indicate, or control. It provides an output which may be a visual display such as a meter or permanent record such as a graph, magnetic tape, etc.

Instrumentation The application of measuring or controlling instruments to a manufacturing or processing operation.

Integral Action An action in which there is a predetermined relationship between some integral function of a controlled variable and the positions of a final control element.

Integral Control Action The mode of control action in which the value of a manipulated variable is changed at a rate proportional to the deviation.

Integrator A device that continuously sums up the quantity being measured over a period of time.

Intensifier A device used to increase the pressure in a system over the source pressure.

Intercooler A device used to cool a gas between compression steps in a multistage compressor.

Intermediate See **Yoke**.

Isolating Valve A hand operated valve located between the packing lubricator assembly and the packing box assembly in a control valve. It is used to shut off the fluid pressure from the lubricator assembly.

Isovel Lines of equal velocity.

Jet A stream, usually high velocity, of fluid emerging from an orifice.

Jet Action A valve design in which the relative position of a nozzle and a receiver control the flow.

Jet Pump A pump that utilizes fluid velocity to force another fluid movement of higher volume, but usually under lower pressure.

Joint A connection of two or more lines. A device which accomplishes the connection of lines.

Joint, Rotary A joint which connects lines which have relative operation rotation.

Joint, Swivel A type of joint which permits variable operational positioning of lines. It is also referred to as a swing joint.

Joule The unit of energy in the SI system of units. One joule equals one Newton-meter.

Joule-Kelvin Effect The cooling effect on a gas as it expands into a low pressure region. It is due to the conversion of internal energy of the gas into work energy during the expansion. The Joule-Kelvin effect is utilized in the liquefaction of gases.

Junket Ring See **Anti-Extrusion Ring**.

K-Factor The head loss factor of a valve or fitting. It is also called resistance coefficient.

$$K = f \frac{L}{D}$$

where

f = friction factor
L = length of pipe
D = diameter of pipe

Labyrinth Seal A type of seal in which the valve seat and poppet contain concentric blades which intermesh to form a seal.

Lag The delay in response of a device to a change in some variable.

Lantern Ring A ring that provides venting and support for adjacent sealing surfaces.

Lantern Ring Type Gland A chamber with a lantern spacer and rings of packing below to wipe the stem clean before it passes into the sealing rings above.

Lapping-In Polishing a surface such as a valve seat to obtain a smooth mating surface.

Latent Heat The heat required to cause a material to undergo a change of phase.

Latent Heat of Condensation The heat given up when a unit mass of vapor condenses back to a liquid.

Latent Heat of Vaporization The heat required to change a unit mass of a liquid to a vapor.

LBH Pounds per hour.

Leakage The amount of fluid passing through a valve when it is off. It is usually expressed in units of volume/time at a given pressure and temperature.

Ledoux Bell A measuring device consisting of a bell floating semi-immersed in some suitable liquid. The liquid surface inside the bell is exposed to the high pressure side of a differential pressure primary element. The surface outside the bell is exposed to the low pressure side. Changing the differential pressure will displace the bell correspondingly.

Level Regulator: A valve which has a positioning actuator which uses a self-generating power signal for moving a closure member relative to the ports in response and in proportion to the changes in the level of the controlled fluid.

Level Regulator, Self-Operated: A self-operated controller in which the energy needed to position the valve closure member is supplied by the changes of level of the controlled fluid.

Lever and Weight: A lever with weights attached to it which is used instead of air pressure or spring loading in a regulating valve. It has the advantage of minimizing droop but is very bulky (See Fig. 45).

Lift: The height of a column of fluid below a given point measured in inches, meters, etc. It is often used to describe vacuum or pressures less than atmospheric. See also: **STEM LIFT**.

Lift, Static Suction: The lift from the centerline of a pump to the surface of the supply source.

Lift Check Valve A type of check valve in which the vertically rising flow opens the gate and reverse flow causes the gate to drop back into the closed position. It is dependent on the

Fig. 45. Lever and Weight. (*Courtesy of Atlas Valve Company.*)

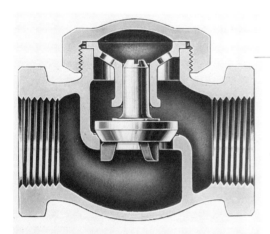

Fig. 46. Lift Check Valve. (*Courtesy of Lunkenheimer Co.*)

presence of gravitational force for its action. (See Fig. 46)

Line A pipe, tube or hose for the conduction of a fluid from a high to low pressure region.

Line, Drain A line used to return leakage to a reservoir or manifold.

Line, Exhaust A line returning a fluid back to a reservoir or to the atmosphere.

Line, Pilot A line used to carry a control fluid.

Line, Suction A supply line at below atmospheric pressure to a pump or compressor.

Line, Working A line which conducts a fluid used for transmitting power.

Linear A straight line relationship between two variables.

Linear Flow Characteristic One in which equal increments of movement of the valve stem will produce equal changes in flow with either a constant pressure drop for incompressible fluids or with constant upstream and downstream pressures for compressible fluids.

Lines Two or more pipes, tubes or hoses carrying fluid for transmitting power.

Lines, Joining Lines which are connected in a curcuit.

Lines, Passing Lines which cross in a circuit but which do not connect.

Load Change The change in demand for a controlled medium in a fluid system.

Load Error See **Accuracy of Regulation**.

Load Pressure Drop The drop in pressure measured between valve control ports.

Loading Spring See **Actuator Spring**.

Logger An instrument that automatically measures parameters such as temperature and pressure and records them on a chart.

Logic Devices The category of components which perform a logic function. They can permit or inhibit signal transmission with prescribed combinations of control signals.

Long Sweep Fitting A fitting which has a long radius turn.

Low Pressure Composition Disc A nonmetallic disc used in low pressure fluid service to provide tight shut off.

Lower Stem See **Valve Plug Stem**.

Lox Liquid Oxygen.

Lubricant Ring See **Lantern Ring**.

Lubricant Spreader See **Lantern Ring**.

Lubricated Plug Valve A valve with a groove which allows a lubricant to seal and lubricate the valve. It also provides a hydraulic jacking force to lift the plug within the valve body. (See Fig. 47)

Fig. 47. Lubricated Plug Valve. (*Courtesy of W-K-M Valve Division, ACF Industries, Inc.*)

Lubricator A device used to add lubricants into a fluid power system.

Mach Number The speed of travel of a system divided by the local acoustic velocity.

Main Body Assembly See **Valve Body Assembly**.

Main Valve The same as **Valve Plug**. Also it is the main valve element of a pilot operated valve.

Male and Female Joint-Bonnet A connection in which the bonnet inserts into the body thus assuring correct alignment and gasket compression. It also eliminates the possibility of the gasket blowing out. (See Fig. 3)

Male Thread The external thread on pipe, bolts, fittings, etc. for making screwed connections.

Malleable Fitting A pipe fitting that is made out of malleable iron.

Malleable Iron A form of iron widely used in the manufacture of valves.

Manifold A fluid conductor that has provisions for multiple connections to it.

Manifold, Vented A manifold that is open or vented to the atmosphere.

Manipulated Variable The quantity which is varied by an automatic controller in order to change the value of the controlled variable.

Manometer A device used for measuring pressure. It generally is in the form of a U-shaped tube partly filled with mercury and one side open to the atmosphere. The pressure on a fixed volume of gas in one arm is given by the difference in height of the mercury columns in the two arms.

Manual Control Hand control of changes in a controlled variable. The opposite of automatic control.

Manual Controller A controller in which all of its basic functions are performed by devices which are hand operated.

Maximum Operating Pressure Differential The maximum difference between the pressure upstream of a valve and the pressure downstream when measured at specific locations.

Measurement A determination of the instantaneous value of a variable.

Measuring Element The part of a control system that senses or measures the direction and/or amount of change of a variable.

Measuring Means That which measures a condition in a fluid system.

Metal-To-Metal Seal A seal effected by very smooth finishes on mating metal parts.

Micron A unit of length in the metric system being one-millionth of a meter.

Miniaturization Reducing the size of a component to minimize space requirements and to reduce its weight.

Minimum Controllable Flow The smallest flow at which control of the condition of the fluid can be maintained or guaranteed.

Modulus of Elasticity The rate of change of unit tensile stress to unit tensile strain for uniaxial stress within the elastic limit. It is also called Young's Modulus.

Modulus of Rigidity The rate of change of unit shear stress to unit shear strain under a condition of pure shear within the proportional limit.

Mollier Chart A family of curves giving enthalpy versus entropy for various pressures and temperatures. It is useful in computing gas flows which depend on the enthalpy of the gas.

Motor A machine which transforms electrical or fluid energy into rotary mechanical motion.

Motor, Fixed Displacement A motor in which the displacement per cycle is constant and cannot be varied.

Motor, Linear See **Cylinder**.

Motor Operator The part of a system that actually opens and closes a valve. See also **Actuator**.

Motor, Rotary A motor which is capable of continuous rotary motion.

Motor, Rotary, Limited A rotary motor which has a limited motion, i.e., does not rotate through a complete revolution.

Motor, Variable Displacement A motor in which the displacement per cycle is variable.

Muffler A device used for reducing noise from flowing gases by controlling the back pressure of the expanding gas.

Multiple Action An action in which two or more individual controller actions are combined.

Multiport Plug Valve A plug valve which has two or more ports. (See Fig. 48)

Multiposition Action An action in which a panel control element is moved to one of several predetermined positions, each of which corresponds to a definite range of values of the controlled variable.

Multiposition Control The type of control response which selects one of several rates of corrective action depending on the deviation of a process variable from a specified value.

Fig. 48. Multiport Plug Valve. (*Courtesy of W-K-M Valve Division, ACF Industries, Inc.*)

Multiposition Controller Action An action in which there are several predetermined positions of a final control element which corresponds to definite values of the variable.

Multispeed Floating Action An action in which a final control element is moved in two or more rates each of which corresponds to a definite range of values of the controlled variable.

National Pipe Thread A standard type of tapered thread used on pipes and pipe fittings to provide a strong joint and tight seal.

Needle Point Valve A type of valve which has a needle point plug and a small seat orifice for metering low flows. (See Fig. 49)

Needle Valve See **Needle Point Valve**.

Neutral Zone The predetermined range of values of a controlled variable within which no control action occurs.

Neutralization Number The measure of total acidity or alkalinity of a fluid including the effects of both inorganic and organic acids and bases.

Newt A unit of kinematic viscosity in the British system of units. One Newt is one square inch per second.

46 VALVES AND VALVE TERMINOLOGY

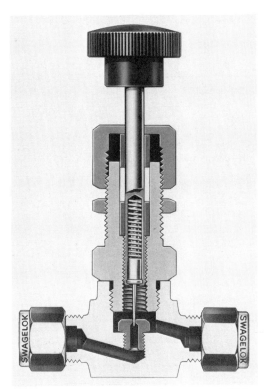

Fig. 49. Needle Point Valve. (*Courtesy of Whitey Valve Co.*) Copyright 1974 by Markad Service Company. All Rights Reserved.

Nipple A short length of tubing or pipe used for joining piping elements.

Noise Stray or random signals in a control system.

Noncorresponding Control See **Floating Controller Action.**

Nonrising Stem A type of valve stem which turns but does not rise as the valve is operated. (See Fig. 3)

Normally Closed A condition of no flow through a valve or other system when there is no input signal.

Normally Closed Solenoid Valve A valve in which the inlet port is closed off when the solenoid coil is de-energized.

Normally Open A valve or other device which allows fluid flow when there is no input signal. An input action must be applied to close the valve.

Normally Open Solenoid Valve A valve in which the inlet port is open when the solenoid coil is de-energized.

Nozzle A device used to shape a stream of fluid emerging from a line. It is used to convert pressure energy into velocity energy.

Nuclear Valve A valve usually of stainless steel or other special alloy that is resistant to radiation damage and corrosion from materials used in nuclear power plants and similar installations. (See Fig. 50)

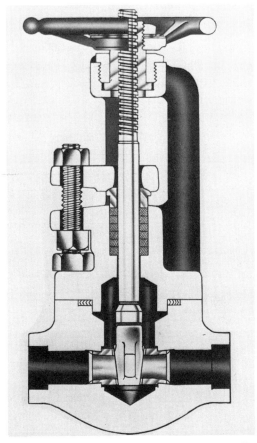

Fig. 50. Nuclear Valve. (*Courtesy of Velan Engineering Companies.*)

Off and On Response The controller response in which the final control element is immediately moved from one extreme to another as the result of a change in a controlled variable.

Offset The difference between the condition or value desired and that actually obtained. See also **Droop.**

On–Off Control A system of control in which the final control element has only two positions from which to choose. It is also known as two-position control.

Operating Life The useful life of a device or system expressed in terms of cycles of operation, hours or similar units.

Operating Pressure The nominal or average pressure in a fluid system.

Operating Pressure Span The difference between the high and low values of fluid pressure on a diaphragm to produce the required value of plug travel.

Operating Temperature The nominal or average temperature of a fluid in a system.

O-Ring A type of seal consisting of an elastomer in the shape of a toroid.

Orifice A restriction whose length is short as compared to its diameter.

Orifice Coefficient A number less than unity which characterizes the flow from an orifice. It gives the fraction of the ideal maximum flow expected.

O S & Y A type of valve stem screw, called outside screw and yoke. The packing is located between the stem screw and valve body isolating the fluid from the stem threads. It is used for abrasive and corrosive applications. (See Fig. 51)

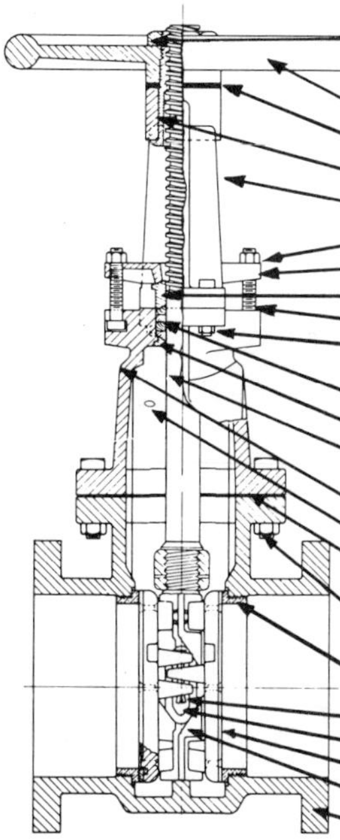

G-1 CAP NUT
G-2 KEY FOR HANDWHEEL
G-3 HANDWHEEL
G-4 BRASS WASHER
G-5 BUSH NUT
G-6 YOKE
G-7 GLAND NUT
G-8 GLAND FLANGE
G-9 GLAND
G-10 GLAND BOLT
G-11 YOKE BOLT AND NUT
G-23 STEM PACKING
G-12 BONNET BUSHING
G-13 STEM AND TOP WEDGE
G-14 BONNET
G-24 TEST PLUG
G-15 GASKET FOR BONNET
G-16 BOLTS AND NUTS FOR BONNET
G-18 SEAT RING
G-20 DISC PIN
G-21 SIDE SPREADER
G-19 DISC
G-22 BOTTOM WEDGE
G-47 BODY

AWWA O.S.&Y. GATE VALVE 4" AND LARGER

Fig. 51. O S & Y. (*Courtesy of Mueller Company.*)

Outgassing The release of absorbed gases into a vacuum chamber.

Outside Screw, Rising Stem See **O S & Y**.

Output The outgoing signal of a control unit or operation.

Overshoot The exceeding or surpassing of a target value as operating conditions change.

Oxidation A chemical process in which oxygen joins with metal atoms to form an oxide coating.

Pack To install packing into a valve.

Packing A device used to seal a valve or other components. It consists of a deformable material or deformable mating elements. (See Fig. 3)

Packing Assembly The part of a valve that contains the packing, packing gland, packing nut, etc.

Packing Box Assembly The part of a valve bonnet assembly used to seal against leakage around the valve stem.

Packing Coil Valve packing in the form of a coil.

Packing, Cup A fluid sealing device which seals primarily on its outside diameter.

Packing, Flange A packing design used for outside packed installations when there is insufficient room for a U or V-packing. It is the reverse of the cup packing design.

Packing Follower A ring shaped device that is installed on top of the packing to hold it in place. It may also be used to adjust the pressure on the packing.

Packing Gland See **Packing Follower**.

Packing Gland Flange See **Packing Flange**.

Packing Lantern See **Lantern Ring**.

Packing Lubricator Assembly An air assembly that is used for lubricating a valve.

Packing, Jamb See **Gland**.

Packing Nut with Gland A packing nut having a gland which compresses the packing.

Packing Nut without Gland A packing nut which compresses the packing in the stuffing box of a valve.

Packing, U A packing in which the deformable element has a U-shaped cross section.

Packing, Y A packing in which the deformable element has a Y-shaped cross section.

Packing, W A packing in which the deformable element has a W-shaped cross section.

Panel A surface or wall for mounting circuit components and instruments.

Paper Stock Valve A valve used in the paper industry in which the wedge has a knifelike closing action against the fluid line. (See Fig. 52)

Fig. 52. Paper Stock Valve. (*Courtesy of Fabri-Valve, A Dillingham Company.*)

Parallax The apparent change in position of an object which results from a change in the direction from which it is viewed.

Pascal The primary unit of pressure in the SI (metric) system. It is one Newton per square meter.

Pascal's Law A law of physics expressing conservation of momentum. It states that a pressure (force) applied to a confined fluid is transmitted equally in all directions through the fluid.

Passage A connection or path within a hydraulic component which acts as a conductor of the fluid.

Passivation A process used to protect stainless steel from corrosion by dipping in a nitric acid solution.

PCV Valve Pollution control valve used on automobiles to allow unburned gases to return to the carburetor.

Percentage Flow Characteristic The characteristic of a valve in which equal increments of plug travel give equal percentage changes in the rated flow.

Pet Cock See **Cock**.

pH The logarithm to the base ten of the reciprocal of the concentration of hydrogen ions in an aqueous solution.

Phase A specific functional operation during a cycle.

Phase, Neutral The phase of a cycle from which the work sequence begins.

Phase Shift The time difference between the input and output signals of a control device or system.

Piezometer A pressure measuring device utilizing the Bernoulli effect to indicate pressure.

Pilot Operated A device in which energy transmitted through a primary element is amplified by energy from some other source.

Pilot Valve A device for controlling the flow of an auxiliary fluid used to amplify the power of a controller in a larger system. It is a small valve requiring little power which is used to operate a larger valve. (See Fig. 52A)

Pinch Valve A valve in which a flexible element, usually a hose, is pinched to close off fluid flow. (See Fig. 53)

Pipe Dope See **Pipe Thread Lubricant**.

Pipe Scale A hard material frequently found in pipe as a result of heating operations used in manufacturing the pipe. The scale material consists of metal oxides and similar compounds.

TERMINOLOGY OF THE VALVE INDUSTRY 49

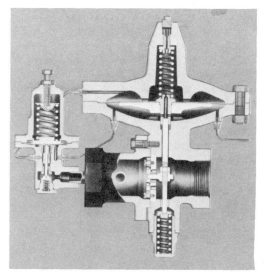

Fig. 52A. Pilot Valve. (*Courtesy of Jordan Valve Co.*)

Fig. 53. Pinch Valve. (*Courtesy of Flexible Valve Corp.*)

Pipe Strap A device used to hold pipe to a wall or from the ceiling. It generally is an adjustable metal strip.

Pipe Support Any device used for supporting pipe lines.

Pipe Threads A type of screw thread used for joining pipe sections. They generally are tapered threads which provide a reasonably tight seal. See also **National Pipe Thread**.

Pipe Thread Lubricant A material used to lubricate threads prior to assembly. It may be a dry lubricant or a greaselike material.

Piping A term for pipe and fittings used in fluid circuits. It also is used to denote a complete fluid system.

Piston Check Valve A type of check valve in which the control element is a movable piston. (See Fig. 54)

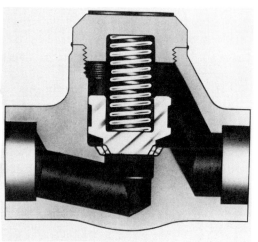

Fig. 54. Piston Check Valve. (*Courtesy of Velan Engineering Companies.*)

Plenum Chamber An auxiliary chamber connected by capillaries to a main actuator chamber used as a source of gas to provide actuator damping.

Plug The part of a valve which actually closes the orifice to stop the flow. See also **Disc**.

Plug Cock A valve in which the fluid passage is a hole in a rotating plug fitted in the valve body. It is also called a stopcock.

Plug Stem See **Valve Plug Stem**.

Plug Type Disc A type of valve disc usually in the form of a tapered plug. It is used in conjunction with a cone shaped seat having a wide bearing seating surface.

Plunger See **Valve Plug**.

Pneumatic Actuator A device used to convert pneumatic energy into mechanical energy. See also **Air Motor**.

Pneumatic Controller A controller which operates by pneumatic means, i.e., an air or gas operated controller. See also **Controller** and **Controller, Hydraulic**.

Pneumatic Motor Actuator A device in which the rotation of a pneumatic motor is translated into mechnical action. (See Fig. 9)

Pneumatic Operator See **Pneumatic Actuator**.

Pneumatic Positioning Relay A device connected to a diaphragm motor or other power unit which responds to a pneumatic pressure signal and to the position of the final control element. Its function is to control the energy applied to the power unit.

Pneumatic Transmission A system in which the control variable is converted into air pressure at a transmitting element. The air pressure variations are conducted through a tube to a receiver where they are converted to another control variable, usually position or force, for operation of the controller. Values of control variables may be transmitted over large distances by these means.

Pneumatics The engineering science dealing with gas flow and power transmission.

Pointer The indicator needle in an instrument.

Poise The unit of absolute viscosity in the metric (CGS) system of units.

Pop Valve A spring loaded, fast opening safety valve which opens automatically when the pressure exceeds a preset value. (See Fig. 55)

Poppet Valve A valve in which the closure member is a poppet usually of conical, spherical or flat shape and actuated by a spring or other mechanical means. (See Fig. 56)

Port The inlet or outlet area of a valve or other device. It may also refer to the valve seat opening.

Port, Bias The port to which a bias signal is applied.

Port, Bleed A port which provides a passage for purging gas from a system or component.

Port, Control A port which provides passage for control signals.

Port, Cylinder A port which provides a passage to an actuator.

Port, Differential Pressure A port which provides a passage to the upstream and downstream sides of a device.

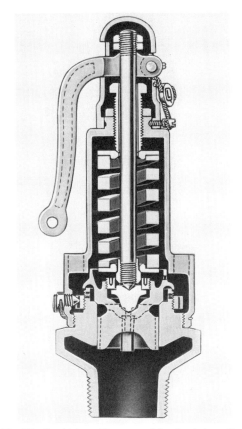

Fig. 55. Pop Valve. (*Courtesy of Lunkenheimer Co.*)

Fig. 56. Poppet Valve. (*Courtesy of Bloomfield Valve Corp.*)

Port, Discharge A port which provides a passage for fluid power to a system.

Port, Drain A port used for removal of fluid from a component.

Port, Exhaust A port which provides a passage to the atmosphere or exhaust reservoir.

Port, Fill A port which provides for filling a fluid system.

Port Guided Control Valve A type of valve in which a plug is aligned by means of the body port only.

Port, Inlet A port which provides a passage for the incoming fluid.

Port Opening The opening of a valve that is connected to a pipe or tube.

Port, Outlet A port which provides a passage for the outgoing fluid.

Port, Pipe A port which is designed for connection by means of pipe threads.

Port, Plain O-Ring A port which utilizes an O ring in a groove machined into the port face.

Port, Pressure A port which provides a passage from the fluid source.

Port, Suction A port which provides a passage for atmospheric charging of a compressor or pump.

Port, Supply The port at which power is provided to a device.

Port, Tank A port which provides a passage back to the fluid source.

Port Valving A controllable opening between two passages which can be varied from open to closed.

Port, Vent A port which provides a passage to the atmosphere.

Portable Air Motor Operator A device used to operate handwheel valves by means of a pinion ring gear.

Position Action An action in which there is a specified relationship between the value of a control variable and the position of the final control element.

Position Actuator An actuator that produces a mechnical motion proportional to the duration or magnitude of a power signal.

Position Control See **Proportional Control**.

Positioning Diaphragm Motor See **Diaphragm Actuator**.

Positive Position Stop A structural member that limits the motion of a moving member in a fixed definite position.

Positive Safety Stop A structural member used to confine travel within the design limits of the device.

Pour Point The lowest temperature at which a liquid will flow under specified conditions.

Power Actuated Valve A valve having an actuator which uses electrical, mechanical, fluid or thermal power to open or close the valve.

Power Signal The energy used to actuate a valve.

Power Unit The part of the control system which supplies the power for operating the final control element.

Prandtl Number A number relating fluid properties of a flowing fluid to heat conduction.

$$N_p = \frac{C_p \mu}{K}$$

where

C_p = specific heat (constant pressure)
μ = absolute viscosity
K = coefficient of thermal conductivity

Preact Response An output pressure change, additive to the proportional response, whose magnitude depends on the rate of the proportional response changes.

Precision The preciseness or exactness with which a measurement can be made. Precision is a function of the fineness of graduation of the measuring instrument.

Precooler A device which cools a gas before compression.

Pressure The force exerted per unit area on a body. The fundamental unit of pressure in the SI (metric) system is the Pascal.

Pressure, Absolute See **Absolute Pressure**.

Pressure, Atmospheric See **Atmospheric Pressure**.

Pressure, Back See **Back Pressure**.

Pressure, Breakloose The lowest pressure that will initiate movement.

Pressure, Cracking The lowest pressure at which a pressure operated valve begins to pass fluid.

Pressure Differential The difference in pressure between any two points in a fluid system. It is also called pressure drop.

Pressure Drop See **Pressure Differential**.

Pressure Element Lag The delay between the moment a change in pressure occurs in a measured medium and the moment it appears on an instrument. The lag is caused by viscosity, instrument friction, etc.

Pressure Gauge Any device used for measuring the pressure of a fluid. It may be a Bourdon gauge, mercury manometer, aneroid gauge, or one of several other types.

Pressure, Gauge See **Gauge Pressure**.

Pressure, Head The pressure exerted due to the weight of a column of fluid. It is normally expressed in terms of the height of the column in feet or inches.

Pressure, Operating The nominal or average pressure at which a system operates.

Pressure, Peak The maximum pressure occurring in the operation of a valve or other device.

Pressure, Pilot The pressure in a pilot circuit.

Pressure, Proof The nondestructive test pressure, in excess of the maximum rated operating pressure, which will cause no permanent deformation.

Pressure Regulator A valve used to automatically reduce and maintain pressure below that of a source. (See Fig. 57)

Pressure Reversal A reversal in direction of movement initiated by a rise in pressure.

Pressure Sealed Bonnet A valve bonnet which seals tighter as the pressure increases rather than being dependent on such things as threads and gaskets.

Pressure, Shock The pressure existing in a pressure pulse or wave.

Pressure, Static The pressure existing in a fluid at rest.

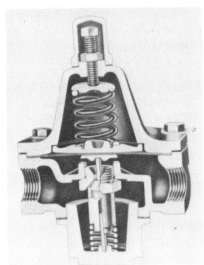

Fig. 57. Pressure Regulator. (*Courtesy of A. W. Cash Valve Mfg. Corporation.*)

Pressure, Suction The absolute pressure of a fluid at the inlet of a pump.

Pressure, Surge The pressure resulting from surge conditions.

Pressure Switch A switch that is operated by a change in the applied pressure.

Pressure, Vapor The pressure, at a given temperature, in which the liquid and gaseous phases coexist in equilibrium.

Pressure Vessel A container designed to hold fluid under pressure.

Primary Control Element The part of the controller which causes a motion or variation of a measuring element to actuate the controller system.

Primary Element The part of the measuring system which first transforms energy from the controlled medium to yield a change in value of the controlled variable.

Primary Feedback A signal which is a function of the controlled medium which is compared with the reference input signal to obtain an actuating signal.

Primary Sensitive Element A device which senses a change in the process medium and determines the size of the change. It is the sensing element, but not the indicating element.

Process A controlled system of events involving physical and/or chemical changes in a controlled medium.

Process Changes Changes in the conditions governing a process such as temperature, pressure, flow rate, etc.

Process Characteristics The physical characteristics such as pressure, temperature, viscosity, density, etc., on which automatic controls depend.

Process Lag The interval of time between movement of a control valve and a subsequent change of the controlled variable.

Process Time Lag The elapsed time between a change in the final control element and the first effect of that change on the measuring element.

Process Variable See **Variable**.

Program Control A control system that automatically holds or changes a target value on the basis of time to follow a previously specified program for the process.

Proportional A relationship of one variable to another for which there is a constant ratio.

Proportional Band See **Throttling Range**.

Proportional Control A controller function in which an output pressure control signal is provided which is proportional to the variable being measured.

Proportional Derivative Control A combination of derivative control action and proportional control action.

Proportional Integral Control A mixture of integral control action and proportional control action.

Proportional-Plus-Floating Action A combination of floating control action and proportional position action.

Proportional-Plus-Floating Controller Action A combination of speed floating action and proportional position action.

Proportional - Plus - Floating - Plus - Derivative Action A combination of derivative action, speed floating action and proportional position action.

Proportional-Plus-Reset Action A control action which will prevent excessive offset from the target value under large load changes or frequent load changes, and which will not allow rapid or drastic changes in the control action.

Proportional-Plus-Reset-Plus-Rate Action A combination of rate action, proportional speed floating action and proportional position action.

Proportional Position Action An action in which a continuous linear relationship exists between the value of the controlled variable and the position of a final control element.

Proportional Position Control See **Proportional Control**.

Proportional Response See **Proportional Control**.

Proportional Position Controller Action An action in which a continuous linear relationship exists between the position of the final control element and the value of the control variable.

Proportional Speed Floating Action An action in which a continuous linear relationship exists between the value of the controlled variable and the rate of motion of a final control element.

Proportional Speed Floating Controller Action An action in which a continuous linear relationship exists between the rate of motion of a final control element and the deviation of the controlled variable from a mean.

PSI The abbreviation for pounds per square inch, the unit of pressure in the British Engineering System.

Pulsometer A steam pump in which an automatic ball valve, the only moving part, admits steam alternately to a pair of chambers, forcing out water which has been sucked in by condensation of the steam after the previous stroke.

Pump Any device used to force a fluid through a circuit usually by increasing its pressure.

Pump, Axial Piston A pump which has multiple pistons arranged with their axes parallel.

Pump, Centrifugal A pump which imparts rotational velocity to a fluid and converts it to a pressure head.

54 VALVES AND VALVE TERMINOLOGY

Pump, Fixed Displacement A pump in which the displacement per cycle is fixed and cannot be varied.

Pump, Gear A pump having two or more intermeshed gears or similar rotating members enclosed in a housing.

Pump, Hand A manually operated pump.

Pump, Multiple Stage A pump having two or more pumping units in series.

Pump, Radial Piston A pump with multiple pistons arranged radially and actuated by an eccentric element.

Pump, Reciprocating Duplex A pump with two reciprocating pistons.

Pump, Reciprocating Single Piston A pump with a single reciprocating piston.

Pump, Screw A pump with one or more screws rotating in a housing.

Pump Section A pump having one or more rotating screws in a housing which impart energy to the fluid.

Pump, Vane A pump with multiple radial vanes within a supporting rotor.

Pump, Variable Displacement A pump in which the displacement per cycle is variable.

Pyrometer An instrument used for measuring very high temperatures beyond the range of ordinary thermometers.

Quick Disconnect Coupling A coupling usually containing a check valve which can be rapidly disconnected with no leakage. (See Fig. 58)

(a)

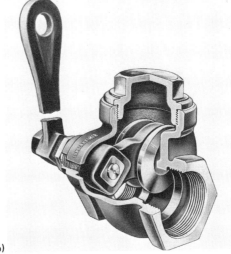

(b)

Fig. 59a,b. Quick Opening Valve. (59a *Courtesy of Canada Valve Ltd.*, 59b *Courtesy of Lunkenheimer Co.*)

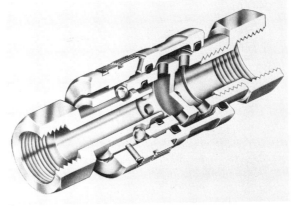

Fig. 58. Quick Disconnect Coupling. (*Courtesy of Chemetron Corporation, Fluid Controls Division.*)

Quick Disconnect Coupling, Break-A-Way A quick disconnect which automatically separates coupling halves when a specified axial force is applied.

Quick Disconnect Coupling, One Valve A quick disconnect which has a shutoff valve in only one side.

Quick Disconnect, Valved A quick disconnect with a shutoff valve in each side.

Quick Opening A characteristic of a fluid component in which there is a maximum flow with a minimum travel of the control element.

Quick Opening Flow Characteristic An inherent flow characteristic for which there is maximum flow with minimum travel.

Quick Opening Valve A gate valve which has a sliding stem, fulcrum and a lever which opens and closes very rapidly. (See Fig. 59)

Quiescent Current A direct current present in a servovalve coil when using a differential coil connection, the polarity of the current in the coils being in opposition such that no electrical control power exists.

Radiation Fin Bonnet A valve bonnet with fins for reducing heat transfer between the valve body and packing box assembly. (See Fig. 60)

Railroad Union See **Union**.

Rangeability The ratio of maximum to minimum flow within which all flow characteristics are maintained within prescribed limits.

Rate Action An action in which there is a continuous linear relationship between the rate of change of a controlled variable and the position of a final control element.

Rate of Flow The volume of fluid flowing past a given point per unit of time. Units in-

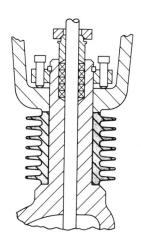

(a)

(b)

Fig. 60a,b. Radiation Fin Bonnet. (60a *Courtesy of High Pressure Equipment Co.*, 60b *Courtesy of ITT Hamel-Dahl-Conoflow.*)

56 VALVES AND VALVE TERMINOLOGY

clude gallons per minute, cubic feet per second, kilograms per second, etc.

Rate Response A controller response that is proportional to the rate of change of a controlled variable.

Rate Time-Derivative Time The unit of rate response measurement. The time interval by which a final control element actuated by rate response anticipates a subsequent position to a proportional response.

Rated Capacity The guaranteed rate of flow through a valve under specified conditions.

Rated C_v The C_v of a valve at its rated full-open position.

Rated Travel The linear movement of a valve plug from the closed position to the rated full-open position.

Ratio Control A control based on a proportion or ratio of two variables.

Reaction Lag A process delay caused by the time necessary to complete a reaction before the result of that reaction can be measured.

Recovery The decrease of aftereffect in a diaphragm with time under zero load.

Reducer A fitting or connector having a smaller line size at one end than at the other end.

Reducing Valve A valve used to reduce the pressure in a system. (See Fig. 61)

Regulator Capacity See **Rated Capacity**.

Regulator Flow Coefficient See **Flow Coefficient**.

Regulator, Valve A valve used to regulate pressure output to a constant value. (See Fig. 62)

Relative Humidity The amount of water vapor in the air as compared to the amount the air could hold at a given temperature. It is expressed as a percentage.

Relay Operated Controller A controller in which the force or motion developed by the measuring means is used to operate an amplifying relay whose output operates the final control element either through additional relays or directly.

Relay Operated Controlling Means The method of controlling wherein the energy transmitted from the measuring means is either

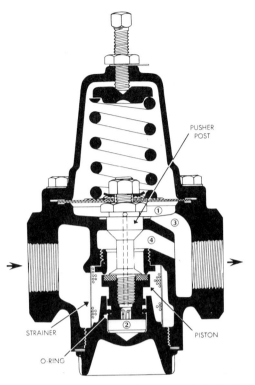

Fig. 61. Reducing Valve. *(Courtesy of A. W. Cash Valve Mfg. Corporation.)*

Fig. 62. Regulator Valve. *(Courtesy of A. W. Cash Valve Mfg. Corporation.)*

supplemented or amplified for operating the final control element.

Relay Operated Measuring Means A method of measuring in which the energy transmitted through a primary element is amplified or supplemented for actuating an automatic controller.

Relay Output The part of a relay supply that is transmitted to a power unit or to another relay.

Relay Supply The energy which is supplied to a relay.

Relief Valve A self-operated, fast acting valve which is used to bleed off excessive pressure in a fluid system. (See Fig. 63)

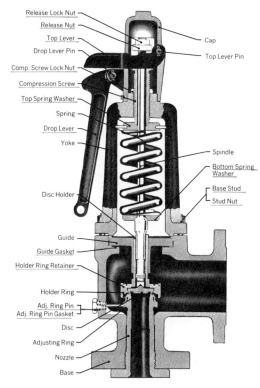

Fig. 63. Relief Valve. (*Courtesy of Dresser Industrial Valve & Instrument Division.*)

Remote Control A system used to control remotely located valves and other devices.

Repack To change the packing in a valve or other component.

Reproducibility The precision with which the measurement of a given quantity can be duplicated.

Reservoir A container used to store the liquid in a fluid power system.

Reservoir, Atmospheric A reservoir which is used to store a fluid at atmospheric conditions.

Reservoir, Sealed A reservoir which is used for storing a fluid isolated from atmospheric conditions.

Reservoir, Sealed, Pressure A sealed reservoir used for storing of fluid under pressure.

Reset Action A type of control response, if automatic reset, giving a rate of value movement proportional to the deviation of the variables. See also **Proportional-Plus-Reset Action**.

Reset Rate The number of connections made per minute by the control system. It is usually expressed as the number of repeats per minute.

Reset Response A response in which the output rate of the controller is proportional to the deviation of the controlled variable.

Reset Time The time for a reset system to again reach equilibrium after an upset.

Residual Unbalance The open circuit output of the pickup at the calibration excitation specified with zero pressure applied and with no external electrical balancing circuitry. It is expressed as a percentage of full range sensitivity.

Resilience The ability of a seal to return to its original shape after deformation.

Resistance An opposition to the flow of a fluid through a device or system. A measure of the effect of friction.

Restrictor A device used to produce a deliberate pressure drop or a resistance in a line by reducing the cross sectional flow area.

Restrictor, Choke A device used to reduce the cross sectional flow area in a fluid component or system.

Restrictor, Orifice A restrictor whose length is small compared to its cross sectional area. It may be a fixed orifice or a variable orifice, which may be compensated by pressure and/or temperature changes.

Restrictor Valve A valve which allows only a prescribed flow or pressure drop in a given direction.

Return Bend A fitting used for reversing the direction of a pipe run.

Reverse Acting Controller An air-operated controller in which the output pressure decreases as the controlled medium increases.

Reverse Acting Valve A valve that is normally closed and which requires an increase of fluid pressure or other mechanical means to open. (See Fig. 64)

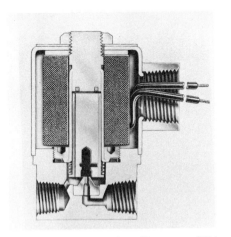

Fig. 64. Reverse Acting Valve. (*Courtesy of Skinner Precision Industries, Inc.*)

Reverse Actuator The actuator in a reverse acting valve.

Reverse Diaphragm Motor See **Reverse Actuator**.

Reyn The standard unit of absolute viscosity in the British system of units, expressed in pound seconds per square foot.

Reynolds Number The ratio of dynamic force due to mass flow to the shear stress due to viscosity.

$$R = \frac{vD\rho}{\mu}$$

where

ρ = the mass density
v = the flow velocity
D = the pipe diameter
μ = the absolute viscosity

Ring, O See **O-Ring**.

Ring, Piston A ring used to seal the space between a cylinder wall and a piston.

Ring, Scraper A ring used to remove material from a surface by means of a scraping action.

Ring, U A ring with a U-shaped cross section.

Ring, V A ring with a V-shaped cross section.

Ring, W A ring with a W-shaped cross section.

Ring, Wiper A ring used to remove material from a surface by a wiping action.

Ring Joint Assembly An assembly in which a metal ring of octagonal or oval cross section fits into machined grooves in mating flange faces. It is generally used in high pressure high temperature service.

Ripple A periodic variation of pressure above and below the operating pressure.

Rising Stem A valve stem that turns and rises when the valve is open. (See Fig. 3)

RMS Root mean square.

Rubber Valve, Check A check valve having a rubber poppet closure member. (See Fig. 65)

Rupture Disc A disc used to relieve pressure as a relief or safety valve. (See Fig. 66)

SAE Society of Automotive Engineers.

Sabot A fixture used for holding an object in a particular attitude.

Safety Valve A self-actuated, fast opening valve used for quick relief of excessive pressure. (See Fig. 67)

  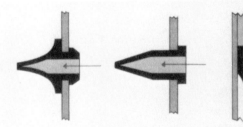

Fig. 65. Rubber Check Valve. (*Courtesy of Vernay Labs, Inc.*)

TERMINOLOGY OF THE VALVE INDUSTRY

Fig. 66. Rupture Disc. (Courtesy of Continental Disc Corp.)

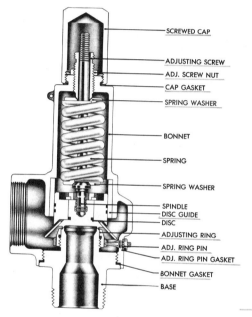

Fig. 67. Safety Valve. (Courtesy of Dresser Industrial Valve & Instrument Division.)

Scale Error The difference between the actual and indicated values of a variable.

Scanner An instrument that automatically checks a number of measuring points and notes which ones have wandered from their specified values.

SCFH Standard cubic feet per hour.

SCFM Standard cubic feet per minute.

Schedule Number A number assigned to a pipe denoting its wall thickness per ASA B36.10-1950.

Scoring Scratches in the direction of motion of mechanical valve parts caused by abrasive contaminants in fluids.

Screwed Bonnet A bonnet which screws into the body of the valve.

Screwed End A type of end fitting on a valve or other fluid component which is joined by threaded connections.

Screwed Flange A flange which attaches to a pipe by means of a threaded connection.

Screwed-In Bonnet See **Screwed Bonnet**.

Screwed Stuffing Box See **Packing Box Assembly**.

Seal, Diaphragm A thin flat sealing device which is clamped and sealed along its periphery with the inner part free to move.

Seal, Dynamic A seal used between parts in relative motion.

Seal, Mechanical A seal obtained by mechanical force.

Seal, Oil A sealing device used to contain oil, usually in a bearing assembly.

Seal, Pressure A seal obtained by means of fluid pressure.

Seal, Sliding A sealing element such as an O-ring used to prevent leakage past a moving member such as a piston. The O-ring is located in a groove in the piston and moves with it, sliding along the cylinder wall.

Seal, Water A seal which uses water as a barrier.

Seat The part of a valve against which the closure element presses to effect a seal.

Seat Bushing See **Seat Ring**.

Seat Insert See **Seat Ring**.

Seat Ring A ring inserted in the valve body to form a valve body port. (See Fig. 4)

Seating Action A type of valve action in which the flow is stopped by a seated obstruction in the flow path.

Seating Action, Ball A type of valve seating action in which a spherical ball is used to obstruct the flow path. (See Fig. 1)

Seating Action, Diaphragm A type of valve seating action in which a diaphragm is used to obstruct the flow path. (See Fig. 5)

Seating Action, Disc A type of valve seating action in which a disc is used to obstruct the flow path. (See Fig. 4)

Seating Action, Disc, Swing Check A type of valve seating action in which a hinged disc is used to obstruct the flow path. (See Fig. 8)

Seating Action, Gate A type of valve seating action in which a wedge is used to obstruct the flow path. (See Fig. 3)

Seating Action, Gate, Spreader A type of valve seating action in which two discs are seated by spreaders to obstruct the flow path.

Seating Action, Gate, Wedge A type of valve seating action in which a solid wedge shaped gate is used to obstruct the flow path. (See Fig. 3)

Seating Action, Globe See **Seating Action, Poppet.**

Seating Action, Needle A type of valve seating action in which a sharply tapered plug or needle is used to obstruct the flow path. (See Fig. 49)

Seating Action, Plug A type of valve seating action in which a plug is used to obstruct the flow path.

Seating Action, Poppet A type of valve seating action in which a poppet is used to control fluid flow. (See Fig. 6)

Second Derivative Control A control response in which the rate of final control element movement is proportional to the acceleration of the deviation from a mean.

Secondary Flow Rotating motion at right angles to pipe axis superimposed on main flow. Caused from friction and centrifugal force when fluid flows around a bend.

Self-Acting Controller A type of controller which employs the energy of the measuring system without amplification to effect necessary corrective action.

Self-Actuated Controller A controller in which all the energy needed to operate the final control element is supplied by the measuring element.

Self-Operating Controller See **Self-Acting Controller.**

Self-Operated Controlling Means A system in which the energy needed to operate the final control element is supplied by the measuring means.

Self-Operated Measuring Means A system in which the energy needed to actuate the controlling means of an automatic controller is supplied by the controlled medium through the primary element.

Semi-Cone Plug Disc A tapered plug used for fine noncharacteristic flow regulation.

Sensible Heat The heat content of a liquid in Btu's measured from a zero heat reference point of 32°F.

Sensitivity The measure of the response of an instrument or control unit to a change in the incoming signal.

Separator A device used to separate out foreign matter held in suspension in a flowing stream.

Servomechanism A device which is activated by electrical or mechanical impulses to automatically operate an instrument or machine.

Servomotor An amplifying device acting under the control of the measuring element operated by an auxiliary power source, used in control systems to position the final control element.

Servo Techniques Techniques or methods devised to study the performance of servomechanisms or control systems.

Servovalve A valve which modulates the output as a function of an input signal. (See Fig. 68)

Fig. 68. Servovalve. (*Courtesy of Hydraulic Research & Mfg. Co.*)

Servovalve, Electrohydraulic A servovalve which controls hydraulic output as a function of an electrical input signal.

Servovalve, Electrohydraulic, Flow Control An electrohydraulic servovalve whose function is the control of output flow.

Servovalve, Four-Way A multi-orifice flow control valve with supply, return and two control ports arranged so that the valve action in one direction opens the supply to control port #1 and opens control port #2 to the return. Reversed valve action opens the supply port to the control port #2 and opens control port #1 to return.

Servovalve, Three-Way A multi-orifice flow control valve with supply, return and one control port arranged so that valve action in one direction opens the supply to the control port and reversed valve action opens the control port to return.

Servovalve, Two-Way A single orifice flow control valve with a supply port and one control port arranged so that action is in only one direction, from the supply to the control port.

Set Point The target value which an automatic control device attempts to reach or to hold.

Set Value The value of a regulated variable at minimum controllable flow.

Shear Action A type of valve action in which the flow is controlled by an element which slides across the flow path. (See Fig. 3)

Shear Action, Ball A type of valve action in which the flow is controlled by a ported ball which rotates on an axis perpendicular to the flow path.

Shear Action, Plug A type of valve action in which the flow is controlled by a ported plug which rotates on an axis perpendicular to the flow path. (See Fig. 7)

Shear Action, Plunger See **Shear Action, Spool** and **Spool Valve**.

Shear Action, Sliding Plate A type of valve action in which the flow is controlled by a plate that slides across the flow path.

Shear Action, Sliding Plate, Linear A sliding plate shear action in which the movement of the plate is linear.

Shear Action, Sliding Plate, Rotary A sliding plate shear action in which the plate movement is rotary.

Shear Action, Spool A type of valve action in which the flow is controlled by a spool sliding through the flow path.

Shock Wave A pressure wave which travels at a supersonic velocity.

Side Outlet A tee or ell fitting which has an outlet on the side.

Signal Information that is transmitted from one point in a system to another.

Silencer See **Muffler**.

Simmer The difference between cracking pressure and the pressure where leakage of a gas begins. See also **Dribble Range**.

Single Speed Floating Action A control action in which the final control element is moved at a single fixed rate.

Single Speed Floating Control A control in which the dependent variable changes at a constant rate in one direction when the deviation is positive and in the opposite direction when the deviation is negative.

Single Speed Floating Controller Action A controller action in which the final control element has a single rate of motion.

Single Wedge Gate A type of valve design in which a wedge shaped plug is inserted between angled seats to obstruct the flow path. (See Fig. 3)

Sleeve See **Bushing**.

Sliding Gate Valve A type of valve in which a sliding disc closes off the orifices in a stationary plate to control the flow. (See Fig. 69)

Fig. 69. Sliding Gate Valve. (*Courtesy of Jordan Valve Co.*)

62 VALVES AND VALVE TERMINOLOGY

Fig. 70. Slurry Valve. (*Courtesy of De Zurik, A Unit of General Signal.*)

Slip on Flange A flange that slips onto a pipe and is welded in place.

Slurry Valve A valve with a knife edged disc which scrapes off solids preventing tight sealing from the valve seat. (See Fig. 70)

Snubber Valve A unit used to isolate shock or dampen surge pressures. (See Fig. 71)

Socket End A type of end fitting used on valves for mating with pipe which is then welded into place.

Socket Welded A valve or other fitting which slips over the end of a pipe and is made pressure tight by welding.

Solder Ends A type of valve end into which pipe is soldered or silver soldered. This type end is commonly used in the plumbing field. See **Brazing Ends**.

Solenoid Operated Valve A valve which is operated by means of an electrical solenoid. (See Fig. 72)

Solid Wedge Disc A wedge of single piece construction.

S.P. Steam pressure.

Spacer See **Lantern Ring**.

Specific Gravity The ratio of the density of a given liquid to the density of water at 4°C. In the British system the comparison is made to water at 60°F.

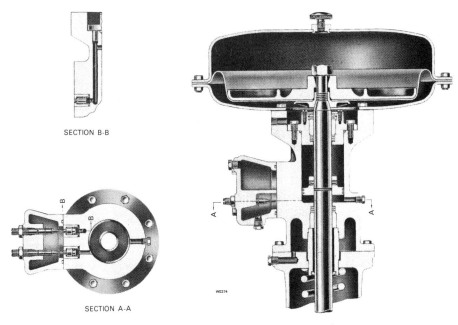

Fig. 71. Snubber Valve. (*Courtesy of Fisher Controls Company.*)

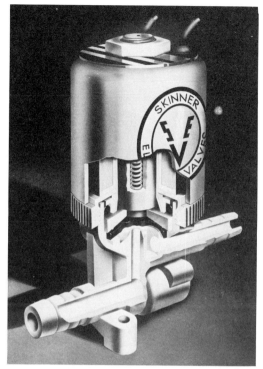

Fig. 72. Solenoid Operated Valve. (*Courtesy of Skinner Precision Industries, Inc.*)

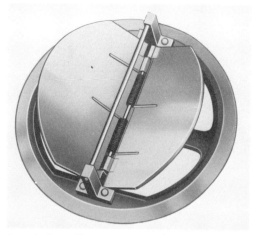

Fig. 73. Split Flapper Valve. (*Courtesy of Essex Cryogenics Industries, Inc.*)

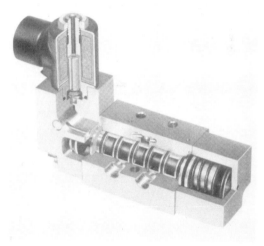

Fig. 74. Spool Valve. (*Courtesy of Automatic Valve Corporation.*)

Specific Heat The amount of heat required to raise the temperature of a unit mass of a substance one degree under specified conditions.

Specific Heat Ratio The ratio of specific heat at constant pressure to specific heat at constant volume. It is usually designated by the lower case Greek letter gamma and has the approximate value 1.4 for air.

Specific Volume The volume per unit mass of a substance, usually the volume per mole.

Specific Weight The weight per unit volume of a substance.

Split Flapper A two-piece flapper disc hinged along a diameter. (See Fig. 73)

Spool An element in a valve used for changing the flow from one port to another.

Spool Valve A spool valve is generally used in servo applications. The name is derived by the shape of the metering or moving element. The spool valve may have two, three, four or more lands. (See Fig. 74)

Spring Any device which makes use of elastic restoring forces to return to its original shape after being deformed. Generally it refers to helical compression springs.

Spring Adjustor A device used to adjust the compression or torsion of a valve spring.

Spring Adjusting Button See **Spring Seat**.

Spring Button See **Spring Seat**.

Spring Case The housing which holds the spring in a valve.

Spring Guide A bearing surface, usually ring shaped, used to guide spring movement.

Spring Holder A device used to align and hold a spring in position.

Spring Nut See **Spring Seat**.

64 VALVES AND VALVE TERMINOLOGY

Spring Range The pressure range over which the spring is operable.

Spring Rate The ratio of the force applied on a spring to the deflection of the spring from its equilibrium position. The units of spring rate are Newtons per meter (metric) or pounds per inch (British).

Spring Retainer See **Spring Seat**.

Spring Seat The device to which a spring is affixed or upon which it sits. The seat may be fixed or it may be adjustable to provide for change in the spring characteristics.

Spring Stem See **Actuator Stem**.

Springless Diaphragm Control Valve A type of valve having a diaphragm actuator that does not utilize a spring for loading. (See Fig. 75)

Fig. 75A. Sprinkler Valve. (*Courtesy of The Reliable Automatic Sprinkler Co.*)

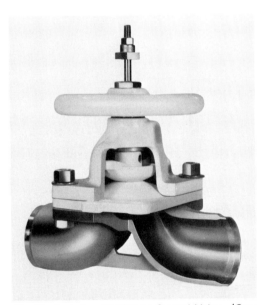

Fig. 75. Springless Diaphragm Control Valve. (*Courtesy of Hills-McCanna Company.*)

Sprinkler Valve A valve normally used for fire protection. It is a valve that opens when heat dissolves a leaded or plasticized substance. (See Fig. 75A)

Squib An explosive device used to initiate an action.

Stable Control Control in which the value of a controlled variable is held within desired limits without sustained oscillation.

Stability The tendency of a material to resist permanent changes in its physical or chemical properties.

Standard Air See **Air, Standard**.

Static Behavior The behavior of a control system under fixed stationary conditions.

Static Error The error that occurs when the set point is changed from one value to another and the controlled variable does not follow exactly.

Static Unbalance The net force exerted on a valve plug in its closed position by the fluid pressure acting on it.

Steam Bronze A type of bronze alloy used in valves and fittings.

Steam, Dry Steam with no water particles mixed in.

Steam, Saturated Steam in contact with liquid water at the boiling point.

Steam Trap A device which automatically allows the discharge of condensation from steam lines without allowing the escape of steam.

Steam, Wet Steam containing particles of water.

Stem Connector A device used to connect the stem of a valve to the spring in the operator.

Stem Guided A type of valve design in which the stem is guided mechanically to assure plug and seat alignment.

Stem Lift The travel of a valve stem upon valve actuation.

Step Change The change from one value of a variable to another in a single step.

Stoke The unit of kinematic viscosity in the cgs metric system. One stoke is one centimeter squared per second.

Stop Check Valve A valve which will automatically close when flow reverses and which can be screwed down into a closed position. (See Fig. 76)

Strain Gauge A transducer which converts a force into an electrical signal.

Street El A pipe elbow which has male threads on one end and female threads on the other.

Stress The internal force exerted by either of two adjacent parts of a body upon the other across an imaginary plane of separation.

Stroke See **Rated Travel**.

Stuffing Box Body See **Bonnet Assembly**.

Stuffing Box Flange See **Packing Flange**.

Stuffing Box Nut See **Packing Nut**.

Stuffing Box Spacer See **Lantern Ring**.

Sump A reservoir serving as a drain for fluids in a system.

Superheated Steam Steam at any pressure which is heated to a temperature above the steam temperature at that pressure.

Superstructure The part of a valve located above the body. It may include the stem, handle, bonnet and other parts.

Supervisory Control A control system which will furnish information to a central location to be utilized by an operator in supervising the control of a process or operation.

Supply Side of Process The part of a process which supplies the material or energy to control the process.

Surface Tension The force within the surface of a contained liquid due to the molecules in the surface being in a higher energy state.

Surge The momentary rise in pressure in a fluid circuit.

Surge Suppressor A device used to filter out pump ripple in a piping system.

Swell The increased volume of a seal caused by immersion in a fluid.

Swing Check Valve A type of valve in which a swinging disc is used to control the flow. The disc opens with increased pressure and closes automatically with decreased pressure. See also **Flap Valve**. (See Fig. 77)

Systems Engineering An engineering method which takes into account all of the elements in a control system and also the process itself.

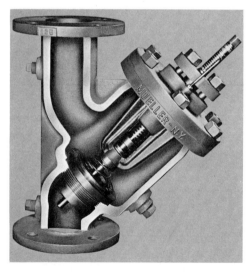

Fig. 76. Stop Check Valve. (*Courtesy of Mueller Steam Specialty Division of SOS Consolidated.*)

Straight Line See **Linear**.

Straight Proportional See **Linear**.

Strainer A device through which a fluid is passed to remove insoluble materials. See also **Filter**.

66 VALVES AND VALVE TERMINOLOGY

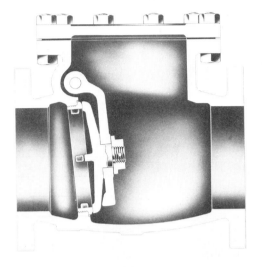

(a)

NAME OF PARTS
1. Bonnet Stud
2. Bonnet Stud Nut
3. Bonnet Cap
4. Bonnet Gasket
5. Hinge Pin
6. Hinge
7. Disc Nut
8. Disc Nut Pin
9. Disc Washer
10. Disc
11. Seat Ring
12. Body
13. Pipe Plug

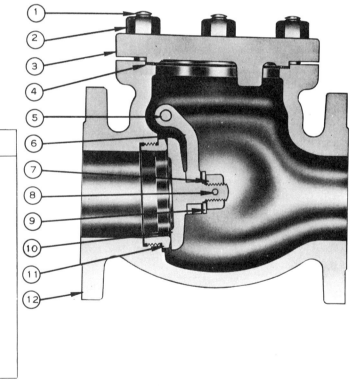

(b)

Fig. 77a,b. Swing Check Valve. (77a *Courtesy of Jenkins Bros.*, 77b *Courtesy of Pacific Valve Co.*)

Tee A three-port fitting used to join one pipe at right angles to two other pipes.

Temperature Pilot Actuator An actuator which translates temperature changes into movement of a pilot stem.

Temperature Regulator A self-operated valve which operates by changes in temperature of the process medium.

Thermal Actuator A device which converts heat energy into mechanical energy.

Thermal Coefficient of Sensitivity The fractional or percentage change in sensitivity of a measuring device with temperature.

Thermally Actuated Valve A valve which utilizes the thermal expansion of metal as an actuating mechanism. (See Fig. 78)

Thermistor A material, generally a semiconductor, the resistance of which is temperature dependent and yields a relatively large electrical signal from a small temperature change.

Thermocouple A temperature sensitive device consisting of two dissimilar metals, such as iron and constantan, between which an EMF is established by the difference in temperature between their junction and a reference junction kept at a zero point (such as the ice point).

Three-Position Control See **Multiposition Control**.

Three-Way Valve A multiple orifice valve with its supply return and control ports arranged such that the valve stem action in one direction opens the supply-to-control port and reversed valve action opens the control port to return. (See Fig. 79)

Throttling The regulation of flow through a valve or other device.

Throttling Control A control which is able to position its final control element in any position between the maximum and minimum limits.

Throttling Process An isenthalpic process in which a fluid flows from a region of constant high pressure through a restriction to a region of constant lower pressure such that no heat can flow into or out of the system.

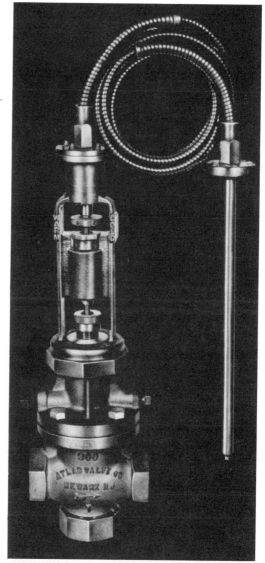

Fig. 78. Thermally Actuated Valve. (*Courtesy of Atlas Valve Co.*)

Fig. 79. Three-Way Valve. (*Courtesy of Manatrol Division, Parker Hannifin Corp.*)

68 VALVES AND VALVE TERMINOLOGY

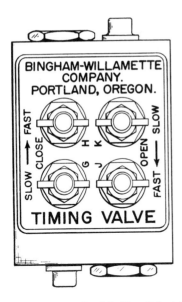

Fig. 80. Time Schedule Controller. (*Courtesy of Bingham-Willamette Company*.)

Throttling Range A fraction of the total range of a variable which will cause a control valve to move from one extreme to the other.

Tight Closing See **Dead End Shutoff**.

Tight Shutoff See **Dead End Shutoff**.

Time Schedule Controller A controller used to regulate batch processes according to some predetermined time schedule. (See Fig. 80)

Tongue-And-Groove Joint A type of joint in which a tongue on a valve bonnet fits into a groove on the body to assure alignment. It is normally used on high pressure, high temperature installations.

Top-And-Bottom Guided A type of valve design in which the plug is aligned by guides in the valve body or in the bonnet and bottom flange. (See Fig. 80A)

Top Bonnet See **Bonnet Assembly**.

Top Diaphragm Case See **Diaphragm Case**.

Top Guided A type of valve design in which the plug is aligned by a single guide in the body adjacent to or in the bonnet.

Topworks See **Superstructure**.

Torricelli's Theorem A statement of the law of conservation of energy relating the velocity of discharge of a liquid column to the pressure head.

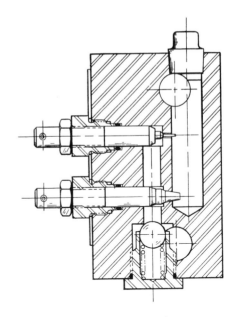

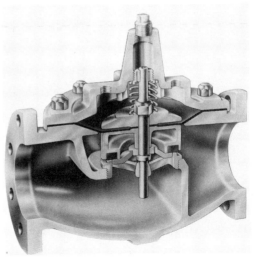

Fig. 80A. Top and Bottom Guided Control Valve. (*Courtesy of Cla-Val Company*.)

$$v = K \sqrt{h}$$

where

v = velocity
h = pressure head
K = proportionality constant

Total Heat The sum of sensible heat plus latent heat.

Transducer An element used to convert one form of energy into another, e.g., pressure into electrical voltage.

Transfer Function A mathematical expression relating the output of a process to the input.

Transient State A short-lived change in the value of a controlled variable.

Trap See **Steam Trap**.

Travel Indicator An indicator used to show the amount of opening or closing of a valve. (See Fig. 30)

Travel Indicator Scale A scale fastened to a valve and graduated to indicate the position of the valve plug.

Travel Scale See **Travel Indicator Scale**.

Trim See **Valve Trim**.

Trip Device A mechanical element used for the actuation of a position control.

Trunnion A pair of short journals supported in bearings projecting coaxially from opposite sides of a cylinder required to pivot about their axis.

Tubing Small diameter, thin-walled pipe, usually used for moderately low pressure work.

Two-Position Action An action in which a final control element is moved from one of two positions to the other. It is also called on-off action.

Two-Position Control A type of control response in which the final control element can be positioned in one of two positions.

Two-Position Differential Gap Action An action in which a final control element is moved from one of two fixed positions to the other when a controlled variable reaches a predetermined value from one direction. Subsequently, the element is moved to the other position only after the variable has passed in the opposite direction through a range of values to the second predetermined value.

Two-Position Differential Gap Controller Action A position action in which a final control element moves in one direction at a predetermined value of the controlled variable, and in the other direction only after the value of the variable has crossed a differential gap to a second specified value.

Two-Position Power Actuated Valve A valve which has only two positions and which is positioned in one mode or the other by means of a power actuator.

Two-Position Single Point Action An action in which the final control element is moved from one of two fixed positions to the other at a given value of the controlled variable.

Two-Way Valve A valve which has one inlet port and one outlet port. (See Fig. 81)

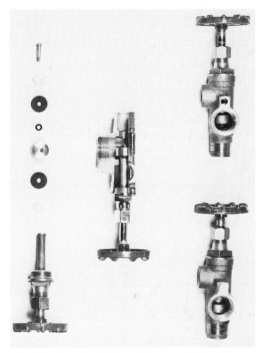

Fig. 81. Two-Way Valve. (*Courtesy of Clack Corp.*)

Ullage The volume of fluid which a vessel lacks of being full.

Ultrasonic Vibration with a frequency higher than that normally audible to the human ear.

Union, Flange A device used for joining two sections of pipe by bolting them together.

Union, Screw A device having threaded ports and a coupling ring used for joining two sections of pipe.

Unloading The release of contaminant that was initially captured by a filter medium.

Unloading Valve A normally closed two-way valve used for unloading excess flow at very low pressure drop to the tank. (See Fig. 82)

70 VALVES AND VALVE TERMINOLOGY

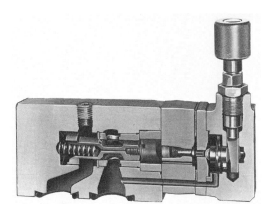

Fig. 82. Unloading Valve. (*Courtesy of Double A Products Co.*)

Vacuum A region in which the air pressure is less than atmospheric pressure.

Vacuum Breaker A valve used to vent a vacuum tank back to atmospheric pressure to relieve stresses in the tank. (See Fig. 83)

Valve Any device which controls the direction or magnitude of fluid flow or its pressure.

Valve, Air A valve used for controlling the flow of air. (See Fig. 84)

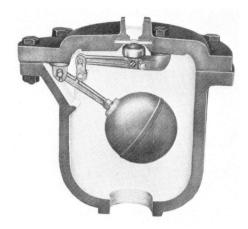

Fig. 84. Air Valve. (*Courtesy of Multiplex Mfg. Co., "Crispin" Valve Div.*)

Valve Body The main part of a valve. It contains the passages for the flow medium, seating surfaces and inlet and outlet fittings.

Valve Body Assembly The valve body, port fittings, seat and bolts for bonnet and/or end fittings.

Valve, Butterfly See **Butterfly Valve**.

Valve, Cam Operated A valve in which the spool is positioned by a cam. (See Fig. 85)

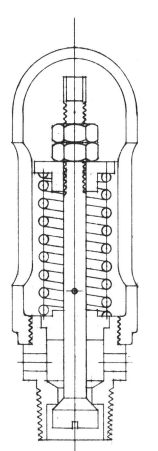

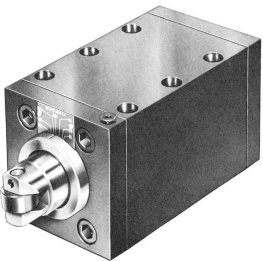

Fig. 83. Vacuum Breaker. (*Courtesy of Gadren Machine Co., Inc.*)

Fig. 85. Cam Operated Valve. (*Courtesy of Manatrol Division, Parker Hannifin Corp.*)

Valve, Cavitating Venturi A valve in which flow is controlled by allowing the fluid to cavitate thereby keeping the throat pressure constant at the vapor pressure independent of the downstream pressure.

Valve, Center Pressure A valve which in the center position connects the supply to the working ports.

Valve, Check See **Check Valve.**

Valve Closure Member The part of a valve such as the plug, disc, etc., used to control the amount of flow.

Valve, Control, Pressure Compensated A control valve that will pass a constant flow independent of change in upstream pressure. (See Fig. 86)

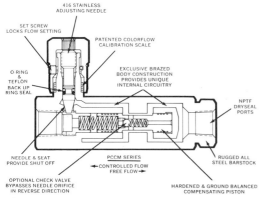

Fig. 86. Pressure Compensated Control Valve. (*Courtesy of Manatrol Division, Parker Hannifin Corp.*)

Valve, Counterbalance A valve which permits free flow in one direction, but provides resistance to the flow in the other. (See Fig. 87)

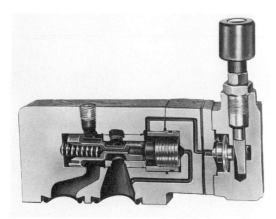

Fig. 87. Counterbalance Valve. (*Courtesy of Double A Products Co.*)

Valve, Directional Control A valve used to direct or prevent fluid flow through specific paths. (See Fig. 88)

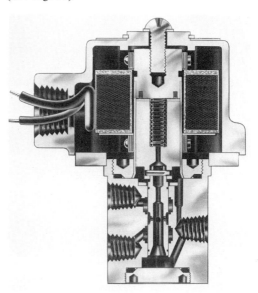

Fig. 88. Directional Control Valve. (*Courtesy of Skinner Precision Industries, Inc.*)

Valve, Directional Control, Check A directional control valve which permits the flow of fluid in only one direction.

Valve, Directional Control, Four-Way A directional control valve whose main function is to pressurize and exhaust two working ports. (See Fig. 89)

Valve, Directional Control, Selector A directional control valve used mainly to selectively interconnect two or more ports. (See Fig. 90)

Valve, Direction Control, Straight-Way A directional control valve with two in-line ports.

Valve, Directional Control, Three-Way A directional control valve used to alternately pressurize and exhaust a working port.

Valve Disc A disc shaped valve closure member.

Valve Disc Guide A guide used to align the valve disc and seat.

Valve Disc Stem The part of the stem connected to the disc.

Valve, Flow Control A valve used to control the rate of fluid flow.

Valve, Flow Control, Deceleration A flow control valve used to gradually reduce a flow rate.

72 VALVES AND VALVE TERMINOLOGY

Fig. 89. Four-Way Directional Control Valve. (*Courtesy of Skinner Precision Industries, Inc.*)

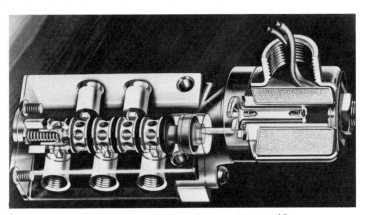

Fig. 90. Directional Control Selector Valve. (*Courtesy of Skinner Precision Industries, Inc.*)

Valve, Flow Control, Pressure Compensated A flow control valve used to control the flow rate independently of the system pressure. (See Fig. 91)

Valve, Flow Control, Pressure Temperature Compensated A flow control valve used to control flow rate independently of both system pressure and temperature.

Valve, Flow Dividing A valve used to divide the flow from a single source into two or more branches. (See Fig. 92)

Valve, Flow Dividing, Pressure Compensating A flow dividing valve that divides the flow into a constant ratio regardless of the changes in pressure in the branches.

Valve, Flow Metering See **Valve, Flow Control**.

Valve, Four-Way A valve having four passages, usually one inlet and three outlet ports.

Valve, Gate See **Gate Valve**.

Valve, Globe See **Globe Valve**.

Valve Guide See **Guide Bushing**.

Valve, Hydraulic A valve used for controlling the flow of a liquid.

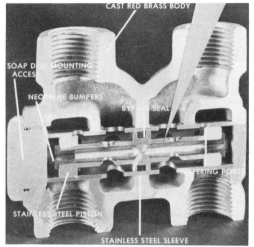

Fig. 91. Pressure Compensated Flow Control Valve. (*Courtesy of Precision Plumbing Products, Inc.*)

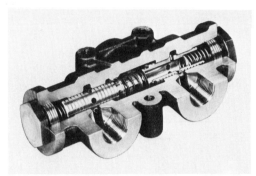

Fig. 92. Valve, Flow Dividing. (*Courtesy of Waterman Hydraulics.*)

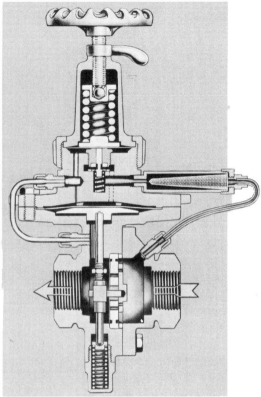

Fig. 92A. Pilot-Operated Valve. (*Courtesy of Jordan Valve Co.*)

Valve, Needle See **Needle Point Valve**.

Valve, Normally Closed A valve which normally does not allow flow and which must be actuated to allow flow.

Valve, Normally Open A valve which normally allows flow and must be actuated to close.

Valve Operator See **Actuator**.

Valve Operator, Manual A valve operator consisting of a hand operated mechanical device such as a lever, button, etc.

Valve Operator, Mechanical A valve operator consisting of mechanical elements such as a lever, gear, etc.

Valve, Pilot A valve used to operate another valve.

Valve, Pilot Check A check valve having a piston to unseat the check poppet when pilot pressure is applied.

Valve, Pilot-Operated A valve which is actuated by pilot fluid pressure. (See Fig. 92A)

Valve Plug The part of a valve which moves to restrict the area through which the fluid travels.

Valve Plug Guide A guide located in the valve bonnet or body used to align the plug with the seat to assure positive seating.

Valve Plug Stem The part of the stem connected to the valve plug.

Valve Plunger Operator See **Diaphragm Actuator**.

Valve, Pneumatic A valve used to control the flow of air or other gases. See **Valve, Air**.

Valve, Poppet See **Poppet Valve**.

Valve Port An opening between valve passages which can be varied continuously, or open and closed. The inlet and outlet openings of a valve are often called ports.

Valve, Power Control A valve which controls fluid power devices.

74 VALVES AND VALVE TERMINOLOGY

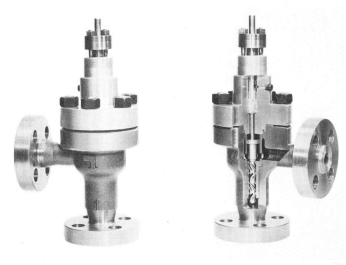

Fig. 93. Pressure Control Valve. (*Courtesy of Yarway Corporation.*)

Valve, Prefill A valve that allows full flow from a tank to a cylinder during the advance part of a cycle, permits the operating pressure to be applied to the cylinder during the pressing portion of the cycle, and allows free flow from the cylinder during the return part of the cycle.

Valve, Pressure Control A valve used to control the pressure of a fluid system. (See Fig. 93)

Valve, Pressure Control, Relief Safety A pressure control valve used to limit pressure in a system after a malfunction.

Valve, Pressure Control, Unloading A pressure control valve used to allow a compressor to operate at minimum load.

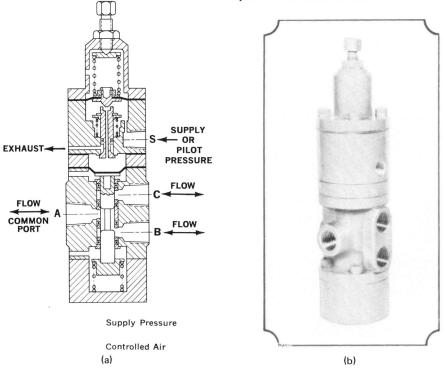

Fig. 94a,b. Relay Valve. (*Courtesy of ITT Hamel-Dahl-Conoflow.*)

(a)

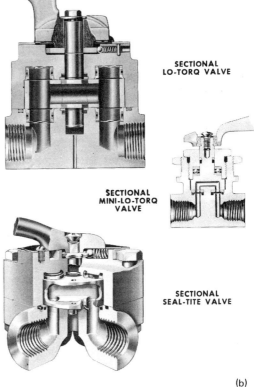

(b)

Fig. 95a,b. Rotary Selector Valve. (*Courtesy of Republic Mfg. Co.*)

Valve, Pressure Reducing A valve used to limit the outlet pressure from a source. See **Pressure Regulator**.

Valve, Pressure Relief A valve whose function is to limit system pressure to some preset value.

Valve, Relay A logic device which receives control signals and changes flow conditions in controlled flow passages. (See Fig. 94)

Valve, Rotary Selector A valve which utilizes rotary actuation to connect the inlet to an outlet port. (See Fig. 95)

Valve, Safety See **Safety Valve**.

Valve Seat Ring A ring inserted in the valve body to form a port.

Valve, Sequence A valve whose function is to direct flow in a predetermined sequence.

Valve, Shutoff A valve designed to operate fully open or fully closed, but not at an intermediate position. (See Fig. 96)

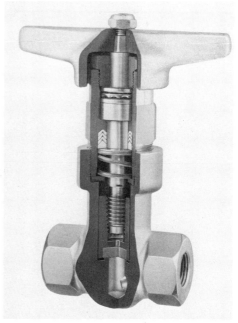

Fig. 96. Shutoff Valve. (*Courtesy of Rego, Division of Golconda Corp.*)

76 VALVES AND VALVE TERMINOLOGY

Valve, Shuttle A valve in which flow or pressure differences between two or more circuits causes a shuttle to move and select one of the circuits. (See Fig. 97)

Valve, Spool A valve in which the closure member is a spool with machined recesses or undercuts which moves in a bore with annular undercuts. Movement of the spool connects ports uncovered by the spool undercuts.

Valve, Standard Action A valve that is positioned by mechanical, manual or pilot means without springs or detents.

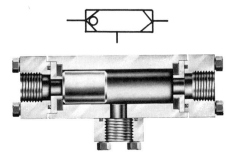

1-INCH OR 1½-INCH SHUTTLE VALVE
Fig. 97. Shuttle Valve. (*Courtesy of Cameron Iron Works.*)

1. INLET AND BODY
2. TWO-PIECE SEAT
3. BONNET
4. STEM
5. GUIDE
6. GUIDE
7. BODY
8. VENTURI SEAT OUTLET

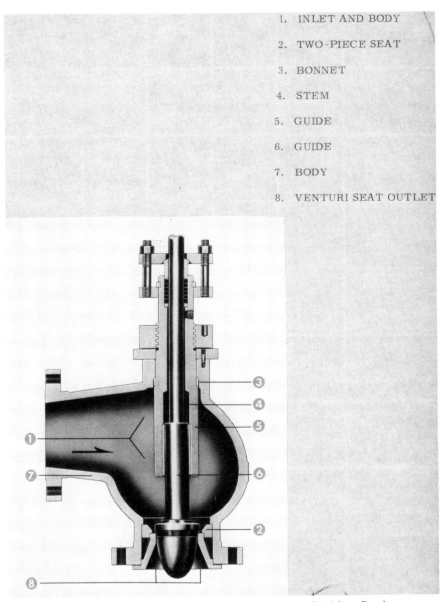

Fig. 98. Venturi Throat Valve. (*Courtesy of ITT Hamel-Dahl-Conoflow.*)

Valve Stem Extension See **Actuator Stem**.

Valve, Surge Dampening A valve that reduces system shock by restricting the acceleration of fluid flow.

Valve, Time Delay A valve in which a change in flow occurs only after the elapse of a specified time interval.

Valve, Three-Position A directional control valve which has three possible choices of flow direction.

Valve Trim The internal components of a valve which are exposed to the flowing fluid.

Valve, Two-Position A directional control valve which has two possible choices of flow directions.

Variable Any quantity that can be measured or controlled. Common variables in fluid systems are pressure, temperature, flow rate, etc.

Velocity The rate of change of position with time. Common units are meters per second (SI metric) and feet per second (British).

Velocity Head The velocity equivalent of pressure.

$$h_v = \frac{v^2}{2g}$$

where

v = velocity
g = acceleration of gravity

Vena Contracta The region of smallest cross section in a fluid stream. As fluid emerges from an orifice, it tends to contract in cross section reaching a minimum and then expanding back to fill the pipe.

Venturi Throat Valve A valve with a smaller opening across the seat than the port diameter. (See Fig. 98)

Viscosity, Dynamic See **Absolute Viscosity**.

Viscosity, Kinematic The ratio of absolute viscosity to the density of the fluid. See also **Stoke**.

Viscosity Index A measure of the viscosity versus temperature characteristics of a fluid as compared to that of a chosen standard fluid.

V-Port Ball Valve A valve with an orifice in the shape of the letter V. It is noted for its throttling and sealing ability. (See Fig. 99)

Fig. 99. V-Port Ball. (*Courtesy of De Zurik, A Unit of General Signal.*)

V-Port Plug A valve plug with an orifice in the shape of the letter V noted for its throttling characteristics. (See Fig. 100)

Water Hammer Vibration in a fluid system due to a rapid decrease in the velocity of a liquid from closing a valve.

Weber Number The ratio of inertia forces to surface tension

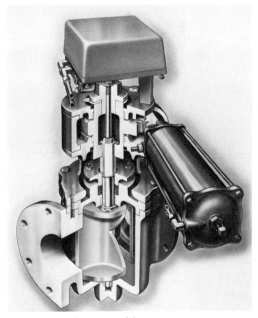

(a)

Fig. 100a. V-Port Plug. (*Courtesy of De Zurik, A Unit of General Signal.*)

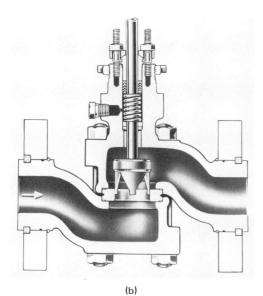

Fig. 100b. V-Port Plug. (*Courtesy of Fisher Controls Co.*)

$$N_w = \frac{V}{\sqrt{\frac{\sigma}{\rho}L}}$$

where

V = velocity
σ = surface tension
ρ = density
L = characteristic length

Weir A device for measuring the flow of a liquid in an open channel.

Welding Ends A type of end fitting on valves or other components suitable for welding to pipe or other fittings.

Wet Bulb Temperature The temperature read by a thermometer whose bulb is enclosed in a wet wicking. It is used in conjunction with dry bulb temperature to determine relative humidity.

Wetted Perimeter The portion of the perimeter of a pipe or channel covered by flowing fluid.

Whistle Valve A valve which whistles as a result of steam velocity and pressure. It is used on trains and boats and in factories. (See Fig. 101)

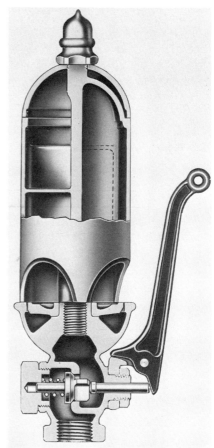

Fig. 101. Whistle Valve. (*Courtesy of Lunkenheimer Co.*)

Wire Drawing The erosion of a valve seat under high velocity flow whereby thin wirelike gullys are eroded away.

W.O.G. Water, Oil, Gas.

W.O.G. Rating The operating pressure rating of a valve as identified by valve markings for Water, Oil or Gas.

W.S.P. Working Steam Pressure.

WYE A fitting with three ports resembling the letter Y.

Y See **WYE**.

Y-Valve A type of globe valve in the shape of the letter Y. (See Fig. 102)

TERMINOLOGY OF THE VALVE INDUSTRY 79

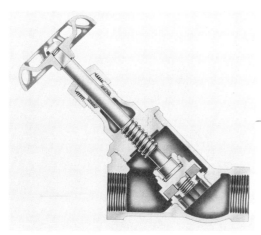

Fig. 102. Y-Valve. (*Courtesy of Jenkins Bros.*)

Y-Factor The expansion factor of a compressible fluid.

$$Y = \frac{W}{W^1}$$

where

W = true flow rate
W^1 = measured flow rate

Yield Point The lowest stress for which strain increases without an increase in stress.

Yoke The part of a valve which connects the valve actuator to the valve body.

Zerk Fitting A type of check valve that allows grease to be inserted in a member for lubrication.

Zero Shift The permanent deformation of a diaphragm, spring, etc., due to loads exceeding the elastic limit.

Zero Shift With Temperature The change in bridge balance, measured with zero pressure applied under stable conditions, which is a function of the change in temperature only. It is expressed as a percent of full scale sensitivity per degree.

OTHER VALVES

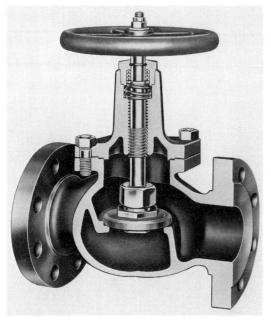

Flange Type Globe Valve. (*Courtesy of Rego, Division of Golconda Corp.*)

Knife Gate With Chain Pulley Actuator. (*Courtesy of De Zurik, A Unit of General Signal.*)

80 VALVES AND VALVE TERMINOLOGY

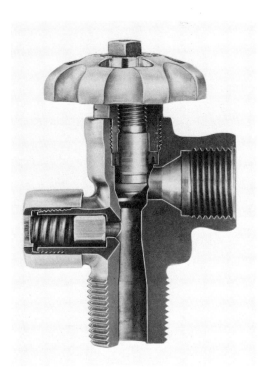

Cylinder Valve. (*Courtesy of Rego, Division of Golconda Corp.*)

An 18-foot fabricated steel, custom designed, metal seated butterfly valve. The valve is shown at shop assembly complete with hydraulic operator consisting of 4 cylinders, an accumulator and pressure tank all mounted and supported on a common base.

156-inch diameter fabricated steel butterfly valve being readied for shipment. The valve as shown is complete with operator linkage and mounting pads or feet.

Butterfly Valves. (*Courtesy of Allis-Chalmers.*)

TERMINOLOGY OF THE VALVE INDUSTRY   81

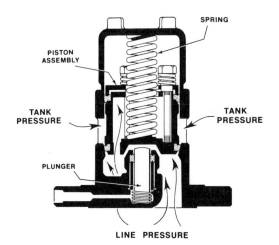

Internal Safety and Operating Valve. (*Courtesy of G.P.E. Controls.*)

Four Inch Stem-Ball Control Valve, With Cylinder Actuator & Positioner. (*Courtesy of Kamyr Inc.*)

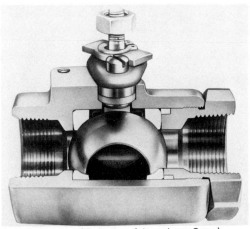

Ball Valve. (*Courtesy of Jamesbury Corp.*)

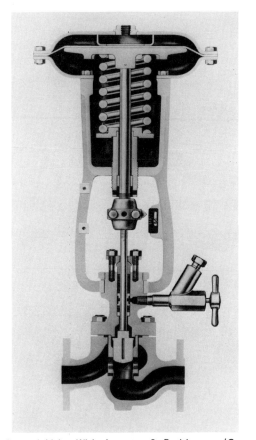

Control Valve With Actuator & Positioner. (*Courtesy of Honeywell Inc.*)

82 VALVES AND VALVE TERMINOLOGY

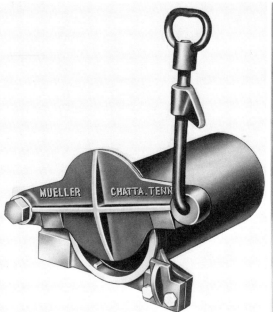

Shear Gate Valve. (*Courtesy of Mueller Co.*)

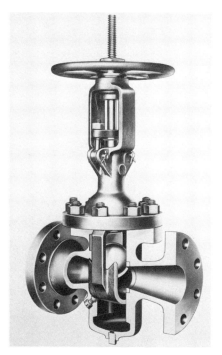

Ball Type Control Valve. (*Courtesy of Bell & Howell.*)

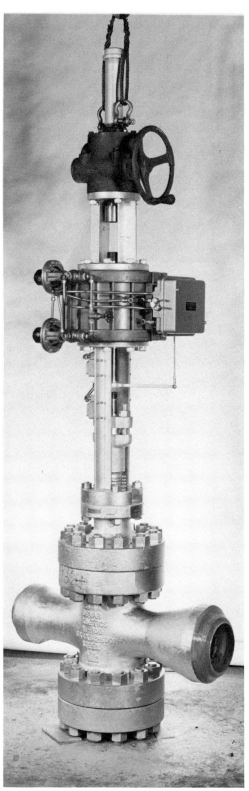

Control Valve. (*Courtesy of Bell & Howell.*)

TERMINOLOGY OF THE VALVE INDUSTRY 83

Swing Check Valve. (*Courtesy of Essex Cryogenics Industries, Inc.*)

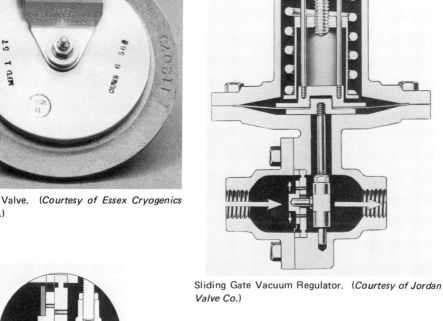

Sliding Gate Vacuum Regulator. (*Courtesy of Jordan Valve Co.*)

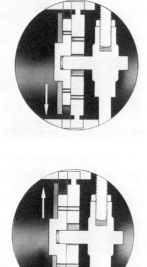

Operation of a Sliding Gate Valve. (*Courtesy of Jordan Valve Co.*)

Valves for Aerospace. (*Courtesy of Essex Cryogenics Industries, Inc.*)

84 VALVES AND VALVE TERMINOLOGY

Cryogenic Valves. (*Courtesy of Essex Cryogenics Industries, Inc.*)

PART TWO

VALVE ENGINEERING AND DESIGN DATA

Design Factors

In this section we consider some of the important factors pertaining to the design of a valve for a specified service. In many applications there are several types of valves that will function equally well and the choice may be based solely on such things as cost, availability, etc. For other applications it may be that no off-the-shelf valve is available and a custom design may be necessary. The proper choice or design may be made easier by proper attention to the factors discussed below.

FACTORS RELATED TO USAGE

In many cases only one or two of the basic valve types are suitable for a specified use because of the design of the valve. Under other conditions, several types of valves are suitable and other factors such as materials may be of prime importance.

Contamination. For control of fluids which may cause contamination buildup, a valve with minimum obstruction to flow is needed. Ball, gate, globe and pinch valves are suitable for this purpose.

Directional Control. For control of the direction of fluid flow, a check valve which blocks flow in one direction and allows full flow in the other, or a restrictor valve which allows only a specified flow in one or more directions, is required. Poppet and swing valves are widely used for this function.

High Pressure. The control of high pressure flow generally calls for a ball or globe valve although gate and poppet valves are occasionally used. The selection of a valve to be used in a high pressure application, particularly pneumatic, should be approached with extreme caution. The design of such valves should be left to qualified engineers.

High Temperature. In most situations the same considerations that apply to high pressure valves apply to high temperature valves. In addition, care must be taken to ensure that thermal expansion does not cause binding or deformation in the valve.

Low Leakage. All of the basic valves can be made leaktight, but often with high cost and complexity. Generally, for tight shutoff one should consider ball, gate, globe and plug valves.

Relief and Safety. For rapid opening response to overpressure and large flow to vent, one should nearly always consider a spring loaded poppet valve. The other types of valves are not normally considered for these functions.

Shutoff. For normal on-off control the best choices are ball, gate, globe and plug valves. The ball and plug valves normally open faster than gate or globe valves.

Steam Service. The control of steam under pressure generally calls for a ball or globe valve.

Throttling. To control the amount of flow by varying the amount of opening, a globe valve is usually recommended. In contrast to most ball and gate valves, it does not tend to vibrate under flow.

FACTORS RELATED TO CONSTRUCTION

Actuator. The means of operating the valve will depend on the type of valve, its location and function in a system, size of the valve, frequency of operation, and degree of control desired. Some common means of actuation are hand, gear, chainwheel, lever, spring, motor, solenoid, servo, gravity, and pressure and flow rate of the fluid media. Generally, a particular type of valve is limited to one of a few types of actuators, for example, relief and safety valves are spring actuated, check valves are spring or gravity actuated, and high pressure globe valves are usually actuated by chainwheels, motors, etc. Automatic process control calls for servo-valves, solenoid and spring actuators.

Table 1.

	Ball	Butterfly	Gate	Globe	Pinch	Plug	Poppet	Swing
CHECK VALVE	P	P	P	P	P	P	G	G
CONTAMINATION FREE	G	F	G	G	G	G	F	P
CORROSIVE FLUIDS	G	P	F-G	F	F-G	P-F	G	G
CRYOGENIC FLUIDS	G	P	P	G	P	P	G	P
GASES	G	G	G	G	G	G	G	G
HIGH ΔP	P	P	F	G	P	F	F	P
HIGH FLOW	G	G	G	G	G	G	G	G
HIGH PRESSURE	G	P	P	G	P	P	G	P
HIGH TEMPERATURE	G	G	G	G	P	P-G	G	G
LEAKTIGHT	G	P	G	G	G	G	G	P
LIGHTWEIGHT	G	G	F	P	F	G	G	G°
LIQUIDS	G	G	G	G	G	G	G	G
LOW ACTUATION FORCE	P	P	P	P	P	P	G	G
LOW COST	G	F-G	G	F-G	G	G	G	G
LOW ΔP	G	G	G	P	G	G	G	G
LOW FLOW CONTROL	G	G	P	G	F	G	G	G
RAPID OPENING	G	G	P	F-P	P	G	G	G
RELIEF	P	P	P	P	P	P	G	P

Table 1. (*Continued*)

SAFETY VALVE	P	P	P	P	P	P	G	P
SEAT EROSION RESISTANCE	F	P	P	G	P	F	F-G	P
SLURRIES	F-G	P	P	P	G	P	P	P
SMALL PHYSICAL SIZE	G	G	P	P	P	G	G	G
STEAM SERVICE	G	P	P-F	G	P	P	G	P
THROTTLING	P	P	P	G	F-P	P	G	P
VIBRATION FREE	F	P	P	G	G	G	P	P

P = Poor, Not Recommended
F = Fair, Better Choices Available
G = Good, Recommend For Use Under Normal Conditions

Closure Member. The type of closure member desired or required will normally determine the type of valve to be used. Conversely, the choice of a valve type will usually determine the type of closure member. The common closure members are the ball, disc, gate, plug and poppet.

End Fittings. The type of end fittings to be specified for a valve is normally determined by the nature of the piping system into which it must be inserted. Some common fittings are brazing end, butt weld ends, compression, flange, flared, hose ends, hub ends, pipe thread, quick disconnect, socket and solder ends. For high pressure and/or high temperature conditions, one of the various types of flange ends or a threaded fitting should be considered.

Material. The material selected for valve trim will depend on the nature of the fluid to be carried, the operating pressure and temperature, the type of closure member and seat, and factors such as cost, weight, etc. Control of corrosive liquids and gases calls for stainless steels, nickel alloys, various plastics, and ceramic materials. For high pressure service and/or high temperatures consider various steels, nickel alloys, titanium alloys and similar high strength materials. For steam service consider cast iron steel, bronze and similar metals. Nuclear valves call for special steels, titanium and other alloys developed especially for this type of service. In all cases of severe use conditions, manufacturers literature should be consulted to determine the suitability of a particular valve.

Packing/Seals. In most valves consideration must be given to possible leakage around the stem or actuator. In common valves a packing material is provided as a seal. However, this tends to wear out with use. If replacement is undesirable or unfeasible, consideration should be given to bellows seal valves, diaphragm seals, etc.

Seat. There are many styles of valve seats available in most of the types of valves discussed. They may differ in geometry, material, rigidity, etc. Conical valve seats can provide a wide sealing surface which minimizes eroding or "wire drawing" of the surface. A conical seat may also be designed with a narrow sealing surface to provide very tight sealing at low pressures. Spherical or ball seats have much the same characteristics as the conical seat. However, they are more costly to produce. Flat seats are used in valves that need not be leaktight since they generally do not seal completely at low pressures.

11 Sizing a Valve

Once the type of valve has been chosen for a particular application, the next step is to determine the correct size, though sometimes the size of valve required will help determine the type. The following data is presented to aid in sizing valves for required flow, pressure drop, etc. The curves and data presented in this section are based on a sampling of data from 10-20 manufacturer's catalogs and are intended only as representative values.

C_v **Method.** The flow coefficient C_v is defined as the flow in gallons per minute of water at 60°F with a pressure drop of one psi (see Table 1 for definition of symbols.

$$Q = C_v \sqrt{\frac{\Delta P}{S}}$$

Note that this equation is valid only for water flow or liquids with viscosities near that of water. If another liquid is to be used, the expected pressure drop must be corrected by

$$\frac{\Delta P_1}{\Delta P_2} = \left(\frac{\mu_1 \rho_2}{\mu_2 \rho_1}\right)^{1/4} \frac{\rho_1}{\rho_2}$$

where subscripts 1 and 2 denote the liquid being considered and water, respectively.

Once the C_v of a value is known, the amount of flow at a given pressure drop can be found, or, conversely, the pressure drop can be determined for a specified flow. On the other hand, for specified flow and pressure drop, the flow coefficient can be computed and the type/size of valve to be used can be determined from published C_v data.

Table 1

A = area (in.^{2})
C = valve coefficient
C_f = orifice coefficient
C_v = valve flow coefficient
d = orifice diameter (in.)
d_{EO} = equivalent orifice diameter (in.)
d_L = line diameter (in.)
F = NBS flow factor
g = acceleration of gravity (ft/sec^2)
K = head loss factor
P = pressure (psia)
ΔP = pressure drop (psi)
Q = flow rate (gpm)
R = gas constant
S = specific gravity
T = temperature (°R)
W = flow rate (lb/sec)
γ = specific heat ratio
μ = absolute viscosity (centipoise)
ν = kinematic viscosity (centistoke)
π = Pi (3.1415926 . . .)
ρ = density (lb/ft^3)
Subscript 1 = value for other liquid
Subscript 2 = value for water

EXAMPLE PROBLEM #1:

A 2 in. full port ball value is to carry water at a rate of 556 gpm. What pressure drop is expected?

Solution:

From Table 2 find C_v = 228, thus:

$$\sqrt{\Delta P} = \frac{556}{228}$$

$$= 2.44$$

$$\Delta P = 5.95 \text{ psi}$$

SIZING A VALVE

Table 2 Typical C_v Values for 2 in. Valves.

Type	C_v
Angle Valve	64.0
Ball Check Valve	154.5
Ball Valve (Full Port)	228
Ball Valve (Standard Port)	120
Butterfly Valve	145
Coaxial Valve	154.5
Cone Poppet Check Valve	166
Flat Poppet Check Valve	133
Gate Valve	210
Globe Valve	44.34
Pinch Valve	181
Plug, Taper Valve	70
Swing Check Valve	138.2
Y-Valve	
45° Angle	72.0
60° Angle	70.8

EXAMPLE PROBLEM #2:

A 2-in. globe valve is allowed a pressure drop of 64 psi when carrying water. How many gallons/minute of oil (specific gravity 0.8, kinematic viscosity 0.82) will it pass and what is the expected pressure drop? The kinematic viscosity of water is 0.93.

Solution:

From Table 2 find $C_v = 44.34$, thus:

$$\Delta P_2 = \frac{\rho_2}{\rho_1}\left(\frac{\mu_2 \rho_1}{\mu_1 \rho_2}\right)^{\frac{1}{4}} \Delta P_1$$

$$= \frac{1}{S}\left(\frac{\nu_2}{\nu_1}\right)^{\frac{1}{4}} \Delta P_1$$

$$= \frac{1}{.8}\left(\frac{.93}{.82}\right)^{\frac{1}{4}} 64$$

$$= 82.56 \text{ psi}$$

$$Q = 44.34 \sqrt{\frac{82.56}{.8}}$$

$$Q = 450.43 \text{ gpm}$$

EXAMPLE PROBLEM #3:

Determine an appropriate type of 2 in. valve that will carry 100 gpm water and have a pressure drop of 1.98 psi.

Solution:

$$C_v = \frac{Q}{\sqrt{\Delta P}}$$

$$= \frac{100}{\sqrt{1.98}}$$

$$= 71$$

From Table 2 we see that a 60° angle valve will meet the requirements.

Table 3 Equations for Valve Sizing.

$$Q = 29.81\, C_f d^2 \sqrt{\frac{\Delta P}{S}}$$

$$Q = C_v \sqrt{\frac{\Delta P}{S}}$$

$$Q = 7.9\, C_v \sqrt{\frac{\Delta P}{\rho}}$$

$$Q = C_v \sqrt{\Delta P \left(\frac{62.4}{\rho}\right)}$$

$$Q = 236\, d_L^2 \sqrt{\frac{\Delta P}{\rho K}}$$

$$d_{EO} = 1.292 \frac{d_L}{\sqrt{K}}$$

$$d_{EO} = 0.2365 \sqrt{C_v}$$

$$d_{EO} = C\, (d_L)^{1.07}$$

$$d = \left(\frac{Q}{29.81\, C_f}\right)^{1/2} \left(\frac{S}{\Delta P}\right)^{1/4}$$

$$\Delta P = \left(\frac{Q}{C_v}\right)^2 \frac{\rho}{62.4}$$

$$\Delta P = 1.8 \times 10^{-5} \frac{K Q^2}{d_L^4}$$

$$C_v = \frac{29.9\, d_L^2}{\sqrt{K}}$$

$$C_v = Q \sqrt{\frac{\rho}{62.4\, \Delta P}}$$

$$W = \frac{C_f A P_1}{\sqrt{R T_1}} \sqrt{g\gamma \left(\frac{2}{\gamma+1}\right)^{(\gamma+1)/(\gamma-1)}}$$

$$W = \frac{C_f A P_1}{\sqrt{R T_1}} \sqrt{\frac{2g\gamma}{\gamma-1}\left[\left(\frac{P_2}{P_1}\right)^{2/\gamma} - \left(\frac{P_2}{P_1}\right)^{(\gamma+1)/\gamma}\right]}$$

EQUIVALENT ORIFICE METHOD: Since the flow through a sharp edge orifice can be conveniently calculated with good accuracy, it would be desirable to relate flow through a valve to that through a sharp edge orifice. This can be done to high accuracy with the use of the equivalent orifice method.

This procedure consists of three basic steps. The first step is to compute the sharp edge orifice diameter. This can be found from one of the following equations (See Table 3 for alternate forms of equations)

$$Q = 29.81\, C_f\, d^2\, \sqrt{\frac{\Delta P}{S}}$$

for liquids,

$$W = \frac{C_f P_1 \pi d^2}{\sqrt[4]{RT_1}} \sqrt{g\gamma \left(\frac{2}{\gamma + 1}\right)^{(\gamma+1)/(\gamma-1)}}$$

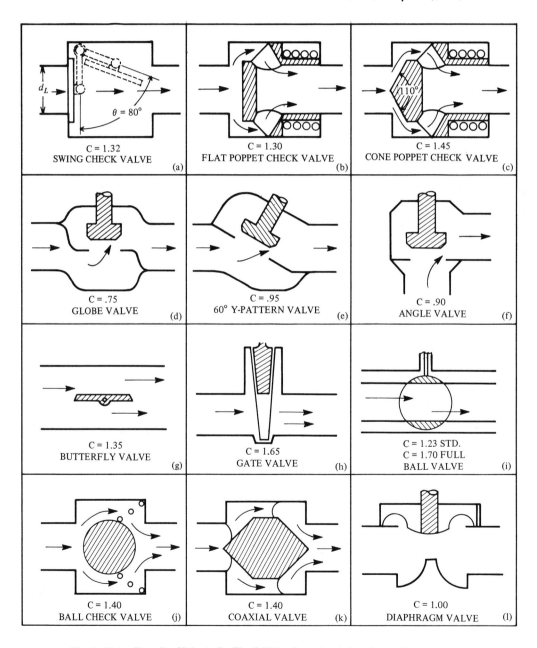

Fig. 1. Valve Flow Coefficient, C. The C-Value for valves is based on full ported design.

for gases under sonic flow conditions, or

$$W = \frac{C_f P_1 \pi d^2}{\sqrt[4]{RT_1}} \sqrt{2g\gamma \left[\left(\frac{P_2}{P_1}\right)^{2/\gamma} - \left(\frac{P_2}{P_1}\right)^{(\gamma+1)/\gamma}\right]}$$

for gases under subsonic flow conditions.

The next step is to determine the equivalent orifice diameter from one of the following equations

$$d_{EO} = \frac{1.292 \, d_L}{\sqrt[4]{K}}$$

$$= 0.2365 \sqrt{C_v}$$

$$= 0.3162 \sqrt{F}$$

if data are available. If d_{EO} can be computed, an adjustment is made in one or more parameters until d_{EO} is made euqal to d. This can be done by changing d_L for a value of known K or by changing K to meet a required line size. If neither is known, K may be computed from

$$K = 1459.24 \frac{A^2}{C_v^2}$$

or

$$K = 449.44 \frac{A^2}{F^2}$$

and the line size chosen to meet this K-factor. Alternatively, a new valve and K-factor may be chosen to fit a line size.

If d_{EO} cannot be computed from known information, a trial and error procedure can be used. Set d_{EO} equal to d and use Figs. 1 and 3 to choose a valve. From its coefficient C and d_{EO} find the proper line diameter, or conversely, from a required line diameter find the valve type from the coefficient C. Once the valve coefficient and line diameter are known, the K-factor for the valve may be found from Fig. 2.

In summary, to fully size a valve for a given flow and pressure drop, determine the coefficient C, the K-factor of the valve, and its line diameter. Note that one or two, but not all of these may be specified beforehand. All three, however, may be specified if the flow and/or pressure drop are not specified. The following problems will illustrate the procedure.

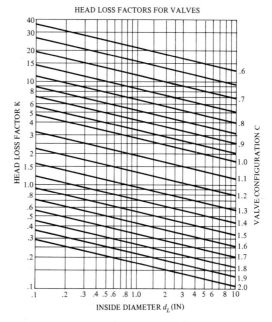

Fig. 2. Headloss Factors for Valves.

EXAMPLE PROBLEM #4:

A $1\frac{1}{2}$-in. full port ball valve has a valve flow coefficient C of 1.7. What is the number of gallons per minute of water it will pass with a pressure drop (ΔP) of 1.2 psi?

Solution:

From Fig. 1, find $C = 1.70$. Then from Fig. 3 find $d_{EO} = 2.62$ in. Hence

$$Q = 29.81 \, C_f d^2 \sqrt{\frac{\Delta P}{S}}$$

$$= 29.81 \, (0.6) \, (2.62)^2 \sqrt{1.2}$$

$$Q = 134.5 \text{ gpm}$$

EXAMPLE PROBLEM #5:

A globe valve is to carry 77 gpm of water with a pressure drop of 3.0 psi. What size valve is required?

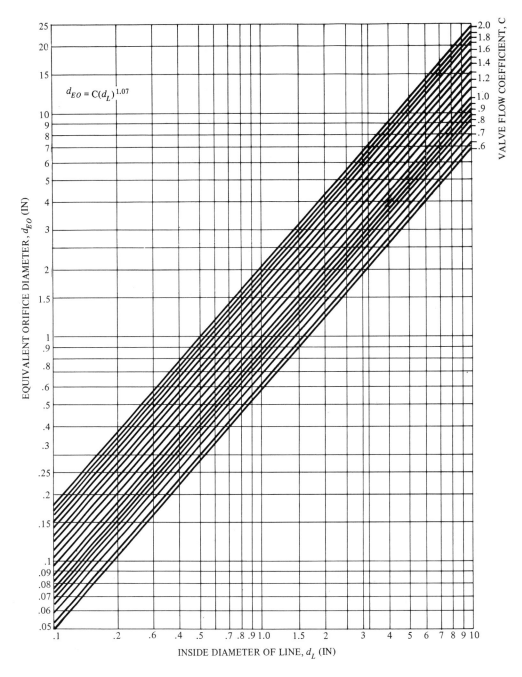

Fig. 3.

Solution:

From the equation of problem 4, find d_{EO}

$$d^2 = \frac{Q}{29.81\, C_f \sqrt{\dfrac{\Delta P}{S}}}$$

$$= \frac{77}{29.81\,(0.6)\sqrt{3.0}}$$

$$d = 1.6$$

From Fig. 1 find C for a globe valve is 0.75. Then use Fig. 3 to find a line diameter, d_L of 2 in. Thus a 2 in. globe valve is required.

EXAMPLE PROBLEM #6:

A gate valve is to have a line diameter of 2 in. and a pressure drop of 9 psi. Determine the K-factor and flow rate.

Solution:

From Fig. 1 find $C = 1.65$. Then from Fig. 2, $K = 0.33$. Figure 3 yields the equivalent orifice diameter 3.75 and thus

$$Q = 29.81\,(0.6)\,(3.50)^2\,\sqrt{9}$$
$$= 657.31 \text{ gpm}$$

For further data on liquid and gas flow, refer to Appendix C.

Valve Spring Design

The proper operation of many check, safety, relief, and various other valves depends upon the proper design of the valve spring. The spring will determine the cracking pressure, the distance the poppet will move, and the stability of the seal in the closed position. Proper design of the spring will help reduce the tendency of the valve to chatter or vibrate and will increase its stability under acceleration in aerospace applications.

Since most spring operated valves utilize either a helical compression spring or a torsion spring made of circular cross-section wire, this discussion is limited to these two cases.

COMPRESSION SPRINGS

The design of a helical compression spring is based primarily on two equations,

$$R = \frac{Gd^4}{8D^3 N}$$

and

$$S = \frac{2.55 \, FDK}{d^3}$$

where symbols have the meanings given in the legend (Table 1).

The first equation expresses the spring rate in terms of geometric factors and material properties and is valid up to the elastic limit of the material. The second equation gives the maximum stress the spring will withstand under a given load. The Wahl factor, K, is defined by

$$K = \frac{4C - 1}{4C - 4} + \frac{0.615}{C}$$

where C is the diameter ratio D/d. This equation is plotted in Fig. 1. The Wahl factor

Table 1. Legend.

A = area (in.^{2})
$C = D/d$
d = wire diameter (in.)
D = mean coil diameter (in.)
ΔD = change in coil diameter (in.)
E = elastic modulus (psi)
f = frequency of vibration (Hz)
F = load on spring (lb)
F_c = cracking force (lb)
F_d = dynamic force (lb)
g = acceleration of gravity (ft/sec^2)
G = torsion modulus (psi)
K = Wahl factor
l = lever arm (in.)
L_f = free length (in.)
L_i = installed length (in.)
L_s = solid height (in.)
M = moment (in.-lb)
M_c = static moment
M_d = dynamic moment
M_n = moment per coil (in.-lb)
N = number of active coils
P = pitch (in.)
P_c = cracking pressure (psi)
q = flow (ft^2/sec)
R = spring rate (lb/in.)
S = stress (psi)
α = angular deflection (turns)
δ = stroke (in.)
$\eta = D_{ball}/d_{seat}$
θ = poppet half-angle (degrees)
π = pi
ρ = density (lb/in.^{3})

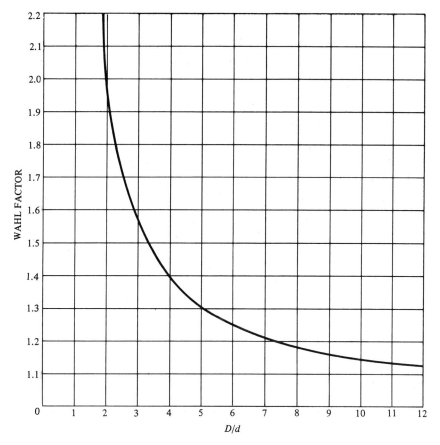

Fig. 1.

should always be taken into account for springs subject to cyclic operation, vibration, impact loading and in situations where the computed stress is close to the rated stress limit of the material. Table 2 gives the maximum uncorrected stress for some of the more common spring materials. Whenever possible the spring should be designed to a stress level under the rated maximum.

To size a compression spring for a check valve with a predetermined cracking pressure, the following procedure should be followed:

(a) From the specified cracking pressure and the inlet area required to pass the flow, determine the cracking force on the poppet,

$$F_c = P_c A$$

(b) From the inlet area and the poppet geometry determine the stroke necessary to give full-port opening. For a conical poppet of half-angle θ the stroke is given by

$$\delta = d \frac{1 - \sqrt{1 - \cos\theta}}{\sin 2\theta}$$

which reduces to

$$\delta = \frac{d}{4}$$

for a flat poppet. For a ball check use

$$\delta = \frac{d}{2} \{\tfrac{1}{4} [1 + \sqrt{1 + \eta^2}]^2 - 1\}^{1/2} - (\eta^2 - 1)^{1/2}$$

(c) From specified flow requirements compute the dynamic force on the poppet in

Table 2. Properties of Spring Materials

MATERIAL	CONDITION	TORSION MODULUS G (psi)	MAXIMUM UNCORRECTED STRESS (psi)	ELASTIC MODULUS E (psi)
CORROSION RESISTANT ALLOYS				
302 CRES	Spring temper	10,000,000	100,000	28,000,000
302 CRES	Cold drawn	9,500,000	80,000	28,000,000
316 CRES	Spring temper	10,000,000	70,000	28,000,000
316 CRES	Cold drawn	9,500,000	70,000	28,000,000
321 CRES	Spring temper	10,000,000	70,000	28,000,000
410 CRES	Cold drawn	10,000,000	72,000	28,000,000
Carpenter 20	Spring temper	10,000,000	80,000	
17-4 pH	Cold drawn	10,500,000	80,000	29,000,000
17-7 pH	Cold drawn	11,000,000	140,000	29,000,000
Hastelloy C	40% Cold reduced	10,500,000	100,000	
18-8 CRES	Cold drawn	10,000,000	140,000	28,000,000
INCONEL-600	Cold drawn	11,000,000	85,000	31,000,000
INCONEL-625	Spring temper	11,000,000	85,000	31,000,000
K-MONEL-500	Cold drawn	9,200,000	80,000	26,000,000
MONEL-400	Spring temper	9,200,000	65,000	26,000,000
NI-SPAN-C	Spring temper	9,500,000	75,000	27,500,000
NS-355	Spring temper	11,000,000	150,000	
NONFERROUS ALLOYS				
Beryllium copper	Cold drawn	7,300,000	80,000	16,000,000
Beryllium copper	Pretempered	7,300,000	85,000	18,500,000
Phosphor bronze	Hard drawn	6,000,000	70,000	15,000,000
Spring brass	Cold drawn	5,500,000	60,000	15,000,000
LOW ALLOY STEELS				
AISI-1095	Hot rolled	10,500,000	90,000	30,000,000
AISI-5160	Hot rolled	10,500,000	105,000	30,000,000
Chrome vanadium	Oil tempered	11,500,000	110,000	30,000,000
Chrome silicon	Oil tempered	11,500,000	110,000	30,000,000
Music wire	Hard drawn	11,500,000	120,000	30,000,000
Spring wire	Oil tempered	11,500,000	105,000	30,000,000

Table 2. (*Continued*)

MATERIAL	CONDITION	TORSION MODULUS G (psi)	MAXIMUM UNCORRECTED STRESS (psi)	ELASTIC MODULUS E (psi)
LOW ALLOY STEELS				
Valve wire	Hard drawn	11,500,000	105,000	30,000,000
Rocket wire	Hard drawn	11,500,000	150,000	30,000,000
HIGH ALLOY STEELS				
A-286	Spring temper	10,000,000	100,000	
AISI-S1	Cold drawn	11,600,000	90,000	
AISI-H21	Cold drawn	11,000,000	100,000	
AISI-T1	Cold drawn	11,000,000	100,000	
INCONEL X-750	Cold drawn	11,500,000	90,000	31,000,000
INCONEL X-750	Spring temper	11,500,000	120,000	31,000,000
RENE 41	Spring temper	12,000,000	130,000	
S-816	Spring temper	11,600,000	130,000	
250 Maraging steel	Centerless ground	10,000,000	160,000	

the full-open position,

$$F_d = \frac{144 \, \rho q^2}{gA}$$

(d) Determine the required spring rate from

$$R = \frac{F_d}{\delta}$$

(e) Considering any restrictions placed on the physical size and shape of the valve, choose an appropriate mean diameter and wire diameter which will yield a stress within safe limits for the spring material chosen. If the mean and wire diameters are restricted the material may be changed to give a stress within safe limits. Be sure to consider the Wahl factor in the stress equation

$$S = \frac{2.55 \, FD}{d^3} K$$

if the computed stress is high. This step may need to be repeated several times before all requirements are satisfied.

In choosing a material to meet the stress limits imposed, be sure to consider any other applicable conditions, such as elevated temperatures, that may affect the stress in the spring. Determine also if the intended use calls for a corrosion resistant material.

(f) With the mean coil diameter, wire diameter and material chosen from the preceding step, determine the number of active coils required from

$$N = \frac{Gd^4}{8D^3R}$$

The total number of coils in the spring will be $N + 2$.

(g) Determine the solid height of the spring from

$$L_s = d(N + 2)$$

If this is less than the length available for installing the spring, proceed to the next step. Otherwise a new spring must be sized with fewer coils and/or a smaller wire diameter. The last two steps may have to be repeated several times.

(h) Determine the free length of the spring from

$$L_f = L_i + \frac{F_c}{R}$$

(i) If the ratio of free length to mean diameter is greater than approximately four, the spring is subject to buckling and should be redesigned.

(j) If the spring is to fit inside a tube or close fitting bore, it may be necessary to allow for the increase in mean coil diameter due to compression to ensure that the spring will not bind in the valve. The increase in diameter is given by

$$\Delta D \approx \frac{0.05 \, (P^2 - d^2)}{D}.$$

Generally this problem can be avoided by a proper choice of mean diameter.

(k) If the intended use of the valve justifies the effort, determine the natural vibration frequency of the coil from

$$f = \frac{48.2 \, G}{\rho} \frac{d}{\pi N D^2}$$

If the valve encounters frequencies near this in operation it may resonate and be damaged.

The determination of the relevant parameters described above will sufficiently describe a spring to be wound or purchased. In addition, it is customary to call for squared and ground ends and to specify either right or left hand winding.

NESTED SPRINGS

In some applications it is not possible to size a single spring to meet all of the requirements imposed on it. Generally, the stress on the spring is far too large for any existing material. In this case it is often possible to nest two or more springs to meet the requirements.

For two compression springs the optimum situation is to size the two springs to equal stress levels such that the outer spring carries two-thirds of the load and the inner spring carries one-third. Since each spring must compress the same amount, the rate of the outer should be two-thirds that of a single spring and the inner should be one-third. The wire and mean diameters must be chosen to fit the geometrical restrictions imposed, and the springs should be wound in opposite directions to prevent interlocking.

EXAMPLE PROBLEM #1:

A flat, poppet check valve has an inlet diameter of 0.391 in. and is to open at 2 psig and allow a flow of water of .01 ft^3/sec. Size a spring of 0.5 in. mean diameter and 0.6 in. installed length for this valve.

Solution:

(a) $F_c = P_c A$

$$= (2) (\pi) \frac{(.391)^2}{4}$$

$$= 0.240 \text{ lb}$$

(b) $\delta = \frac{d}{4}$

$$= \frac{.391}{4}$$

$$= 0.098 \text{ in.}$$

(c) $F_d = \frac{144 \, \rho q^2}{gA}$

$$= \frac{144 \, (62.4) \, (.01)^2}{32.2 \, \pi \, (.391)^2 / 4}$$

$$= 0.232 \text{ lb}$$

(d) $R = \frac{F_d}{\delta}$

$$= 2.37 \text{ lb/in.}$$

(e) $S = \frac{2.55 \, FD}{d^3} K$

$$= \frac{2.55 \, (.240 + .232) \, (0.5)}{d^3} K$$

Try a wire diameter 0.026 in.

$S = 34,389 \, K$

This is low enough stress so we need not consider the Wahl factor. If we had chosen $d = 0.020$ we would have

$$S = 75{,}225\,K$$

and the Wahl factor would be important unless a very high strength wire was chosen.

(f) Choose 302 stainless wire and find for N

$$N = \frac{(10^7)(.026)^4}{8(.5)^3(2.37)}$$

$$= 1.9 \text{ active coils}$$

(g) $L_s = 0.026\,(1.9 + 2)$

$$= 0.102 \text{ in.}$$

This is less than the installed length of 0.6, so the spring will not go solid under load.

(h) $L_f = L_i + \dfrac{F_c}{R}$

$$= 0.6 + \frac{0.240}{0.232}$$

$$= 1.634 \text{ in.}$$

(i) $\dfrac{L_f}{D} = \dfrac{1.634}{0.5}$

$$= 3.27$$

This is less than four so the spring should be stable.

TORSION SPRINGS

Torsion springs are widely used on swing check valves and flap valves. The basic equations used in the design of a torsion spring are

$$S = \frac{10.8\,M}{d^3}$$

$$M_n = \frac{Ed^4}{10.8\,DN}$$

$$\alpha = \frac{10.8\,MND}{Ed^4}$$

The first equation gives the stress on the spring in terms of the applied moment and wire diameter. The second gives the moment per turn in terms of geometrical and material parameters, and the third gives the angular deflection of the spring under load. The moment per turn of a torsion spring is the analog of the rate of a compression spring.

To size a torsion spring for a swing check valve with a predetermined cracking pressure, use the following procedure:

(a) Determine the required force and moment or torque from the specified cracking pressure and the inlet area,

$$F_c = P_c A$$

$$M_c = F_c l.$$

(b) For a swing check valve, full-port opening is defined as an opening angle θ of 80 degrees (1.396 radians). If some other angle is required, use it in place of 80 degrees in succeeding steps.

(c) From specified flow conditions determine the dynamic force and its moment against the disc in the open position

$$F_d = \frac{144\,\rho q^2}{gA}\sqrt{2(1-\cos\theta)}$$

$$M_d = F_d l$$

(d) Determine the moment per turn (spring rate) from

$$M_n = \frac{360\,M_d}{\theta}$$

(e) Considering any physical restrictions placed on the design, choose a coil diameter and a wire diameter that will yield a stress within safe limits for the spring material

$$S = \frac{10.8\,(M_c + M_d)}{d^3}$$

Some trial and error and/or trade-offs may be needed at this point. Suitable allowance must be made for high temperature operation or impact loading, etc.

(f) Using the moment per turn, wire diameter and coil diameter determined above for the chosen spring material, compute the number of active coils required from

$$N = \frac{Ed^4}{M_n D\,10.2}$$

(g) If there is a restriction on the length of the coiled spring, compute the solid height from

$$L_s = d(N + 2)$$

If this is longer than the allowed length, a new spring must be sized by changing the wire diameter and/or a number of the coils, which will change the stress level.

(h) If the spring is to be installed around an arbor, sufficient space must be allowed for a decrease in mean coil diameter as the applied moment is increased. The change in diameter can be estimated from

$$D = 0.05 \frac{P^2 - d^2}{D}$$

(i) Determine the required angle between ends in the free position and the change in angle giving the specified installed load from

$$\alpha = \frac{10.8 \, M_c N D}{E d^4}$$

For most applications this completes the required calculations. For design purposes it is also customary to specify the nature of the spring ends, the direction of coiling and any special requirements such as finish, heat treatment, etc.

EXAMPLE PROBLEM #2:

A swing check valve has a 3-in. disc which is to open at a cracking pressure of 0.1 psig and pass 0.05 ft³/sec. of jet fuel ($S = 0.7752$). Size a torsion spring with a coil length not to exceed 2 in.

Solution:

(a) $M_c = F_c l$

$$= (0.1)(\pi)\frac{(3.0)^2}{4}(1.5)$$

$$= 1.06 \text{ in-lb}$$

(b) Require the door to open 80° from vertical.

(c) $M_d = \dfrac{144 \, \rho q^2}{gA} \sqrt{2(1 - \cos 10°)} \, 1.5$

$$= \frac{144(.7752)(62.4)(0.5)^2\sqrt{2(1 - .98481)}(1.5)}{32.2(\pi)(3)^2/4}$$

$$= 2.000 \text{ in-lb}$$

(d) $M_n = \dfrac{360 \, M_d}{\theta}$

$$= 9.00 \text{ in-lb/turn}$$

(e) $S = \dfrac{10.8 \, (M_c + M_d)}{d^3}$

Try a coil with $D = 0.5$ and $d = 0.072$.

$$S = \frac{10.8 \, (1.06 + 2.00)}{(.072)^3}$$

$$= 88{,}542 \text{ psi}$$

Depending on expected use, 302 stainless wire should do the job.

(f) $N = \dfrac{E d^4}{10.2 \, M_n D}$

$$= \frac{3 \times 10^7 \, (.072)^4}{10.2 \, (9.0)(0.5)}$$

$$= 17.5 \text{ coils}$$

(g) $L_s = d(N + 2)$

$$= (.072)(19.5)$$

$$= 1.405 \text{ in.}$$

This would easily fit in the allowed space of 2 in.

(h) Since no arbor diameter is specified, choose one sufficiently small, such as 0.4, to allow coiling under load without binding.

(i) Require a 90 degree angle between spring ends when the spring is installed under a moment M_c. In the free position

$$\alpha = \frac{10.8 \, M_c N D}{E d^4}$$

$$= \frac{10.8 \, (1.06)(17.5)(0.5)}{3 \times 10^7 \, (0.072)^4}$$

$$\alpha = .1242 \text{ turns}$$

IV Some Useful Formulas

In this section we present a collection of formulas that may be useful in the design of valves and piping systems. The formulas are often given in several sets of units to minimize the need for conversion factors. Symbols used in these equations have the meanings given in the accompanying legend. Most of the numerical constants used in the equations are expressed in scientific notation. Most of the equations are given in British Engineering units although many have the same form in SI units with appropriate changes in the quantities involved. A few equations are given directly in SI units.

For convenience we have tried to group equations according to their basic nature, application, etc. Where this may cause overlap, we have usually repeated the equation under each name or category. A word of caution, many of these equations are valid only in a restricted range of some parameter. For example, an equation may be valid for a low Reynolds number only. Reference to a textbook on fluid mechanics or other source should be made whenever there is doubt about the applicability of an equation.

LEGEND

A = cross sectional area (ft^2)
a = cross sectional area (in.^{2})
B = flow rate (barrels/hr)
C = flow coefficient for orifice or nozzle
C_v = flow coefficient for valves
D = diameter (ft)
D_s = mean coil diameter (in.)
d = diameter (in.)
d_w = wire diameter (in.)
E = elastic modulus
e = base of natural logarithims (2.718)
F = force (lb)
f = friction factor
G = torsion modulus
g = acceleration of gravity (32.2 ft/sec^2)
H = total pressure head (ft of fluid)
h_s = total heat of steam (btu/lb)
h_L = static head loss (ft of fluid)
h_w = static pressure head (in. of water)
K = head loss factor
K' = Wahl factor
L = length (ft)
L_m = length (miles)
L/D = equivalent length in pipe diameters
M = molecular weight
M_o = torque (in.-lb)
M_t = moment per turn (in.-lb)
N = number of active spring coils
n = number of moles of gas
P = pressure (lb/in.^{2} gauge)
P' = pressure (lb/in.^{2} absolute)
p = pressure (lb/ft^2)
Q = flow rate (gal/min)
Q = flow rate (lb/hr)
q = flow rate (ft^3/sec)
q_h = flow rate (ft^3/hr standard)
q_m = flow rate (ft^3/min)
q'_m = flow rate (ft^3/min standard)
R = gas constant (155/M)
R' = displacement (in.)
R_e = Reynolds number
S = specific gravity relative to water
S' = stress (psi)
S_g = specific gravity relative to air
T = absolute temperature (°R)
T_t = stagnation temperature (°R)
t_s = time (sec)
V = mean flow velocity (ft/min)
v = mean flow velocity (ft/sec)
V' = volume (ft^3)

$\overline{V}$ = specific volume (ft³/lb)
v_S = sonic velocity (ft/sec)
W = gas flow rate (lb/hr)
w = gas flow rate (lb/sec)
Y = expansion factor
Z = potential head (ft)
γ = specific heat ratio
Δ = differential between reference points
δ = displacement or movement (in.)
ρ = density (lb/ft³)
ρ' = density (g/cm³)
μ = absolute viscosity (centipoise)
μ_a = absolute viscosity (lb/ft sec)
μ_a' = absolute viscosity (slug/ft sec)
ν = kinematic viscosity (centistokes)
ν' = kinematic viscosity (ft²/sec)

Subscripts
1 = upstream condition
2 = downstream condition
c = critical point value

BABCOCK FORMULA:

$$\Delta P = 3.63 \times 10^{-8} \left(\frac{d + 3.6}{d^6}\right) W^2 L \overline{V}$$

$$\Delta P = 0.470 \left(\frac{d + 3.6}{d^6}\right) w^2 L \overline{V}$$

BERNOULLI EQUATION:

$$H = \frac{v^2}{2g} + \frac{144P}{\rho} + Z$$

$$H = \frac{v^2}{2g} + \frac{p}{\rho} + Z$$

$$H = 1.38 \times 10^{-4} \frac{V^2}{g} + \frac{144P}{\rho} + Z$$

BLASIUS EQUATION:

$$f = \frac{0.316}{\sqrt[4]{R_e}}$$

CRITICAL POINT EQUATIONS:

$$T_c = \left(\frac{2}{\gamma + 1}\right) T_t$$

$$P_c = P_1 \left(\frac{2}{\gamma + 1}\right)^{\gamma/(\gamma-1)}$$

$$V_c' = \sqrt{\frac{2g\gamma R T_t}{\gamma + 1}}$$

$$V_c' = 109.7 \sqrt{T_t} \text{ (air)}$$

DARCY'S EQUATION:

$$h_L = f \frac{L}{D} \frac{v^2}{2g}$$

$$\Delta P = \frac{\rho}{144} \left(f \frac{L}{D}\right) \frac{v^2}{2g}$$

(For other forms and derivatives of Darcy's equation see: Head Loss Equations–Laminar, Head Loss Equations–Turbulent, Liquid Flow Equations, and Gas Flow Equations.)

DYNAMIC FORCE:

$$F = \frac{\rho v^2 A}{g}$$

$$F = \frac{\rho q^2}{gA}$$

$$F = 22.199 \times 10^{-6} \frac{\rho Q^2}{a}$$

$$F = 1.542 \times 10^{-7} \frac{\rho Q^2}{A}$$

$$F = 1.385 \times 10^{-3} \frac{SQ^2}{a}$$

$$F = 5.53 \times 10^{-9} \frac{\overline{Q}^2}{a}$$

$$F = 7.96 \times 10^{-7} \frac{\overline{Q}^2}{A}$$

FLOW OF GASES:

$$w = \frac{CaP_1}{\sqrt{RT_1}} \sqrt{g\gamma \left(\frac{2}{\gamma + 1}\right)^{(\gamma+1)/(\gamma-1)}}$$

$$w = 0.5322 \frac{CaP_1}{\sqrt{T_1}} \text{ (air)}$$

$$w = \frac{CaP_1}{\sqrt{RT_1}} \sqrt{\frac{2g\gamma}{\gamma - 1} \left[\left(\frac{P_2}{P_1}\right)^{2/\gamma} - \left(\frac{P_2}{P_1}\right)^{(\gamma+1)/\gamma}\right]}$$

$$w = 0.525 \, Y d^2 \sqrt{\frac{\Delta P}{K \overline{V}_1}}$$

$$w = 0.525 \, Y C d^2 \sqrt{\frac{\Delta P}{\overline{V}_1}}$$

$$W = 1891 \, Y d^2 \sqrt{\frac{\Delta P}{K \overline{V}_1}}$$

$$W = 1891\, YCd^2 \sqrt{\frac{\Delta P}{V_1}}$$

$$q = YCa \sqrt{\frac{288\, g \Delta P}{\rho}}$$

$$q_h = 4.07 \times 10^4\, Yd^2 \sqrt{\frac{P_1' \Delta P}{KT_1 S_g}}$$

$$q_h = 4.07 \times 10^4\, YCd^2 \sqrt{\frac{P_1' \Delta P}{T_1 S_g}}$$

$$q_h = 2.47 \times 10^4\, \frac{Yd^2}{S_g} \sqrt{\frac{\rho_1 \Delta P}{K}}$$

$$q_h = 2.47 \times 10^4\, \frac{YCd^2}{S_g} \sqrt{\rho_1 \Delta P}$$

$$q_m' = 678\, Yd^2 \sqrt{\frac{P_1' \Delta P}{KT_1 S_g}}$$

$$q_m' = 678\, YCd^2 \sqrt{\frac{P_1 \Delta P}{T_1 S_g}}$$

$$q_m' = 412\, \frac{Yd^2}{S_g} \sqrt{\frac{\rho_1 \Delta P}{K}}$$

$$q_m' = 412\, \frac{YCd^2}{S_g} \sqrt{\rho_1 \Delta P}$$

$$q = 11.3\, Yd^2 \sqrt{\frac{P_1' \Delta P}{KT_1 S_g}}$$

$$q = 6.87\, \frac{Yd^2}{S_g} \sqrt{\frac{\rho_1 \Delta P}{K}}$$

$$q = 6.87\, \frac{YCd^2}{S_g} \sqrt{\rho_1 \Delta P}$$

$$q_m' = 33 d^2\, P_1' \sqrt{\frac{P_2'}{P_1'}\left[\left(\frac{P_2'}{P_1'}\right)^{0.43} - \left(\frac{P_2'}{P_1'}\right)^{0.71}\right]}$$

$$q_m' = 16.1\, C_v \sqrt{\frac{\Delta P'(P_1' + P_2')}{ST_1}}$$

$$q_m' = 22.67\, C_v \sqrt{\frac{P_1' \Delta P'}{ST_1}}$$

$$q_m' = 23.16\, C_v \sqrt{\frac{P_2' \Delta P'}{ST_1}}$$

$$q_m' = 86.33\, C_v \sqrt{\frac{(P_1')^2 - (P_2')^2}{MT_1}}$$

$$q_m' = 16.1\, C_v \sqrt{\frac{(P_1')^2 - (P_2')^2}{ST_1}}$$

FLOW OF LIQUIDS:

$$Q = C_v \sqrt{\frac{\Delta P}{S}}$$

$$Q = 7.9\, C_v \sqrt{\frac{\Delta P}{\rho}}$$

$$Q = 19.65\, d^2 \sqrt{\frac{h_L}{K}}$$

$$Q = 19.65\, Cd^2 \sqrt{\frac{h_L}{K}}$$

$$Q = 236\, d^2 \sqrt{\frac{\Delta P}{K\rho}}$$

$$Q = 236\, Cd^2 \sqrt{\frac{\Delta P}{K\rho}}$$

$$Q = 9.1 \sqrt{\frac{d^5 \Delta P}{fL}}$$

$$Q = 4080\, \frac{d^4 \Delta P}{\nu L}$$

$$Q = 11\, \frac{D^4 \Delta p}{\mu L}$$

$$q = 0.0438\, d^2 \sqrt{\frac{h}{K}}$$

$$q = 0.525\, d^2 \sqrt{\frac{\Delta P}{K\rho}}$$

$$q = CA \sqrt{2gH}$$

$$q = 0.0203 \sqrt{\frac{d^5 \Delta P}{fL}}$$

$$q = 9.1\, \frac{d^4 \Delta P}{\nu L}$$

$$q = 0.0245\, \frac{D^4 \Delta p}{\mu L}$$

$$w = 0.0438\, d^2 \sqrt{\frac{h_L \rho^2}{K}}$$

$$w = 0.0438\, Cd^2 \sqrt{\frac{h_L \rho^2}{K}}$$

$$w = 0.525\, d^2 \sqrt{\frac{\rho \Delta P}{K}}$$

$$w = 0.525\, Cd^2 \sqrt{\rho \Delta P}$$

$$W = 157.6 \, d^2 \sqrt{\frac{h_L \rho^2}{K}}$$

$$W = 157.6 \, Cd^2 \sqrt{h_L \rho^2}$$

$$W = 1891 \, d^2 \sqrt{\frac{\rho \Delta P}{K}}$$

$$W = 1891 \, Cd^2 \sqrt{\rho \Delta P}$$

FRICTION FACTOR EQUATIONS:

$$f = \frac{64}{R_e} \quad \text{(laminar)}$$

$$f = 0.1225 \left(\frac{d}{q_h S_g}\right)^{0.1461} \quad \text{(panhandle)}$$

$$f = \frac{0.316}{\sqrt[4]{R_e}} \quad \text{(turbulent)}$$

$$f = \frac{0.032}{\sqrt[3]{d}} \quad \text{(Weymouth)}$$

HAGEN-POISEVILLE LAW:

$$\Delta P = 6.68 \times 10^{-4} \frac{\mu L}{d^2}$$

HAZEN-WILLIAMS FORMULA:

$$Q = 61.88 \, d^{2.63} \left(\frac{P_1 - P_2}{L}\right)^{0.54}$$

HEAD LOSS EQUATION—LAMINAR FLOW:

$$h_L = 0.0962 \frac{\mu v L}{\rho d^2}$$

$$h_L = 0.0393 \frac{\mu Q L}{\rho d^4}$$

$$h_L = 17.65 \frac{\mu q L}{\rho d^4}$$

$$h_L = 4.9 \times 10^{-3} \frac{\mu W L}{\rho^2 d^4}$$

$$h_L = 0.0275 \frac{\mu B L}{\rho d^4}$$

$$h_L = 3.28 \times 10^{-3} \frac{L}{D} \frac{\mu Q}{\rho d^3}$$

$$h_L = 8.02 \times 10^{-3} \frac{L}{D} \frac{\mu v}{\rho d}$$

$$h_L = 1.47 \frac{L}{D} \frac{\mu q}{\rho d^3}$$

$$h_L = 4.08 \times 10^{-4} \frac{L}{D} \frac{\mu W \overline{V}^2}{d^3}$$

HEAD LOSS EQUATION—TURBULENT FLOW

$$h_L = 0.1863 \frac{fLv^2}{d}$$

$$h_L = 0.0311 \frac{fLQ^2}{d^5}$$

$$h_L = 6260 \frac{fLq^2}{d^5}$$

$$h_L = 4.83 \times 10^{-4} \frac{fLW^2 \overline{V}}{d^5}$$

$$h_L = 0.01524 \frac{fLB^2}{d^5}$$

$$h_L = 2.59 \times 10^{-3} \frac{KQ^2}{d^4}$$

$$h_L = 522 \frac{Kq^2}{d^4}$$

$$h_L = 4.03 \times 10^{-5} \frac{KW^2 \overline{V}^2}{d^4}$$

$$h_L = 1.27 \times 10^{-3} \frac{KB^2}{d^4}$$

HYDROSTATIC FORCE

$$P = .433 \, \rho H$$

IDEAL GAS EQUATION

$$P' = \frac{\rho RT}{144}$$

$$pV' = wRT$$

$$pV' = nMRT$$

$$pV' = 1544 \, nT$$

$$pV' = 1544 \frac{W}{M} T$$

$$p = \rho RT$$

SOME USEFUL FORMULAS

Isothermal Equation

$$w = \left[\left(\frac{144gA^2}{\overline{V}_1\left(\frac{fL}{D} + 2\ln\frac{P'_1}{P'_2}\right)}\right)\left(\frac{(P'_1)^2 - (P'_2)^2}{P'_1}\right)\right]^{1/2}$$

$$w = 0.371\left[\frac{d^4\left(\frac{(P'_1)^2 - (P'_2)^2}{P'_1}\right)}{\overline{V}_1\left(\frac{fL}{D} + 2\ln\frac{P'_1}{P'_2}\right)}\right]^{1/2}$$

Pressure Drop

$$\Delta P = \frac{\rho f L v^2}{144 D 2g}$$

$$\Delta P = 1.294 \times 10^{-4}\, \frac{fL\rho v^2}{d}$$

$$\Delta P = 3.59 \times 10^{-7}\, \frac{fL\rho V^2}{d}$$

$$\Delta P = 3.36 \times 10^{-6}\, \frac{fLW^2\overline{V}}{d^5}$$

$$\Delta P = 1.078 \times 10^{-4}\, K\rho v^2$$

$$\Delta P = 3.00 \times 10^{-8}\, K\rho v^2$$

$$\Delta P = 3.62\, \frac{K\rho q^2}{d^4}$$

$$\Delta P = 1.799 \times 10^{-5}\, \frac{K\rho Q^2}{d^4}$$

$$\Delta P = 2.80 \times 10^{-7}\, \frac{KW^2\overline{V}}{d^4}$$

$$\Delta P = 7.26 \times 10^{-9}\, \frac{fLTq_h^2 S_g}{d^5 P'}$$

$$\Delta P = 1.959 \times 10^{-8}\, \frac{fLq_h^2 S_g^2}{\rho d^5}$$

$$\Delta P = 6.05 \times 10^{-10}\, \frac{Kq_h^2 TS_g}{d^4 P'}$$

$$\Delta P = 1.633 \times 10^{-9}\, \frac{Kq_h^2 S_g^2}{\rho d^4}$$

$$\Delta P = 0.072\, f\rho\, \frac{L}{d}\, v^2$$

$$\Delta P = 0.0234\, \rho f\, \frac{LQ^2}{d^5}$$

$$\Delta P = 6 \times 10^{-4}\, \rho v\, \frac{Lv}{d^2}$$

$$\Delta P = 2.45 \times 10^{-4}\, \rho v\, \frac{L}{d^4}\, Q$$

Reynolds Number

$$R_e = \frac{\rho v D}{\mu_a}$$

$$R_e = 123.9\, \frac{\rho v d}{\mu}$$

$$R_e = \frac{\rho v D}{g\mu'_a}$$

$$R_e = 22{,}700\, \frac{\rho q}{\mu d}$$

$$R_e = 50.6\, \frac{\rho Q}{\mu d}$$

$$R_e = 35.4\, \frac{B\rho}{\mu d}$$

$$R_e = 6.31\, \frac{W}{\mu d}$$

$$R_e = 0.482\, \frac{q_h S_g}{\mu d}$$

$$R_e = \frac{vD}{v'}$$

$$R_e = 7740\, \frac{vD}{v}$$

$$R_e = \frac{vd}{12 v'}$$

$$R_e = 1.419 \times 10^6\, \frac{q}{dv}$$

$$R_e = 3160\, \frac{Q}{dv}$$

$$R_e = 394\, \frac{W\overline{V}}{dv}$$

Sonic Flow Equation

$$w = \frac{CaP_1}{\sqrt{RT_1}} \sqrt{g\gamma\left(\frac{2}{\gamma+1}\right)^{(\gamma+1)/(\gamma-1)}}$$

$$w = 0.5322\, \frac{CaP_1}{\sqrt{T_1}}\ \text{(air)}$$

SPECIFIC GRAVITY

$$S = \frac{\rho}{\rho^{water}}$$

$$S_g = \frac{\rho}{\rho^{air}}$$

$$S = \frac{141.5}{131.5 + °API} \quad \text{(oil)}$$

$$S = \frac{140}{140 + °BAUME} \quad \text{(light liquids)}$$

$$S = \frac{145}{145 - °BAUME} \quad \text{(heavy liquids)}$$

SPITZ GLASS FORMULA

$$q_h = 3550 \sqrt{\frac{\Delta h_w d^5}{S_g L \left(1 + \frac{3.6}{d} + 0.03 d\right)}}$$

SPRING RATE:

$$R' = \frac{F}{\delta}$$

$$R' = \frac{Gd^4}{8D^3 N}$$

$$M_t = \frac{E d_w^4}{10.8 D_s N}$$

SPRING STRESSES:

$$S' = \frac{2.55 F D_s K'}{d_w^3} \quad \text{(compression springs)}$$

$$S' = \frac{10.8 M_o}{d_w^3} \quad \text{(torsion spring)}$$

VELOCITY OF FLOW:

$$v = \frac{q}{A}$$

$$v = \frac{q}{144 a}$$

$$v = 3.73 \sqrt{\frac{d \Delta P}{fL}}$$

$$v = 183.3 \frac{q}{d^2}$$

$$v = 1670 \frac{d^2 \Delta P}{vL}$$

$$v = 0.408 \frac{Q}{d^2}$$

$$v = 183.3 \frac{w \overline{V}}{d^2}$$

$$v = 0.0312 \frac{D^2 \Delta \rho}{\mu L}$$

$$v = 0.0509 \frac{W \overline{V}}{d^2}$$

$$v = 0.286 \frac{B}{d^2}$$

$$v = 1.44 \times 10^{-3} \frac{q_h T}{P' d^2}$$

$$v = 3.89 \times 10^{-3} \frac{q_h S_g}{\rho d^2}$$

$$V = \frac{q_m}{A}$$

$$V = \frac{2.40 \, W \overline{V}}{a}$$

$$V = 3.06 \frac{W \overline{V}}{d^2}$$

$$V = 0.233 \frac{q_h S_g}{\rho d^2}$$

$$V = 0.0865 \frac{q_h T}{P' d^2}$$

VELOCITY, SONIC:

$$v = \sqrt{g \gamma R T}$$

$$v = 49.1 \sqrt{T} \quad \text{(air)}$$

VISCOSITY:

$$v = 0.0022 \, t_s - \frac{1.8}{t_s}$$

$$= \frac{\mu}{\rho'}$$

$$= \frac{\mu_a}{g \rho}$$

WEYMOUTH FORMULA:

$$q_h = 28.0 d^{2.667} \sqrt{\frac{(P'_1)^2 - (P'_2)^2}{S_g L_m} \left(\frac{520}{T}\right)}$$

V Properties of Some Valve Materials

110 VALVE ENGINEERING AND DESIGN DATA

Table 1 Corrosion Resistance of Some Typical Valve Materials. (Courtesy of Jordan Valve, Division of Richards Industries, Inc.)

CHEMICAL 1	Aluminum	Asbestos	Bronze	Carpenter 20	Cast Iron	Carbon Steel	Ductile Iron	Hastelloy B	Jordonite	Monel	Neoprene	Nickel	Plastisol	Polyethylene	303 S.S.	304 S.S.	316 S.S.	Teflon	CHEMICAL 2
Acetaldehyde	1	2	1 2	1		2		1	1 2	1 2	1	1	1 2	1 2	1 2	1 2	1 2	1 2	Zinc Sulfate (acid)
Acetate Solvents (crude)	1 2		1 2	1 2	2	2	1 2	2	1 2	1 2	2 1		2	2	1 2	1 2	1 2	1 2	Zinc Sulfate
Acetate Solvents (pure)	1		1	1		1		1	1	1	1	1	1	1	1	1	1		Zinc Plating Solution
Acetic Acid (crude)	1		1 2	1	1	1		1	1	1	2	1	1	1 2	1 2	1 2	1 2	2	Zinc Hydrosulfite
Acetic Acid (pure)	1 2		1 2	1 2	1 2	1 2	1 2	1 2	1 2	1 2	2 1	1	1 2	1 2	1 2	1 2	1 2	1 2	Zinc Chloride
Acetic Acid Vapors	1		1	1 2	1 2	1	1 2	1	1	1	1	1		1	1	1 2	1		Zinc Cyanide Solution
Acetic Anhydride	1		1 2	1 2	1	1		1	1	1	1	1	1	1	1	1	1		Zinc Carbonate
Acetone	1		1	1	1	1	1	1	1	1 2	2 1	1	1 2	1	1	1	1		Zinc Ammonium Chloride
Acetylene	1	1	1	1	1	1		1	1	1	2 1	1			1	1	1		Zinc Acetate
Alcohol – Amyl	1	1	1	1	1	1		1	1	1	2 1	1			1	1 2	1		Zeolite
Alcohol – Butyl	1 2		1 2	1	2 1 2	2	2	1	1	1 2	2 1	1	1 2	1 2	1 2	1 2	1 2	2	Xylene
Alcohol – Ethyl	1 2	2	1 2	1 2	1 2	1 2	1 2	1	1	1 2	2 1	1	1 2	1 2	1 2	1 2	1 2	2	Whiskey & Wines
Alcohol – Isopropyl	1 2		1 2	1		1		1	1	1 2	1	1	2	1 2	1 2	1 2	1 2	1	White Liquor
Alcohol – Methyl	1		1	1	1 2	1 2		1	1	1 2	2	1		1	1 2	1 2	1 2	1	Wax - molten
Alcohol – Diacetone			2	2 1 2	2	2	2	1	1	1 2	1 2	1			1	1	1		Water - sea
Alcohol – Oleyl	2		2	2	1	2	2	1	1	1 2	2	2	2		2	2	2	2	Water - fresh
Alkaform						2										1			Vinyl Chloride
Alum	1					2	1			1					1 2	1 2	1 2		Vinyl Acetate
Alumina	2	1	1 2	2	2	2	2	2	1	1 2	1 2	2	2	2	2	2	1 2	2	Vinegar
Aluminum Acetate	1 2	2	2				2	1 2	1 2	1 2	2	2	2	2	2	2	1 2	2	Vegetable Oils
Aluminum Chloride	1 2		1 2	1 2	1 2			1 2	1	1 2	1 2	1	1 2	1 2	1 2	1 2	1 2	2	Varnish
Aluminum Fluoride		1	2 1	1				1 2	1 2	1 2	2	1 2		2	2	2	2	1	Uric Acid
Aluminum Hydroxide	2		2	1 2	2	2		1 2	1 2	1 2	1	2	2		1	1 2	1 2	1	Turpentine
Aluminum Sulfate 10% Boil	1		1 2	1		1		1	1 2	1 2	1 2		1 2	1 2	1 2	1 2	1 2	1	Trisodium Phosphate
Aluminum Sulfate Saturate room	1		1 2	1 2	1	1	1	1	1	1	1		1	1	1	1 2	1	1	Triodium Phosphate
Aluminum Sulfate Saturate Boil	1		1 2	1 2	1	1	1	1	1	1	1	1			1	1 2	1	1	Tripotassium Phosphate
Amines			1	2					1	1					1	1 2	1	1	Triphenylphosphite
Ammonia, Dry		1		1					1	1 2		1	1	1	1	1 2	1	1	Triethylamine
Ammonia (gas of liquid)	1	1	1 2	1		1		1	1 2	1 2	2	1	2		1	1 2	1	1	Triethanolamine
Ammonium – Bicarbonate	1			1 2	1			1	1	1	1	1			1	1 2	1	1	Trichlorotrifluroethane
Ammonium Carbonate	1	1	1	1 2	1		1 2	1	1	1	1			1	1	1 2	1	1	Trichloropropane
Ammonium Chloride	1		1 2	1	1	2		1	1	1	1	1	1	1	2	1 2	1 2	1	Trichloromonofluoroethane
Ammonium Dophospate	2			1 2				1	1	1	1 2			1	2	1 2	1 2	1	Trichloroethylene (moist)
Ammonium Hydroxide	1 2	1	1 2	1			1	1 2	1 2	2 1	2 1	1	1 2	1 2	1 2	1 2	1 2	1 2	Trichloroethylene (dry)
Ammonium Hydrozide				1						1 2					1		1		Trichlorobenzene
Ammonium Monosulfate	2		2 1												1 2	1 2		1	Trichloroacetic Acid
Ammonium Nitrate	1		1	1	1	1		1	1 2 1	1	1	1		1	1	1	1	1	Tretolite

KEY: 1 - Chemical in column 1 may be used   ● 1 - Chemical in column 1 may not be used
      2 - Chemical in column 2 may be used   ● 2 - Chemical in column 2 may not be used
      ( ) Conditional or no information

PROPERTIES OF SOME VALVE MATERIALS 111

Table 1 (Continued)

CHEMICAL 1	Aluminum	Asbestos	Bronze	Carpenter 20	Cast Iron	Carbon Steel	Ductile Iron	Hastelloy B	Jordanite	Monel	Neoprene	Nickel	Plastisol	Polyethylene	303 S.S.	304 S.S.	316 S.S.	Teflon	CHEMICAL 2
Ammonium Oxalate	1		2	1	1					1 2		1				1	1 2	1	Transmission Oil
Ammonium Persulfate	1		1		1					1 2	1	2 1				1			Toxaphene
Ammonium Phosphate (mono)	1 2		1 2	1	1	1 2				1 2		1			1 2 1	1 2	1 2		Tomato Juice
Ammonium Phosphate (di)	1 2		1		1	1				1 2		1			1	1		1	Toluene Sulfonic Acid
Ammonium Phosphate (tri)	1 2	2 1	1 2	1	1 2	1 2		2		1 2 1	1 2			2	1 2	1 2	1 2	2	Toluene or Toluol
Ammonium Sulfate	1 2	1	1 2	1	1 2	1		2		1 2 1	1	1		1	1 2	1 2	1	1	Titanium Tetrachloride
Ammonium Thiocyanate										1									Tin Plating Solution
Amyl Acetate	1			1						1 2	1				1	1	1	1	Thiophene
Amyl Chloride	1									2 1									Thiamine Hydrochloride
Aniline	1		1	1						2 1	1			1	1 2	1	1 2	1 2	Tetraphosphoglucosate
Aniline Sulfite	1		1 2					1 2			1 2						1 2	1 2	Tetraphosphoric Acid
Aniline Dyes		1								1								1	Tetramine
Aniline Oils		1																1	Tetrachloroethane
Aniline Hydrochloride	1				2					1	1		1			1 2 1	1 2	1	Terpene Monocyclic
Antimony Chloride									1 2										Tennox
Antimony Trichloride:	2		2 1		1 2	2			1 2	1 2	2 1	1 2 1	2	1	1 2	1 2 1	1 2	2	Tartaric Acid
Antioxidants									1 2	1 2				1				1	Tar Acids
Arochlor	2	2	2	1 2						2	1				2 1	2 1			Tar
Arsenic Acid		1		1	2 1	2		1	1 2 1	1 2 1	1 2	1		1 2	2 1	1 2 1		1	Tannin
Arsenic Trichloride		2		1	1				1	1									Tanning Liquor
Asphalt	1	1 2	2 1	1 2			2		1 2	1 2	2 1	2	2	1 2	1 2	1 2	1 2	1 2	Tannic Acid
Barium Carbonate	1		1	2	1 2	1			1 2	1 2	1	1 2		2 1	2	1	1	1	Tallow, Molten
Barium Chloride		1	1	1 2	1		1 2		1	1	1			1	1 2	1 2	1		Tall Oil
Barium Hydrate	1								1 2	1 2	2 1				1				Talc Slurry
Barium Hydroxide	2	1 2	1 2	1 2	1 2	1 2	1	1	2	2	1	2		1 2	1 2	1 2	1 2	1 2	Sulfurous Acid
Barium Nitrate			1		1					2				2	2 1	1 2 1			Sulfuric Acid (spent)
Barium Sulfate	2		2 1	2		2			1 2	1 2	1 2		2	1 2	1 2	1 2	1 2	1	Sulfuric Acid (95-100%)
Barium Sulfide	2		2	2	1 2	2			1 2	1 2	1 2		2	2	2	2	2	1	Sulfuric Acid (90-95%)
Beer	1 2	1	2	1 2	1 2	1 2	1 2	2	1 2	1 2	1		1 2	1 2	1 2	1 2	1 2	1	Sulfuric Acid (75-90%)
Beet Sugar Liquor	1 2	1	1 2	2	1	1	2	2	1 2	1 2	1		2	2 1	1 2	1 2	1 2	1	Sulfuric Acid (10-75%)
Benzaldehyde	2		2	2		2	2	2	1 2 1	1	2	1		2	1 2	1 2	1 2		Sulfuric Acid (0-10%)
Benzene (Benzol)	1 2		1	2	1 2	1	2			1 2 1	1 2 1	2	1	1 2	1 2	1 2	1 2		Sulfur Trioxide (dry)
Benzenesulfonic Acid	1 2	2	2	1	1 2	2	2		1 2	2	2	1	2	2	2	2	2		Sulfur Dioxide (dry)
Benzine	1 2		1 2	2	1 2	1 2	2		1 2	1 2				1 2	1 2	1 2	1 2	1 2	Sulfur Chloride
Benzoic Acid	1 2	2	2	1 2	1 2	1 2	2	1 2	1 2	1 2	1 2	1		2	1 2	1 2	1 2	1 2	Sulfur (molten)
Black Liquor		1			1							2		2		2	2	2	Sulphonyl Chloride
Blast Furnace Gas					2														Sulfite Liquor

KEY: 1 - Chemical in column 1 may be used • 1 - Chemical in column 1 may not be used
2 - Chemical in column 2 may be used • 2 - Chemical in column 2 may not be used
( ) Conditional or no information

Table 1 (Continued)

CHEMICAL 1	Aluminum	Asbestos	Bronze	Carpenter 20	Cast Iron	Carbon Steel	Ductile Iron	Hastelloy B	Jordanite	Monel	Neoprene	Nickel	Plastisol	Polyethylene	303 S.S.	304 S.S.	316 S.S.	Teflon	CHEMICAL 2
Bleaching Powder, Wet	1									1	1					1	1	1	Sulfate Oils
Boiler Compounds (pH 8.0)	2	2	2	2	2			1 2	1	1 2	1			2	2	2	2		Sulfate Liquor
Boiler Acid-phosphate type	2	2	2						1 2	1 2	1 2	2		1	2	2	2		Sugar Solution
Borax	1	1		1 2		2		2		1 2				1	1 2	1 2	1 2	1	Styrene
Bordeaux Mixture	2		2	2	2	1			1 2	1 2	1			1 2		2	2		Stoddard Solvent
Boric Acid	1 2	2	1 2	1 2	1	1 2		1 2	1 2	1 2 1	1	1	1 2	1 2	1 2	1 2	1 2	1	Stearic Acid
Boron Trichloride					1						2						1		Steam Condensate
Boron Trifluoride			2	2				2		2		2	2	2	2	1 2	1 2	2	Steam
Brine		1 2	1 2		1 2				1 2	1 2	1 2			1 2	2	1 2	1 2		Starch
Bromine (Wet)			1 2	1 2	1	2	1	1				1	1	1	2	1	1 2		Stannous Chloride
Bromine (Dry)	1	1	1				2		1	1		1		1			2		Stannous Bisulfate
Butadiene	1 2	2	1 2	2	1	1		1 2	1	1 2	2	2	1	2	1	1 2	1 2	2	Stannic Chloride
Butane	1 2	1 2	1 2	1 2	2	2	1	2		1 2		2		1		1 2	1 2		Soybean Oil
Buttermilk	1		1	1	2	1			1	1	1				1 2	1 2		1	Sorbitol
Butyl Acetate	1	1	1	1 2		1 2	1	1	1 2	1 2					2	1	1 2		Sodium Triphosphate
Butyl Catechol Tert										2	1				1				Sodium Tetraphosphate
Butyl Cellosolve	1		1	1 2					1	2						2			Sodium Tetraborate
Butyl Chloride	2		2	2 1		2	2		2	2		2	2		2	2	1 2	2	Sodium Sulfite
Butyl-p-Aminopheno	2	2	2	2	2	2	2	2	1 2	2	2	2		1	2	2	2	2	Sodium Sulfide
Butyl-Stearate	1 2	2	1 2	1 2	2	2			1 2	1 2	1 2	2		2	2	2	2	2	Sodium Sulfate
Butylene	1					1			1		1			1		1	1		Sodium Silicofluoride
Butyric Acid	1 2	2	1 2	1 2	1 2	2	1 2	1 2	1 2	1 2	1 2	1 2	2	2	2	1 2	1 2	2	Sodium Silicate
Borax					2				1	1 2	1 2			1	1 2	1 2	1 2		Sodium Salts
Cadmium Sulfate					2	2									2				Sodium Salicylate
Calcium Acetate	1	1	1	1	2	2	1		1						2	1			Sodium Resinate
Calcium Bisulfite	1	1				1			1	1	1	1		1 2	1	1	1		Sodium Pyrophosphate
Calcium Carbonate	1		1	2	1		1		1	1 2	1 2			1 2	1	1	1		Sodium Polyphosphate
Calcium Chlorate		2	1		2	2		2	1		1			1		1 2	1 2		Sodium Phosphate (neutral)
Calcium Chloride	1 2	2	1	1 2	1 2	1 2	1 2	1 2	1	1 2 1	1	1		1 2	1 2	1 2	1 2	1	Sodium Phosphate (acid)
Calcium Nitrate 40%	1								1 2	1 2		1				1			Sodium Plumbite
Calcium Hydroxide	1 2	1 2	1 2	1 2	2	2	1 2		1 2	1 2	2	1 2		2	1 2	1 2	1 2	1 2	Sodium Peroxide
Calcium Hypochlorite	1	1 2	1 2	1	1	1 2 1	1		1 2	1 2	1 2	1 2		2	1	1 2	1 2	1 2	Sodium Perborate
Calcium Sulfate	1		1	1					1	1 2	1 2					1	1		Sodium Orthosilicate
Calgon		2	2	2 1	2	2			1	1	1 2			1 2	2	2	2		Sodium Nitrite
Camphene	2	2			1 2	2	2	2	2	2		2		2	2	2	2	2	Sodium Nitrate
Camphor	1						1		1	1	1		1			1			Sodium Naphthsulfonate
Cane Sugar Liquor	1	1 2		1 2	1	1			1 2	1 2				1	1	1	1 2		Sodium Triphosphate

KEY: 1 - Chemical in column 1 may be used   • 1 - Chemical in column 1 may not be used
2 - Chemical in column 2 may be used   • 2 - Chemical in column 2 may not be used
( ) Conditional or no information

PROPERTIES OF SOME VALVE MATERIALS 113

Table 1 (Continued)

CHEMICAL 1	Aluminum	Asbestos	Bronze	Carpenter 20	Cast Iron	Carbon Steel	Ductile Iron	Hastelloy B	Jordantie	Monel	Neoprene	Nickel	Plastisol	Polyethylene	303 S.S.	304 S.S.	316 S.S.	Teflon	CHEMICAL 2
Carbolic Acid (phenol)	1				2					1 2							1 2	1	Sodium Diphosphate
Carbon Monoxide			2	2	1						2					1	1 2		Sodium Monophosphate
Carbon Dioxide (wet)	1	1	1			1	1			1 2	2 1	1		1	1	1	1		Sodium Methylate
Carbon Dioxide (dry)			1	1	1					1 2	2 1	1					1	1	Sodium-M-Silicate
Carbon Disulfide	1 2	2	1 2	1 2		1 2				1 2	1 2			1 2	1 2	1 2	1 2	1	Sodium-M-Phosphate
Carbon Tetrachloride			1			1		1	1	1 2	1 2	1	1	1	1	1 2	1 2	1	Sodium Diphosphate
Carbonated Beverages	1	1			2 1	2	1				1				1		1		Sodium Oleate
Carbonated Water		1			2 1	2 1				1	1				1 2	1 2	1		Sodium Lactate
Carbolic Acid	1 2			2 1	1 2	1 2	1 2		1	1 2	2	1		1	1 2	1 2	1 2		Sodium Hyposulfite
Castor Oil	2	1 2	1 2	2	2	2	2		1	1 2			2	2	2	2	2	2	Sodium Hypochlorite
Catechol	2	2	2		2	2	2	2		1 2	1 2	2	2	2	2	2	2	2	Sodium Hydroxide 20% or Hot
Caustic Soda	1 2	2	1 2	1 2	2	2	2 1 2	2		1 2	2	2	2	2	1 2	1 2	1 2	1 2	Sodium Hydroxide 0-20%
Cellosolve (butyl or ethyl)	2			2 1						1 2							1		Sodium Hydrosulfite
Cellosolve (methyl)	1	1		1	2										2		1		Sodium Glutamate
China Wood Oil (tung)	1 2	1	2	1 2	1 2		2			1 2	1 2	1 2	2	2 1	1	2	1 2	1 2	Sodium Fluoride
Chloric Acid	2			1 2			2			2		2	2	1					Sodium Ferrocyanide
Chlorinated Water			1		1 2		1		1		1			1	1 2	1 2	1		Sodium Ethylate
Chlorine (wet)	1 2	1 2	1 2		2	2 1	1		1	1		1	1	1 2	1 2	1 2	2 1	1	Sodium Dichromate
Chlorine (dry)	1 2	1 2	1	1 2	1 2	1 2	2	1		1 2	2 1		2	1 2	1 2	1 2	1 2	2	Sodium Cyanide
Chloroacetic Acid	1 2	1	1 2	1			1 2			1 2		1		1	1 2	1 2	1	1	Sodium Citrate
Chlorobenzene			1 2	2 1	2 1		2			1	1			1 2	1 2	1 2	1 2		Sodium Chromate
Chlorobromomethane	2	2	1 2	2	2	2	2	2		1 2	2	2	2	2	2	2	2	2	Sodium Chloride
Chloroethane	2	2	2	2	2		2			1 2	2 1	2	2	2	2	2	2	2	Sodium Carbonate (soda ash)
Chloroethylbenzene	2		2	2			2			2		2	1 2		2	2	2		Sodium Bromide
Chloroform	1 2	1 2	1 2	1 2	1	1	1 2			1 2	1 2 1	2	2	1	1 2	1 2	1 2		Sodium Borate
Chlorex	1		1		2			2		1 2	2	2	2		1	1 2	1 2		Sodium Bisulfite
Chlorosulfonic Acid	1 2	2	2	1 2	1 2	2	2	1 2	1	1 2	2	2 1	1 2	2	2	1 2	1 2	1 2	Sodium Bisulfate
Chlorox	2		2 1	2 1		1	2		1 2	1 2	2		1 2		2	2	2		Sodium Bichromate
Chromic Acid (free of SO3)	1 2	2	1	1		1 2	2	2	1 2	1 2	2 1	1 2	1 2	1 2	2	1 2	1 2	1	Sodium Bicarbonate
Chromic Acid (contains SO3)	1 2	1	1		1	1	1 2	2	1	1 2	1	1	1 2		1	1	1	1	Sodium Benzoate
Chrome Plating Sol.		1 2				2			1	1	2 1			1 2	1 2		1 2		Sodium Aluminate
Chromium Sulfate			1					1			2								Sodium Acid Sulfate
Cider	1 2		1 2		2		1 2			1 2	1 2	1 2	2 1 2	2 1 2	2 1	2 1	1 2	2	Sodium Acetate
Citric Acid	1 2	1	2 1	2 1	1 2	1	1 2	1 2	1 2	1 2	1 2	1 2	1 2	1 2	1 2	1 2	1 2	1	Soap (molten)
Clay Slurries	2				1					2					1		1 2		Sludge Acid
Coal Tar (creosote)	1	1	1	2		2	1		1 2	2							1		Sizing, Alkaline
Coca Cola Syrup					1					2					1		1		Sizing, Acid

KEY: 1 - Chemical in column 1 may be used • 1 - Chemical in column 1 may not be used
2 - Chemical in column 2 may be used • 2 - Chemical in column 2 may not be used
( ) Conditional or no information

114 VALVE ENGINEERING AND DESIGN DATA

Table 1 (Continued)

CHEMICAL 1	Aluminum	Asbestos	Bronze	Carpenter 20	Cast Iron	Carbon Steel	Ductile Iron	Hastelloy B	Jordanite	Monel	Neoprene	Nickel	Plastisol	Polyethylene	303 S.S.	304 S.S.	316 S.S.	Teflon	CHEMICAL 2
Coconut Oil	1 2		1 2	1						1						1 2	1 2	2	Silver Plating Sol.
Coffee	1 2		2	1 2			1 2		1	1 2	1 2		2	2	2	1 2	1 2		Silver Nitrate
Coke Oven Gas	1	1		2 1		1			1 2	1	1		2	1		1	1 2		Silver Cyanide
Cod Liver Oil			2 1		2		2		1 2	1 2					2 1	2 1		2	Silver Chloride
Copal Varnish	1 2		1 2	1 2		2	1 2		1 2	1 2	1				2	1 2	1 2	2	Silver Bromide
Copper Acetate	1		1				1 2								1	1	1		Silicon Tetraiodide
Copper Carbonate	1		1			2			2	2	1				2 1	2 1	1		Silicon Tetrachloride
Copper Chloride	1 2		1 2	1		2 1			1 2	1 2			1 2	1 2	1 2	1 2	1 2		Shellac (bleached)
Copper Cyanide	1 2		2 1	1 2		2 1			1 2	1 2	1				2 1	2 1	1 2		Shellac Orange
Copper Nitrate	1		1	1			1		1 2	1 2	1				1	1	1		Santosite
Copper Sulfate	1	1	1	2	1	1	1		1 2	1 2	1			1	1	1 2	1 2	1	Santophen
Cupric Chloride	1		1	2	1		1		1 2	1 2	1				1 1	1 1	1		Santomerse
Cupric Nitrate	1		1	1			1		1 2	1 2	1				1	1			Santobrite
Core Oil	1 2		1	2 1	2	2 1	2		2	1 2	2	2		1	1 2	1 2	1 2	2	Salicylic Acid
Cornstarch Slurries	2		2	2	2	2	2		1 2	1 2	1 2	2	2	2	2	2	2	2	Sal Ammoniac
Cottonseed Oil	1	1	1	1		1			1 2	1 2					1	1	1		Rustang
Cream of Tartar	2		2	1 2	2	2	2		1 2	1 2	2				1 2	1 2	1 2	2	Rosin (light)
Cresylic Acid	2		2	2 1	1 2	2	2		1 2	1 2 1	1 2			1 2	1 2	1 2	1 2	2	Rosin (dark)
Cyanogen Chloride						1				2					1				Resorcinol
Cyanohydrin						1 2				2					1 2	1 2			Quinine Sulfate
Cyclohexane	1		1	2		2	2									2 1	2 1	2	Quinine Bisulfate
Cyclohexylamine				1						2									Quebracho
DDT	1		1							2	1						1		Quasol 80
Detergents	1	1				2			1	1					2 1	2 1	1		Pyroligneous Acid
Developing Solutions	2		1 2	2	2 1	1 2	1 2		2	2	2	2			2	2 1 2	2		Pyrogallic Acid
Dextrose		1		2	2	2		1						1 2	2 1 2	2 1 2			Pyridine
Diacetone																			Pyrethrum Sol.
Diamylamine						2	2		1 2						2	2	2		Propylene Oxide
Dichloroethane			2	2 1	2 1	1 2			1						1 2	1 2			Propylene Glycol
Dichloropentane				2	2	2			1 2						2	2			Propylene Dichloride
Diesel Oil (light)			2						1	1 2	1 2				2	2		2	Propyl Alcohol
Diethanolamine									1 2 1	2 1 1									Propene, Liquefied
Diethylbenzene	2			2		2			1 2	2 1 2		2	2	2	2	2	2	2	Propane, Liquefied
Diethyl Sulfate	2	2	2		2	2			1 2	1 2	2	2	2		2	2	2	2	Propane Gas
Diethylene Glycol									1 2 1 2	1 2 1 2	1			1	1	1 1 2	1		Producer Gas
Dimethyl Phthalate			1	1					2 1	2 1									Prestone
Dinitrochlorobenzene				2 1					2	2				2	2 1	2 1	2		Potassium Triphosphate

KEY:
- 1 - Chemical in column 1 may be used
- 2 - Chemical in column 2 may be used
- 1 - Chemical in column 1 may not be used
- 2 - Chemical in column 2 may not be used
- ( ) Conditional or no information

Table 1 (Continued)

PROPERTIES OF SOME VALVE MATERIALS

CHEMICAL 1	Aluminum	Asbestos	Bronze	Carpenter 20	Cast Iron	Carbon Steel	Ductile Iron	Hastelloy B	Jordanite	Monel	Neoprene	Nickel	Plastisol	Polyethylene	303 S.S.	304 S.S.	316 S.S.	Teflon	CHEMICAL 2
Dioctyl Phthalate				2	1 2										1 2	1 2	2		Potassium Sulfide
Dioxane	2		2		2	2	2	2 1	1 2	1 2	2	2	2	2	2	2	2	2	Potassium Sulfate
Dipentene					2	2			1 2	1 2	2		2		2	2			Potassium Phosphate (alkaline)
Diphenyl					1 2	2			1 2	1 2	2				1 2	1	2 1		Potassium Phosphate (acid)
Diphenyloxide		1	2 1							1									Potassium Peroxide
Distilled Water	2	1	1	2 1 2	1 2		2		1 2	1 2 1 2	1 2		2		2	2	2 1	2	Potassium Permanganate
Distillery Wort	1 2		2	2	2		2							2	2 1	2 1	2 1		Potassium Oxalate
Doctor Sol.	1 2	1 2	2	2		1			1 2 1	1 2 1		2				2 1	2 1		Potassium Nitrate
Dowtherm					2 1			1	1 2	1 2	2	2		2	2 1	2 1	1 2		Potassium Monophosphates
Dyewood Liquor	2		2		1 2		2		2	2	2	2	2		1 2	1 2	1 2		Potassium Iodide
Embalming Fluid	2		2	2	1 2		2	2	2	2	2	2	2	2	1 2	1 2	1 2	2	Potassium Hypochlorite
Enamel	2		2		2	2	2	2	1 2	1 2	2	2	2		1 2	1 2	1 2	2	Potassium Hydroxide
Ethanolamine	1 2				1			2	1 2 1	1 2 1	2	2	2		2	1 2	1 2		Potassium Hydrate
Ether, Diethyl	1 2	1	1 2	2	1 2	1 2 1	1 2	1 2	1 2	1 2 1 2	1 2	2	2 1 2	1 2	1 2 1 2	1 2 1 2	1 2 1 2	2	Potassium Ferrocyanide
Ether, Dibutyl	1	1	1 2	1 2	1 2 1	1 2 1	1 2	1 2	1 2	1 2	1 2	2	2	1 2	1 2 1 2	1 2 1 2	1 2 1 2	2	Potassium Ferricyanide
Ether, Petroleum	1	1			1	1					2 1				2 1	2 1	2 1		Potassium Diphosphate
Ethyl Acetate	1 2	1	1 2	1 2	1 2	1 2	1 2	1 2	1 2 1	1 2 1		2	2 1		1 2 1	1 2 1	2 1 2	2	Potassium Dichromate
Ethyl Acrylate	2		2 2	2 2	2	2	2	2	2	2	2	2	2 2	2	2	2	2	2	Potassium Cyanide
Ethylbenzene	2		2	2	2	2	2	2	1	1			2	2	1	2	2	2	Potassium Chromate
Ethyl Cellulose	2	1	2	2	2 1	2	2	2	2	2	2	2	2	2	2	2	2	2	Potassium Chloride
Ethyl Chloride	1 2		1 2	1 2	2 1 2	1 2	2	1 2	1 2	1 2	1 2	1 2	2	2	1 2 1	1 2 1	1 2 1	1 2	Potassium Chlorate
Ethyl Mercaptan					2 1				1	1					2 1	2 1	2		Potassium Bicarbonate
Ethyl Sulfate	2	1	2	2	1 2	2 1	2	2	1 2	1 2	2	2	2	2	2 1	2 1	1 2	2	Potassium Carbonate
Ethylene (liquefied)	1 2	1	2 1		2		2	2	1 2	1 2	2			2	2 1	2 1	2 1	2	Potassium Bromide
Ethylene Chloride	1		1							1						1	2		Potassium Bisulfite
Ethylene Chlorohydrin	2		2	2		1		2			2				1 2	1 2	1 2		Potassium Bichromate
Ethylene Dibromide			1														2		Potassium Antimonate
Ethylene Dichloride		1					1	1			2								Potassium Alum
Ethylene Glycol	1	1			1	1			1 2	1 2 1		1	1 2 1	1	1	1	1	1	Plating Solution
Ethylene Oxide		1			1				1 2	1 2		1	2 1		1	1	1	1	Pitch
Esters	2		2				2		1 2	1 2 2		1	1	1	1	1	1	1	Pine Oil
Fatty Acids	1 2	1 2			2	2	2		1 2	1 2	1	1	1	1 2	2 1	2 1	2 1	1 2	Picric Acid, Aqueous Sol.
Ferric Chloride	1 2	1 2	1 2	2	2 1 2	1 2 1	2 1 2	1	1 2 1	1 2 1	1	1	1	1	1 2	1 2	1 2	1 2	Picric Acid, Molten
Ferric Hydroxide	1				2	2			1	2	2		1	1	1 2	1 2	1 2	2	Phthalic Anhydride
Ferric Nitrate	1 2		1		1 2	1	1		1 1	1 1			1	1 1	1 2	1 2	1		Phthalic Acid
Ferric Sulfate	1 2	1	1 2 1	1	1 1	1	1 2	2 1	1 1	1 1	2	1	1	1	1 2	1 2	1 2 1	1	Phosphorous Trichloride
Ferrous Ammonium Citrate	1					2	1		2						2	2			Phosphorous Molten

KEY: 1 - Chemical in column 1 may be used • 1 - Chemical in column 1 may not be used
2 - Chemical in column 2 may be used • 2 - Chemical in column 2 may not be used
( ) Conditional or no information — contact: Jordan Valve for assistance

116 VALVE ENGINEERING AND DESIGN DATA

Table 1 (Continued)

CHEMICAL 1	Aluminum	Asbestos	Bronze	Carpenter 20	Cast Iron	Carbon Steel	Ductile Iron	Hastelloy B	Jordanite	Monel	Neoprene	Nickel	Plastisol	Polyethylene	303 S.S.	304 S.S.	316 S.S.	Teflon	CHEMICAL 2
Ferrous Chloride	1 2	1	1 2	1 2	2		1	1	1	1	1	1	1	1 2	1 2	2	2		Phosphoric Anhydride
Ferrous Sulfate	1 2	2	1 2	1 2	2	2	1 2	1 2	1 2	1 2	1 2	1 2	1 2	1 2	1 2	1 2	1 2	1 2	Phosphoric Acid 45%
Filter Aid		2	2	2	2		2	2	2	1 2	1 2	2	2	2	2	2	1 2	2	Phosphoric Acid 0-45%
Fish Oil	2	2	2	2	1 2	2	2	2	2 1 2	1 2 1	2	2	2	2	1 2	1 2	1 2	2	Phosphoric Acid, Crude
Flue Gases	1		1			2										2 1			Phosgene
Fluoboric Acid							1						1 2	2		2 1	1		Phoscaloid
Fluorine	1		1					1		1 2	1	1			1 2				Phenolic Sulfonate
Fluosilicic Acid				1 2			2		1	1	1	1	1				2		Phenosulfonic Acid
Formaldehyde	1 2	1	1 2		1	1	1 2		1	1	1 2	1 2	1	1	1 2	1 2	1 2	1	Phenolic Resins
Formalin	1 2			2	2		2	2	2	2	1	2	2	2	2	2	2	1	Phenol
Formic Acid	1 2	1 2		1 2	1 2	1 2			1	1 2	1 2		2		1 2	1 2	1 2	1	Petroleum Oils (refined)
Freon (liquefied)	1 2	2 1	2	1 2 1	1 2	1 2	2 1	1	2	1 2	1 2	1			1 2	1 2	1 2	1	Petroleum Oils (sour)
Freon (dry)	1		1	1		2	1			1	1				1	1 2	1 2		Perfume
Fruit Juices	1			1		1	1			1 2	2 1	1				1	1	1	Pentane
Fuel Oil	1	1	1	1	1 2	1 2	1			1 2 1	2 1					1	1		Penicillin, Sol.
Fumeric Acid									2		1								Pelargonic Acid
Furfural	1			1	1 2	1 2	1	1	1	1	1		2	1	1 2	1 2	1	1	Pectin
Gallic Acid	1		2	2 1		1 2	1 2	1 2					2		1	1 2	1 2		Peanut Oil
Gasoline (refined)	1	1 2	1	1	1 2	2			1	1				1	1 2	1 2	1 2		Parez 607
Gasoline (sour)	1		1	1		2	1 2	1	2 1 2		1			1	1	1 2	1 2		Paregoric Compound
Gasoline (antioxident)					1				2	2									Paraldehyde
Ginger Ale										1 2					1	1	1 2		Para-formaldehyde
Gelatin	1	1	1	1		1				1 2	1 2		1	1	1	1 2	1 2		Paraffin Oil
Glauber's Salt	1 2			1 2	2	1 2		1 2	1 2 1	1 2 1	2 1	1			1	1 2	1 2		Paraffin
Glucose	1 2	1 2	1 2		1	1	1		1 2	1 2			1		1	1 2	1 2		Palmitic Acid
Glue	1	1	1	1		1	1		1	2	1 2		1		1	1	1		Palmic Acid
Glycerine	1	1	1 2	1		1	1	1			2					1	1		Palm Oil
Glycerol	1	1	1	1 2		1			1	2	1 2 1	1	1	1	1	1 2	1 2	1	Paint Vehicles (except soya)
Glutamic Acid			1						2								1		Paint
Grease					1	2			1							2 1			Ozone
Geen Sulfate Liquor	2	1 2	2	1 2	2	2		2		2	1 2		2	1 2	2 1	2 1 2	1 2	2	Oxygen
Gypsum	2	2	2	1 2	2	2		2		2	2		2	2 1	1	1 2	1 2		Oxalic Acid
Hagan Solution										1 2									Organic Esters
Heptane (liquefied)			1 2	1 2						1 2					1	1	1 2		Olive Oil
Hexamine	2				2	2	1		2	2	2					1 2	1 2	2	Oleum
Hexane	1 2	2	1 2	2 1	2	2	2	2	2 1 2	2 1 2		2		2 1	2	2 1 2	2 1 2		Oleic Acid
Hydazine Hydrate																			Octyl Alcohol

KEY: 1 - Chemical in column 1 may be used   •   1 - Chemical in column 1 may not be used
      2 - Chemical in column 2 may be used   •   2 - Chemical in column 2 may not be used
      ( ) Conditional or no information

PROPERTIES OF SOME VALVE MATERIALS 117

Table 1 (Continued)

CHEMICAL 1	Aluminum	Asbestos	Bronze	Carpenter 20	Cast Iron	Carbon Steel	Ductile Iron	Hastelloy B	Jordanite	Monel	Neoprene	Nickel	Plastisol	Polyethylene	303 S.S.	304 S.S.	316 S.S.	Teflon	CHEMICAL 2
Hydraulic Oil	1		1							1 2							1		Oakite
Hydrobromic Acid	1		1	1				1		1 2	2 1	1		1	1				Nordihydroguaraetic Acid
Hydrocarbons (chlorinated)	2		1	2	2	2	1 2	2	1	2 1		2				2			Nitrous Oxide
Hydrocarbons (alkylated)					2		1									2			Nitric Acid
Hydrocarbons (H_2SO_4)					2	2										2			Nitropropane
Hydrochloric Acid, Cold	1		1									1		1	1	1 2	1		Nitroethane
Hydrocyanic Acid			1	1	2	1 2	1	1		1 2	1 2	1		1 2	1	1 2	1 2	1	Nitrobenzene
Hydrofluoric Acid 5%	1 2	1 2	1 2	2	1 2	1 2	1 2	1 2	1 2	1 2	1 2	1 2	1	2 1	1	1 2	1 2	2	Nitric Acid 50-100%
Hydrofluoric Acid 50%	1 2	1	1 2	1	1 2	1 2	1 2	1 2	1 2	1 2	1 2	2	1 2	1 2	1	1 2	1 2	2	Nitric Acid 40%
Hydrofluoric Acid 60%	1 2	1	1 2	1	1 2	1 2	1	1 2	1 2	1 2	1 2	2	1 2	1 2	1 2	1 2	1 2	2	Nitric Acid 20%
Hydrofluosilicic Acid	2	1	1 2	1 2	2	2		2	2	1 2	1	1	1 2	1 2	1 2	1 2	1 2	2	Nitric Acid 5%
Hydrogen Chloride (gas)	1 2		2	2	2	2	2	2	2	1 2	2	1 2	1 2	2	2	1 2	1 2	2	Nitric Acid (crude)
Hydrogen Fluoride	2	1	2	2				2	2				2	2	2	1 2	1 2		Nickel Sulfate
Hydrogen Gas											1								Nickel Plating Sol.
Hydrogen Peroxide	1 2		1 2	1 2	1	1	1 2	1 2		1 2	1 2	1 2		1	1 2	1 2	1 2	1	Nickel Nitrate
Hydrogen Sulfide	1 2		1 2	1 2	1 2	1 2		1 2	1 2	1 2	2	1 2	1 2	1 2	1 2	1 2	1 2	2	Nickel Chloride
Hydrogen Sulfide (Wet)	1		1			1				1				1		1 2	1	1	Nickel Acetate
Hydroquinone	2	2	2		2	2				1 2	1 2	1 2	1 2	2	2	1 2	1 2	2	Natural Gas
"HYPO" (hyposulfite soda)				2 1		2				1	1					1 2	1 2		Naphthalenic Acid
Ink	1 2		1 2	1	1 2	1 2	1	2	2	1 2	2		2	1 2	1 2	1 2	1 2		Naphthalene
Iodine	1 2		2	1 2	2 1	1 2	1 2	1 2	1 2	1 2	1	1 2	2	1 2	1 2	1 2	1 2	2	Naphtha
Iodoform	1				1											2 1	2		Nalco Solution
Isobutane			2		1 2	2			1	1 2				1		2	1 2		Mustard
Isobutyl Acetate	1		1						2								1		Monoethanolamine
Isoctane				2					2 1	1						2		1	Monochlorodifluromethane
Isopropyl Acetate						1			1						2	2		1	Monochlorobenzene
Isopropyl Ether								1 2	1	1 2									Monochloroacetic Acid
Jet Fuel	1 2		1 2		1 2	2	2	2	1 2	1 2	2	2	2	2	2	1 2	1 2	2	Molasses
Kerosene	1	1 2	1 2		1 2	1	1 2	1	1 2	1 2	2	1	1	2 1	1 2	1 2	1 2	1	Mineral Oil U.S.P.
Ketchup				1 2	1	1 2	1 2		1 2	1 2	2	2		1	1	1 2	1 2		Mine Water
Ketones	2		1 2	2	2 1	2	2	2	2	2	1	2	2	2	1 2	1 2	1 2	2	Milk
Lacquers and Lacquer Solvents	1 2		1 2	2 1	2 1	2			1 2	1 2				1	1 2	1 2	1 2		Methylene Chloride
Lactic Acid	1	1	1	1	1	1	1		1	1 2	1	1	1	1	1 2	1 2	1		Methyl Methacrylate
Lard								2								1 2			Methyl Ketone
Latex			1							2						1		1	Methyl Isobutyl Ketone
Lead Acetate	1				1	1	1		2	1 2		1	1	1	1	1	1		Methyl Formate
Lead Nitrate	2		2		2 1				2	2 1 2		2			1	2	2	2	Methyl Ethyl Ketone

KEY: 1 - Chemical in column 1 may be used • 1 - Chemical in column 1 may not be used
2 - Chemical in column 2 may be used • 2 - Chemical in column 2 may not be used
( ) Conditional or no information

118 VALVE ENGINEERING AND DESIGN DATA

Table 1 (Continued)

CHEMICAL 1	Aluminum	Asbestos	Bronze	Carpenter 20	Cast Iron	Carbon Steel	Ductile Iron	Hastelloy B	Jordanite	Monel	Neoprene	Nickel	Plastisol	Polyethylene	303 S.S.	304 S.S.	316 S.S.	Teflon	CHEMICAL 2
Lead Sulfamate					2	2				2		2			2		2	2	Methyl Chloride (dry)
Lime Slurry					1					1 2 1 2						1			Methyl Cellosolve
Lime Sulfur	1 2	1 2		1 2	1 2	1				1 2 1				1		1	1		Methyl Benzene
L.P.G.	1									2							2		Methyl Acrylate
Levulinic Acid							1			2	2								Methyl Acetate
Linoleic Acid	1 2			2						1 2						1 2	1 2		Methane
Linseed Oil	1	1		1	1		1			1	1 1	1		1	2 1	2 1	2 1		Mesityl Oxide
Lithium Chloride					1										2 1	1			Mercury Salts
Lithium Hydroxide	2	2		2	2		2 1 2			2	2	2		1	2	1 2	1 2	2	Mercury
Lubricating Oils	2	1		2 1	2					1			1			1 2	1 2		Mercurous Nitrate
Magnesium Carbonate	1 2		1 2	1 2	2		2			1		1 2	2	2	1 2	1 2	1 2		Mercuric Cyanide
Magnesium Chloride	1 2	1 2		1 2	1 2	1 2	1 2			1 2 1	1	1	1 2	1 2	1 2	1 2	1 2	1 2	Mercuric Chloride
Magnesium Hydroxide	1 2	1		1	1		1 2			2 1		1	1	1	1	1 2	1 2	1	Mercuric Bichloride
Magnesium Nitrate	1	1					1			2									Mercaptobenzothiazole
Magnesium Oxide					2					1	1								Mercaptans
Magnesium Oxychloride			2	2 1	2 1					2					2	1 2	1 2		Melamine Resins
Magnesium Sulfate	1 2	1		1 2	1 2	1	1 2			1 2 1 2	1 2	1		1	1 2	1 2	1 2	1	Mayonnaise
Maleic Acid	2	2		1	1		2			1									Mash
Maleic Anhydride									2 1										Manganese Sulfate
Malic Acid	1			1 2 1			1			1 2 1		1 2 1		1	1 2	1 2	1 2		Manganese Chloride
Malt Beverages				2						1 2 1	2	2			2	2	2		Manganese Carbonate

KEY: 1 - Chemical in column 1 may be used • 1 - Chemical in column 1 may not be used
2 - Chemical in column 2 may be used • 2 - Chemical in column 2 may not be used
( ) Conditional or no information

VI Fluid Power Symbols and Standards

A. AMERICAN STANDARD GRAPHICAL SYMBOLS FOR PIPE FITTINGS, VALVES, AND PIPING

(Extracted from American Standard Graphical Symbols for Pipe Fittings, Valves, and Piping (ASA Z32.2.3-1949) with the permission of the publisher, The American Society of Mechanical Engineers, United Engineering Center, 345 E. 47th Street, New York, New York 10017)

	FLANGED	SCREWED	BELL & SPIGOT	WELDED	SOLDERED
1 BUSHING					
2 CAP					
3 CROSS					
3.1 REDUCING					
3.2 STRAIGHT SIZE					
4 CROSSOVER					
5 ELBOW					
5.1 45-DEGREE					
5.2 90-DEGREE					
5.3 TURNED DOWN					
5.4 TURNED UP					

120 VALVE ENGINEERING AND DESIGN DATA

GRAPHICAL SYMBOLS FOR PIPE FITTINGS & VALVES

	FLANGED	SCREWED	BELL & SPIGOT	WELDED	SOLDERED
5.5 BASE					
5.6 DOUBLE BRANCH					
5.7 LONG RADIUS					
5.8 REDUCING					
5.9 SIDE OUTLET (OUTLET DOWN)					
5.10 SIDE OUTLET (OUTLET UP)					
5.11 STREET					
6 JOINT					
6.1 CONNECTING PIPE					
6.2 EXPANSION					
7 LATERAL					
8 ORIFICE FLANGE					

AMERICAN STANDARD					
	FLANGED	SCREWED	BELL & SPIGOT	WELDED	SOLDERED
9 REDUCING FLANGE					
10 PLUGS					
10.1 BULL PLUG					
10.2 PIPE PLUG					
11 REDUCER					
11.1 CONCENTRIC					
11.2 ECCENTRIC					
12 SLEEVE					
13 TEE					
13.1 (STRAIGHT SIZE)					
13.2 (OUTLET UP)					
13.3 (OUTLET DOWN)					
13.4 DOUBLE SWEEP)					
13.5 REDUCING					
13.6 SINGLE SWEEP)					

122 VALVE ENGINEERING AND DESIGN DATA

GRAPHICAL SYMBOLS FOR PIPE FITTINGS & VALVES

	FLANGED	SCREWED	BELL & SPIGOT	WELDED	SOLDERED
13.7 SIDE OUTLET (OUTLET DOWN)					
13.8 SIDE OUTLET (OUTLET UP)					
14 UNION					
15 ANGLE VALVE					
15.1 CHECK					
15.2 GATE (ELEVATION)					
15.3 GATE (PLAN)					
15.4 GLOBE (ELEVATION)					
15.5 GLOBE (PLAN)					
15.6 HOSE ANGLE	SAME AS	SYMBOL	23.1		
16 AUTOMATIC VALVE					
16.1 BY-PASS					

AMERICAN STANDARD

	FLANGED	SCREWED	BELL & SPIGOT	WELDED	SOLDERED
16.2 GOVERNOR-OPERATED					
16.3 REDUCING					
17 CHECK VALVE					
17.1 ANGLE CHECK	SAME AS	SYMBOL	15.1		
17.2 (STRAIGHT WAY)					
18 COCK					
19 DIAPHRAGM VALVE					
20 FLOAT VALVE					
21 GATE VALVE					
*21.1					
21.2 ANGLE GATE	SAME AS	SYMBOLS	15.2 & 15.3		
21.3 HOSE GATE	SAME AS	SYMBOL	23.2		

*ALSO USED FOR GENERAL **STOP VALVE** SYMBOL WHEN AMPLIFIED BY SPECIFICATION

124 VALVE ENGINEERING AND DESIGN DATA

GRAPHICAL SYMBOLS FOR PIPE FITTINGS & VALVES

	FLANGED	SCREWED	BELL & SPIGOT	WELDED	SOLDERED
21.4 MOTOR-OPERATED					
22 GLOBE VALVE					
22.1					
22.2 ANGLE GLOBE	SAME AS	SYMBOLS	15.4 & 15.5		
22.3 HOSE GLOBE	SAME AS	SYMBOL	23.3		
22.4 MOTOR-OPERATED					
23 HOSE VALVE					
23.1 ANGLE					
23.2 GATE					
23.3 GLOBE					
24 LOCKSHIELD VALVE					
25 QUICK OPENING VALVE					
26 SAFETY VALVE					
27 STOP VALVE	SAME AS	SYMBOL	21.1		

AMERICAN STANDARD

AIR CONDITIONING

28	BRINE RETURN	— — —BR— — —
29	BRINE SUPPLY	————B————
30	CIRCULATING CHILLED OR HOT-WATER FLOW	———— CH ————
31	CIRCULATING CHILLED OR HOT-WATER RETURN	— — —CHR— — —
32	CONDENSER WATER FLOW	————C————
33	CONDENSER WATER RETURN	— — —CR— — —
34	DRAIN	————D————
35	HUMIDIFICATION LINE	——·——H——·——
36	MAKE-UP WATER	——·——·——·——
37	REFRIGERANT DISCHARGE	————RD————
38	REFRIGERANT LIQUID	———— RL ————
39	REFRIGERANT SUCTION	— — —RS— — —

HEATING

40	AIR-RELIEF LINE	— —— — ——
41	BOILER BLOW OFF	—— —— ——
42	COMPRESSED AIR	————A————
43	CONDENSATE OR VACUUM PUMP DISCHARGE	—o— —o— —o—
44	FEEDWATER PUMP DISCHARGE	—oo— —oo— —oo—
45	FUEL-OIL FLOW	————FOF————
46	FUEL-OIL RETURN	— — —FOR— — —
47	FUEL-OIL TANK VENT	— — —FOV— — —
48	HIGH-PRESSURE RETURN	—#— —#— —#—
49	HIGH-PRESSURE STEAM	—#——#——#—
50	HOT-WATER HEATING RETURN	— — — —
51	HOT-WATER HEATING SUPPLY	————————

GRAPHICAL SYMBOLS FOR PIPING

52 LOW-PRESSURE RETURN

53 LOW-PRESSURE STEAM

54 MAKE-UP WATER

55 MEDIUM PRESSURE RETURN

56 MEDIUM PRESSURE STEAM

PLUMBING
57 ACID WASTE

58 COLD WATER

59 COMPRESSED AIR

60 DRINKING-WATER FLOW

61 DRINKING-WATER RETURN

62 FIRE LINE

63 GAS

64 HOT WATER

65 HOT-WATER RETURN

66 SOIL, WASTE OR LEADER (ABOVE GRADE)

67 SOIL, WASTE OR LEADER (BELOW GRADE)

68 VACUUM CLEANING

69 VENT

PNEUMATIC TUBES
70 TUBE RUNS

SPRINKLERS
71 BRANCH AND HEAD

72 DRAIN

73 MAIN SUPPLIES

B. USA STANDARD GRAPHIC SYMBOLS FOR FLUID POWER DIAGRAMS

(Extracted from USA Standard Graphic Symbols for Fluid Power Diagrams (USAS Y32.10-1967) with the permission of the publisher, The American Society of Mechanical Engineers, United Engineering Center, 345 E. 47th Street, New York, New York 10017)

1. Introduction

1.1 General

Fluid power systems are those that transmit and control power through use of a pressurized fluid (liquid or gas) within an enclosed circuit.

Types of symbols commonly used in drawing circuit diagrams for fluid power systems are Pictorial, Cutaway, and Graphic. These symbols are fully explained in the USA Standard Drafting Manual (Ref. 2).

1.1.1 *Pictorial symbols* are very useful for showing the interconnection of components. They are difficult to standardize from a functional basis.

1.1.2 *Cutaway symbols* emphasize construction. These symbols are complex to draw and the functions are not readily apparent.

1.1.3 *Graphic symbols* emphasize the function and methods of operation of components. These symbols are simple to draw. Component functions and methods of operation are obvious. Graphic symbols are capable of crossing language barriers, and can promote a universal understanding of fluid power systems.

Graphic symbols for fluid power systems should be used in conjunction with the graphic symbols for other systems published by the USA Standards Institute (Ref. 3–7 inclusive).

1.1.3.1 Complete graphic symbols are those which give symbolic representation of the component and all of its features pertinent to the circuit diagram.

1.1.3.2 Simplified graphic symbols are stylized versions of the complete symbols.

1.1.3.3 Composite graphic symbols are an organization of simplified or complete symbols. Composite symbols usually represent a complex component.

1.2 Scope and Purpose

1.2.1 *Scope*
This standard presents a system of graphic symbols for fluid power diagrams.

1.2.1.1 Elementary forms of symbols are:

Circles	Triangles	Lines
Squares	Arcs	Dots
Rectangles	Arrows	Crosses

1.2.1.2 Symbols using words or their abbreviations are avoided. Symbols capable of crossing language barriers are presented herein.

1.2.1.3 Component function rather than construction is emphasized by the symbol.

1.2.1.4 The means of operating fluid power components are shown as part of the symbol (where applicable).

1.2.1.5 This standard shows the basic symbols, describes the principles on which the symbols are based, and illustrates some representative composite symbols. Composite symbols can be devised for any fluid power component by combining basic symbols.

Simplified symbols are shown for commonly used components.

1.2.1.6 This standard provides basic symbols which differentiate between hydraulic and pneumatic fluid power media.

1.2.2 *Purpose*

1.2.2.1 The purpose of this standard is to provide a system of fluid power graphic symbols for industrial and educational purposes.

1.2.2.2 The purpose of this standard is to simplify design, fabrication, analysis, and service of fluid power circuits.

1.2.2.3 The purpose of this standard is to provide fluid power graphic symbols which are internationally recognized.

1.2.2.4 The purpose of this standard is to promote universal understanding of fluid power systems.

1.3 Terms and Definitions

Terms and corresponding definitions found in this standard are listed in Ref. 8.

2. Symbol Rules (See Section 10)

2.1 Symbols show connections, flow paths, and functions of components represented. They can indicate conditions occurring during transition from one flow path arrangement to another. Symbols do not indicate construction, nor do they indicate values, such as pressure, flow rate, and other component settings.

2.2 Symbols do not indicate locations of ports, direction of shifting of spools, or positions of actuators on actual component.

2.3 Symbols may be rotated or reversed without altering their meaning except in the cases of: a.) Lines to Reservoir, 4.1.1; b.) Vented Manifold, 4.1.2.3; c.) Accumulator, 4.2.

128 VALVE ENGINEERING AND DESIGN DATA

USA STANDARD

2.4 Line Technique (See Ref. 1)
Keep line widths approximately equal. Line width does not alter meaning of symbols.

2.4.1 Solid Line

———————————————

(Main line conductor, outline, and shaft)

2.4.2 Dash Line

— — — — —

(Pilot line for control)

2.4.3 Dotted Line

- - - - - - - - - - - - - - -

(Exhaust or Drain Line)

2.4.4 Center Line

——— — ——— — ———

(Enclosure outline)

2.4.5 Lines Crossing
(The intersection is not necessarily at a 90 deg angle.)

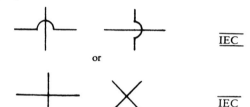

or

IEC

2.4.6 Lines Joining

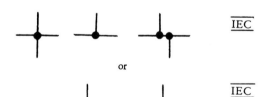

or

IEC

2.5 Basic symbols may be shown any suitable size. Size may be varied for emphasis or clarity. Relative sizes should be maintained. (As in the following example.)

2.5.1 Circle and Semi-Circle

2.5.1.1 Large and small circles may be used to signify that one component is the "main" and the other the auxiliary.

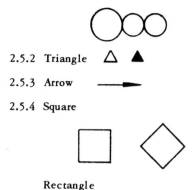

2.5.2 Triangle △ ▲

2.5.3 Arrow ——▶

2.5.4 Square

Rectangle

2.6 Letter combinations used as parts of graphic symbols are not necessarily abbreviations.

2.7 In multiple envelope symbols, the flow condition shown nearest an actuator symbol takes place when that control is caused or permitted to actuate.

2.8 Each symbol is drawn to show normal, at-rest, or neutral condition of component unless multiple diagrams are furnished showing various phases of circuit operation. Show an actuator symbol for each flow path condition possessed by the component.

2.9 An arrow through a symbol at approximately 45 degrees indicates that the component can be adjusted or varied.

2.10 An arrow parallel to the short side of a symbol, within the symbol, indicates that the component is pressure compensated.

2.11 A line terminating in a dot to represent a thermometer is the symbol for temperature cause or effect.

GRAPHIC SYMBOLS FOR FLUID POWER DIAGRAMS

See Temperature Controls 7.9, Temperature Indicators and Recorders 9.1.2, and Temperature Compensation 10.16.3 and 4.

2.12 External ports are located where flow lines connect to basic symbol, except where component enclosure symbol is used.

External ports are located at intersections of flow lines and component enclosure symbol when enclosure is used, see Section 11.

2.13 Rotating shafts are symbolized by an arrow which indicates direction of rotation (assume arrow on near side of shaft).

3. Conductor, Fluid

3.1 Line, Working (main)

3.2 Line, Pilot (for control)

3.3 Line, Exhaust and Liquid Drain

3.4 Line, sensing, etc. such as gage lines shall be drawn the same as the line to which it connects.

3.5 Flow, Direction of

   3.5.1 Pneumatic

   3.5.2 Hydraulic

3.6 Line, Pneumatic
Outlet to Atmosphere

   3.6.1 Plain orifice, unconnectable

   3.6.2 Connectable orifice (e. g. Thread)

3.7 Line with Fixed Restriction

3.8 Line, Flexible

3.9 Station, Testing, measurement, or power take-off

   3.9.1 Plugged port

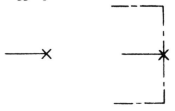

3.10 Quick Disconnect

   3.10.1 Without Checks

      Connected

      Disconnected

   3.10.2 With Two Checks

      Connected

      Disconnected

   3.10.3 With One Check

      Connected

      Disconnected

3.11 Rotating Coupling

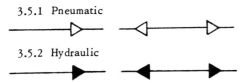

130 VALVE ENGINEERING AND DESIGN DATA

USA STANDARD

4. Energy Storage and Fluid Storage

4.1 Reservoir

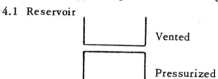

Note: Reservoirs are conventionally drawn in the horizontal plane. All lines enter and leave from above. Examples:

4.1.1 Reservoir with Connecting Lines

Above Fluid Level

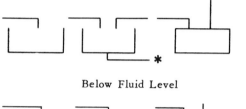

Below Fluid Level

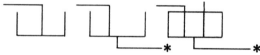

4.1.2 Simplified symbol

The symbols are used as part of a complete circuit. They are analogous to the ground symbol of electrical diagrams. ──┤├── IEC. Several such symbols ⊔ may be used in one diagram to represent the same reservoir.

4.1.2.1 Below Fluid Level

4.1.2.2 Above Fluid Level

(The return line is drawn to terminate at the upright legs of the tank symbol.)

4.1.2.3 Vented Manifold

---

*Show line entering or leaving below reservoir only when such bottom connection is essential to circuit function.

4.2 Accumulator

4.2.1 Accumulator, Spring Loaded

4.2.2 Accumulator, Gas Charged

4.2.3 Accumulator, Weighted

4.3 Receiver, for Air or Other Gases

4.4 Energy Source
(Pump, Compressor, Accumulator, etc.)

This symbol may be used to represent a fluid power source which may be a pump, compressor, or another associated system.

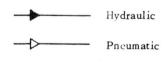

Simplified Symbol

Example:

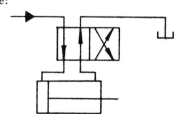

GRAPHIC SYMBOLS FOR FLUID POWER DIAGRAMS

5. Fluid Conditioners

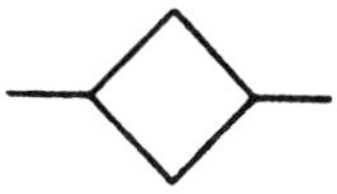

Devices which control the physical characteristics of the fluid.

5.1 Heat Exchanger

5.1.1 Heater

Inside triangles indicate the introduction of heat.

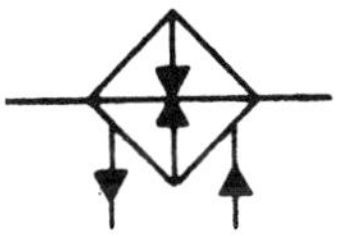

Outside triangles show the heating medium is liquid.

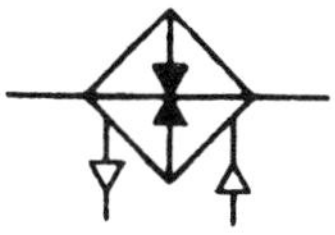

Outside triangles show the heating medium is gaseous.

5.1.2 Cooler

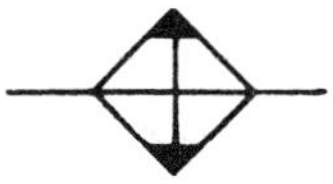

Inside triangles indicate heat dissipation

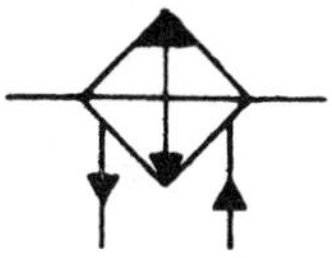

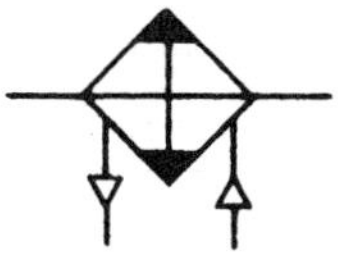

(Corners may be filled in to represent triangles.)

5.1.3 Temperature Controller
(The temperature is to be maintained between two predetermined limits.)

 or

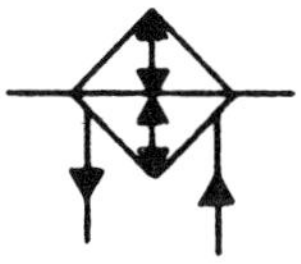

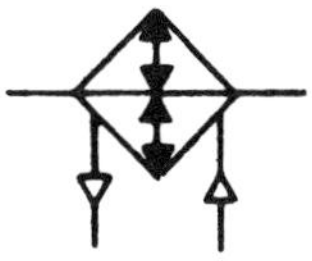

5.2 Filter – Strainer

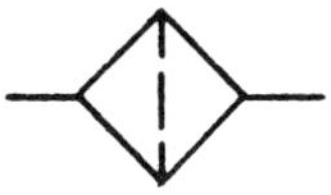

5.3 Separator

5.3.1 With Manual Drain

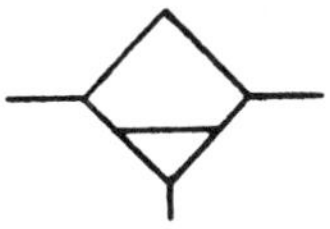

5.3.2 With Automatic Drain

5.4 Filter – Separator

5.4.1 With Manual Drain

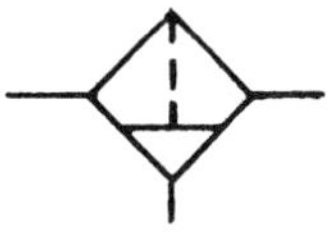

5.4.2 With Automatic Drain

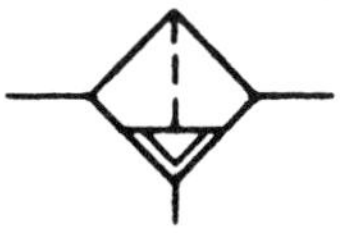

5.5 Dessicator (Chemical Dryer)

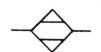

5.6 Lubricator

5.6.1 Less Drain

USA STANDARD

5.6.2 With Manual Drain

6. Linear Devices

6.1 Cylinders, Hydraulic & Pneumatic

6.1.1 Single Acting

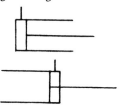

6.1.2 Double Acting

6.1.2.1 Single End Rod

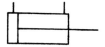

6.1.2.2 Double End Rod

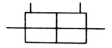

6.1.2.3 Fixed Cushion, Advance & Retract

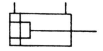

6.1.2.4 Adjustable Cushion, Advance Only

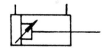

6.1.2.5 Use these symbols when diameter of rod compared to diameter of bore is significant to circuit function.

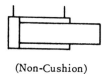

(Non-Cushion)

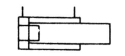

(Cushion, Advance & Retract)

6.2 Pressure Intensifier

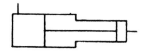

6.3 Servo Positioner (Simplified)

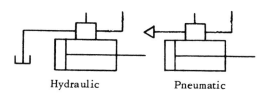

Hydraulic   Pneumatic

6.4 Discrete Positioner

Combine two or more basic cylinder symbols.

7. Actuators and Controls

7.1 Spring

7.2 Manual

(Use as general symbol without indication of specific type; i.e., foot, hand, leg, arm)

7.2.1 Push Button

7.2.2 Lever

FLUID POWER SYMBOLS AND STANDARDS

GRAPHIC SYMBOLS FOR FLUID POWER DIAGRAMS

7.2.3 Pedal or Treadle

7.3 Mechanical

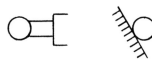

7.4 Detent

(Show a notch for each detent in the actual component being symbolized. A short line indicates which detent is in use.) Detent may, for convenience, be positioned on either end of symbol.

7.5 Pressure Compensated

7.6 Electrical

   7.6.1 Solenoid (Single Winding)

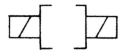

   7.6.2 Reversing Motor

7.7 Pilot Pressure

   7.7.1

       Remote Supply

   7.7.2

       Internal Supply

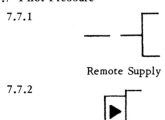

7.7.3 Actuation by Released Pressure

by Remote Exhaust

by Internal Return

7.7.4 Pilot Controlled, Spring Centered

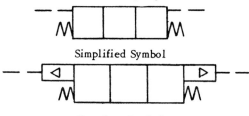

Simplified Symbol

Complete Symbol

7.7.5 Pilot Differential

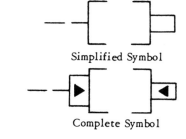

Simplified Symbol

Complete Symbol

7.8 Solenoid Pilot

   7.8.1 Solenoid or Pilot

      External Pilot Supply

      Internal Pilot Supply and Exhaust

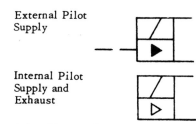

   7.8.2 Solenoid and Pilot

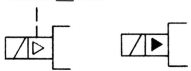

134   VALVE ENGINEERING AND DESIGN DATA

USA STANDARD

7.9 Thermal

A mechanical device responding to thermal change.

   7.9.1 Local Sensing

   7.9.2 With Bulb for Remote Sensing

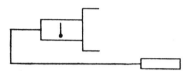

7.10 Servo

(This symbol contains representation for energy input, command input, and resultant output.)

7.11 Composite Actuators (and, or, and/or)

Basic One signal only causes the device to operate.

And One signal and a second signal both cause the device to operate.

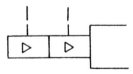

Or One signal or the other signal causes the device to operate

And/Or The solenoid and the pilot or the manual override alone causes the device to operate.

The solenoid and the pilot or the manual override and the pilot

The solenoid and the pilot or a manual override and the pilot or a manual override alone.

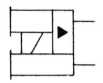

8. Rotary Devices

8.1 Basic Symbol

   8.1.1 With Ports

   8.1.2 With Rotating Shaft, with control, and with Drain

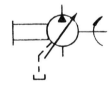

8.2 Hydraulic Pump

   8.2.1 Fixed Displacement.

      8.2.1.1 Unidirectional

      8.2.1.2 Bidirectional

FLUID POWER SYMBOLS AND STANDARDS 135

GRAPHIC SYMBOLS FOR FLUID POWER DIAGRAMS

8.2.2 Variable Displacement, Non-Compensated

8.2.2.1 Unidirectional

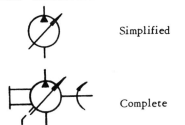

Simplified

Complete

8.2.2.2 Bidirectional

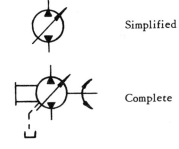

Simplified

Complete

8.2.3 Variable Displacement, Pressure Compensated

8.2.3.1 Unidirectional

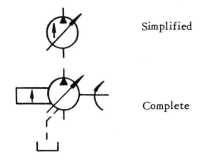

Simplified

Complete

8.2.3.2 Bidirectional

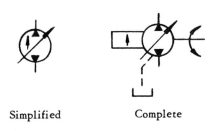

Simplified   Complete

8.3 Hydraulic Motor

8.3.1 Fixed Displacement

8.3.1.2 Bidirectional

8.3.2 Variable Displacement

8.3.2.1 Unidirectional

8.3.2.2 Bidirectional

8.4 Pump-Motor, Hydraulic

8.4.1 Operating in one direction as a pump. Operating in the other direction as a motor.

8.4.1.1 Complete Symbol

8.4.1.2 Simplified Symbol

8.4.2 Operating one direction of flow as either a pump or as a motor.

8.4.2.1 Complete Symbol

136 VALVE ENGINEERING AND DESIGN DATA

USA STANDARD

8.4.2.2 Simplified Symbol

8.4.3 Operating in both directions of flow either as a pump or as a motor.
(Variable displacement, pressure compensated shown)

    8.4.3.1 Complete Symbol

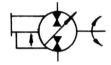

    8.4.3.2 Simplified Symbol

8.5 Pump, Pneumatic

    8.5.1 Compressor, Fixed Displacement

    8.5.2 Vacuum Pump, Fixed Displacement

8.6 Motor, Pneumatic

    8.6.1 Unidirectional

    8.6.2 Bidirectional

8.7 Oscillator

    8.7.1 Hydraulic

    8.7.2 Pneumatic

8.8 Motors, Engines

    8.8.1 Electric Motor

  IEC

    8.8.2 Heat Engine (E. G. internal combustion engine)

9. Instruments and Accessories

9.1 Indicating and Recording

    9.1.1 Pressure

    9.1.2 Temperature

    9.1.3 Flow Meter

        9.1.3.1   Flow Rate

        9.1.3.2   Totalizing

9.2 Sensing

    9.2.1 Venturi

GRAPHIC SYMBOLS FOR FLUID POWER DIAGRAMS

9.2.2 Orifice Plate

9.2.3 Pitot Tube

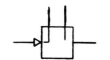

9.2.4 Nozzle

Hydraulic   Pneumatic

9.3 Accessories

9.3.1 Pressure Switch

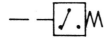

9.3.2 Muffler

10. Valves

A basic valve symbol is composed of one or more envelopes with lines inside the envelope to represent flow paths and flow conditions between ports. Three symbol systems are used to represent valve types: single envelope, both finite and infinite position; multiple envelope, finite position; and multipe envelope, infinite position.

10.1 In infinite position single envelope valves, the envelope is imagined to move to illustrate how pressure or flow conditions are controlled as the valve is actuated.

10.2 Multiple envelopes symbolize valves providing more than one finite flow path option for the fluid. The multiple envelope moves to represent how flow paths change when the valving element within the component is shifted to its finite positions.

10.3 Multiple envelope valves capable of infinite positioning between certain limits are symbolized as in 10.2 above with the addition of horizontal bars which are drawn parallel to the envelope. The horizontal bars are the clues to the infinite positioning function possessed by the valve re-represented.

10.4 Envelopes

10.5 Ports

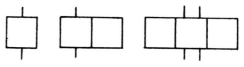

10.6 Ports, Internally Blocked

Symbol System 10.1

Symbol System 10.2

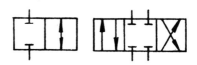

10.7 Flow Paths, Internally Open (Symbol System 10.1 and 10.2)

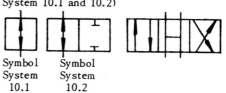

Symbol System 10.1   Symbol System 10.2

10.8 Flow Paths, Internally Open (Symbol System 10.3)

138 VALVE ENGINEERING AND DESIGN DATA

USA STANDARD

10.9 Two-Way Valves (2 Ported Valves)

10.9.1 On-Off (Manual Shut-Off)

Simplified

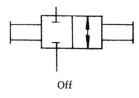

Off

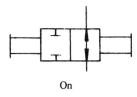

On

10.9.2 Check

 Simplified Symbol

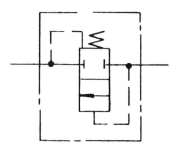

Flow to the right is blocked. Flow to the left is permitted)

(Composite Symbol)

10.9.3 Check, Pilot-Operated to Open

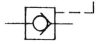

10.9.4 Check, Pilot-Operated to Close

10.9.5 Two-Way Valves

10.9.5.1 Two-Position

Normally Closed   Normally Open

10.9.5.2 Infinite Position

Normally Closed   Normally Open

10.10 Three-Way Valves

10.10.1 Two-Position

10.10.1.1 Normally Open

10.10.1.2 Normally Closed

10.10.1.3 Distributor (Pressure is distributed first to one port, then the other)

10.10.1.4 Two-Pressure

10.10.2 Double Check Valve
Double check valves can be built with and without "cross bleed". Such valves with two

GRAPHIC SYMBOLS FOR FLUID POWER DIAGRAMS

poppets do not usually allow pressure to momentarily "cross bleed" to return during transition. Valves with one poppet may allow "cross bleed" as these symbols illustrate.

10.10.2.1 Without Cross Bleed (One Way Flow)

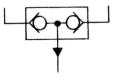

10.10.2.2 With Cross Bleed (Reverse Flow Permitted)

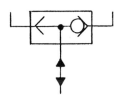

10.11 Four-Way Valves

10.11.1 Two Position

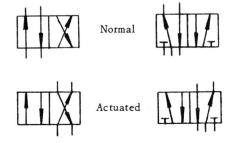

10.11.2 Three Position

(a) Normal

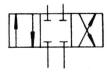

(b) Actuated Left

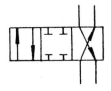

(c) Actuated Right

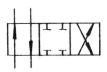

10.11.3 Typical Flow Paths for Center Condition of Three Position Valves

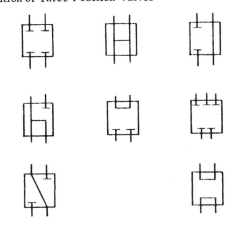

10.11.4 Two-Position, Snap Action with Transition.

As the valve element shifts from one position to the other, it passes through an intermediate position. If it is essential to circuit function to symbolize this "in transit" condition, it can be shown in the center position, enclosed by dashed lines.

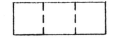

Typical Transition Symbol

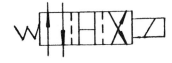

10.12 Infinite Positioning (Between Open & Closed)

10.12.1 Normally Closed

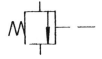

140 VALVE ENGINEERING AND DESIGN DATA

USA STANDARD

10.12.2 Normally Open

10.13 Pressure Control Valves

10.13.1 Pressure Relief

Simplified Symbol Denotes

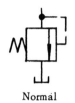 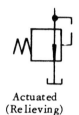

Normal   Actuated (Relieving)

10.13.2 Sequence

10.13.3 Pressure Reducing

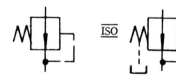

ISO

10.13.4 Pressure Reducing and Relieving

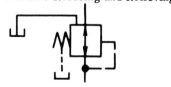

10.13.5 Airline Pressure Regulator (Adjustable, Relieving)

10.14 Infinite Positioning Three-Way Valves

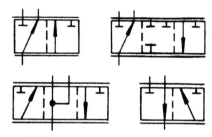

10.15 Infinite Positioning Four-Way Valves

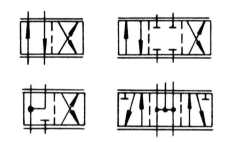

10.16 Flow Control Valves (See 3.7)

10.16.1 Adjustable, Non-Compensated (Flow control in each direction)

10.16.2 Adjustable with Bypass

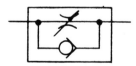

Flow is controlled to the right
Flow to the left bypasses control

10.16.3 Adjustable and Pressure Compensated With Bypass

GRAPHIC SYMBOLS FOR FLUID POWER DIAGRAMS

10.16.4 Adjustable, Temperature & Pressure Compensated

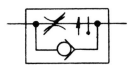

11. Representative Composite Symbols

11.1 Component Enclosure

Component enclosure may surround a complete symbol or a group of symbols to represent an assembly. It is used to convey more information about component connections and functions. Enclosure indicates extremity of component or assembly. External ports are assumed to be on enclosure line and indicate connections to component.

Flow lines shall cross enclosure line without loops or dots.

11.2 Airline Accessories
(Filter, Regulator, and Lubricator)

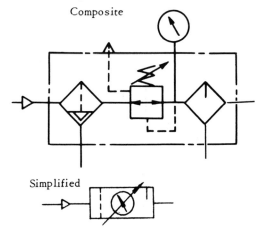

11.3 Pumps and Motors

   11.3.1 Pumps

      11.3.1.1 Double, Fixed Displacement, One Inlet and Two Outlets

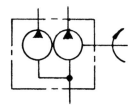

11.3.1.2 Double, with Integral Check Unloading and Two Outlets

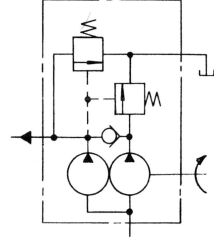

11.3.1.3 Integral Variable Flow Rate Control with Overload Relief

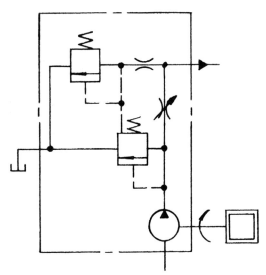

142   VALVE ENGINEERING AND DESIGN DATA

USA STANDARD

11.3.1.4 Variable Displacement with Integral Replenishing Pump and Control Valves

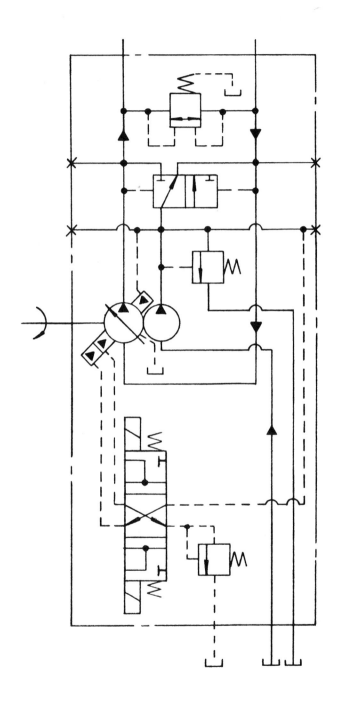

GRAPHIC SYMBOLS FOR FLUID POWER DIAGRAMS

11.3.2 Pump Motor

Variable displacement with manual, electric, pilot, and servo control.

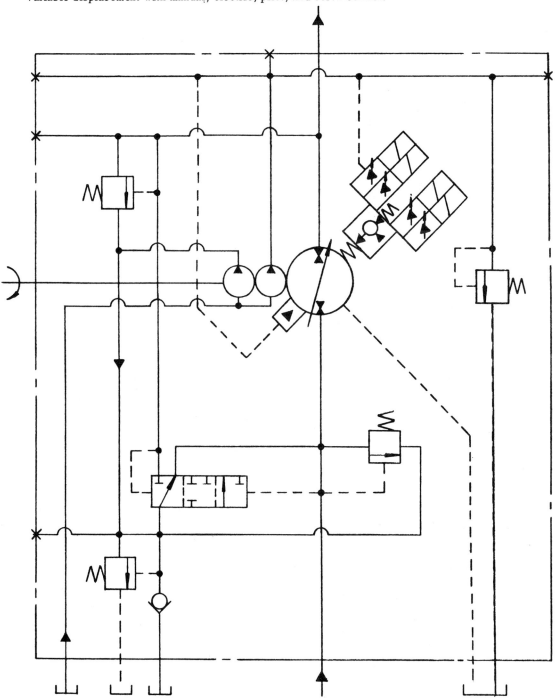

144 VALVE ENGINEERING AND DESIGN DATA

USA STANDARD

11.4 Valves

11.4.1 Relief, Balanced Type

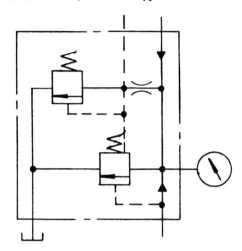

11.4.2 Remote Operated Sequence with Integral Check

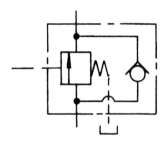

11.4.3 Remote & Direct Operated Sequence with Differential Areas and Integral Check

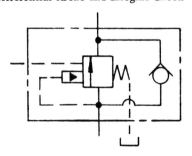

11.4.4 Pressure Reducing with Integral Check

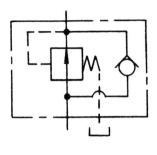

11.4.5 Pilot Operated Check

11.4.5.1 Differential Pilot Opened

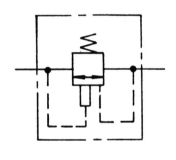

11.4.5.2 Differential Pilot Opened and Closed

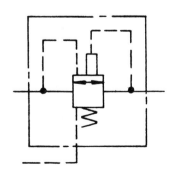

FLUID POWER SYMBOLS AND STANDARDS

GRAPHIC SYMBOLS FOR FLUID POWER DIAGRAMS

11.4.6 Two Positions, Four Connection Solenoid and Pilot Actuated, with Manual Pilot Override.

Simplified Symbol

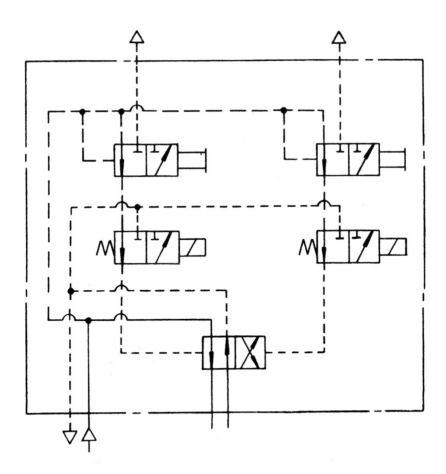

Complete Symbol

146   VALVE ENGINEERING AND DESIGN DATA

USA STANDARD

11.4.7 Two Position, Five Connection, Solenoid Control Pilot Actuated with Detents and Throttle Exhaust

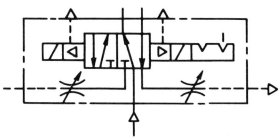

Symplified Symbol

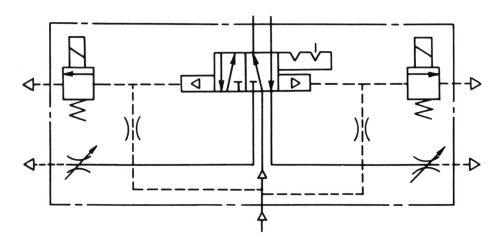

Complete Symbol

11.4.8 Variable Pressure Compensated Flow Control and Overload Relief

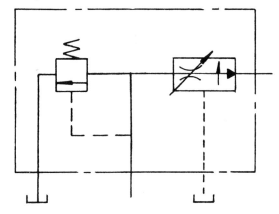

FLUID POWER SYMBOLS AND STANDARDS 147

GRAPHIC SYMBOLS FOR FLUID POWER DIAGRAMS

11.4.9 Multiple, Three Position, Manual Directional Control with Integral Check and Relief Valves

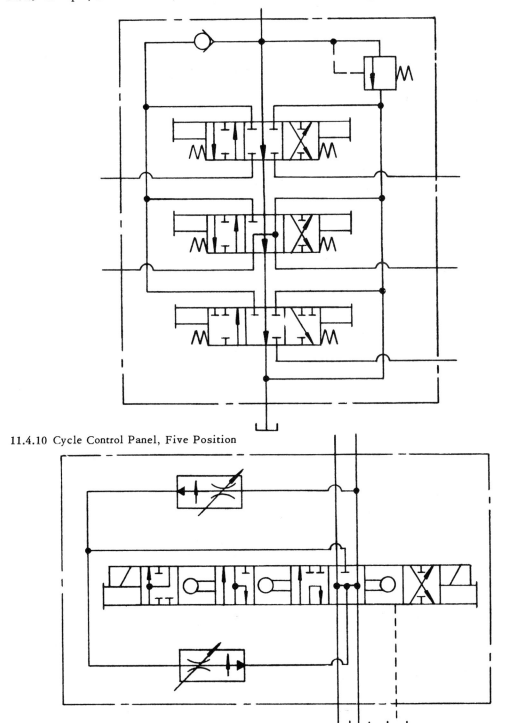

11.4.10 Cycle Control Panel, Five Position

148 VALVE ENGINEERING AND DESIGN DATA

USA STANDARD

11.4.11 Panel Mounted Separate Units Furnished as a Package (Relief, Two Four-Way, Two Check, and Flow Rate Valves)

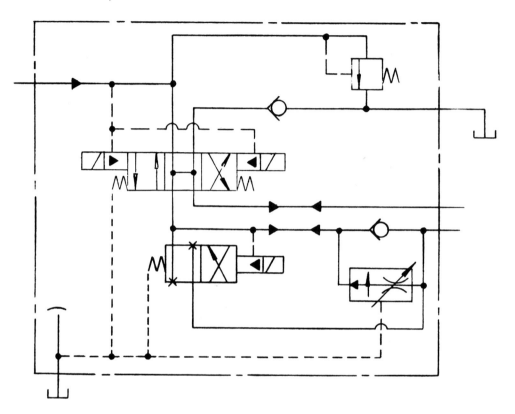

11.4.12 Single Stage Compressor with Electric Motor Drive, Pressure Switch Control of Receiver Tank Pressure

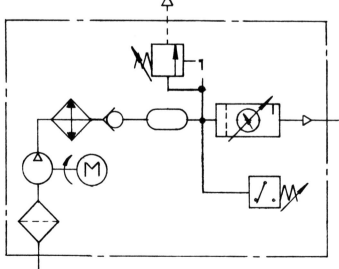

C. AMERICAN NATIONAL STANDARD SYMBOLS FOR MARKING ELECTRICAL LEADS AND PORTS ON FLUID POWER VALVES

(Extracted from American National Standard Symbols for Marking Electrical Leads and Ports on Fluid Power Valves (ANSI B93.9-1969) with the permission of the publisher, the National Fluid Power Association, Inc.)

REFERENCES

1. ANSI Standard Glossary of Terms for Fluid Power, B93.2
2. ANSI Standard Fluid Power Diagrams, Y14.17
3. ANSI Standard Graphic Symbols for Fluid Power Diagrams, Y32.10

1. INTRODUCTION

Fluid power systems are those that transmit and control power through use of a pressurized fluid (liquid or gas) within an enclosed circuit. In such circuits, valves are devices which control fluid direction, pressure, or flow rate. The simplest valve usually has an inlet port and an outlet port. Sometimes one is the other, but often the inlet must be identified and distinguished from the outlet so that piping the valve incorrectly is avoided.

In more complex valves, the importance of port identification is magnified. The valve may be multi-purpose and multi-ported. It may be special-purpose and special-ported. Its proper use will depend on the correct identification of all of its ports. Such identification is usually carried somewhere on the actual component adjacent to the port being named.

2. SCOPE

This recommended standard offers symbols for the identification of valve ports, for the identification of valve pilot and solenoid actuators, and for the identification of valve solenoid leads.

3. PURPOSE

Correct identification assists in providing:
Safety of personnel
Long life of the equipment
Proper function of other system components
Proper service and maintenance
Universal understanding of fluid power valve function

4. TERMS AND DEFINITIONS

For terms, abbreviations, and definitions applicable to fluid power, consult Ref. 1.

5. RULES

5.1 Identify valve pilot and solenoid actuators with the same symbol as the port which is pressurized when the control is caused or permitted to actuate.

5.2 Identify coil leads when the solenoid from which they originate cannot be determined visually.

5.2.1 When solenoid coil leads are identified, use the same symbol as the port which is pressurized when the coil is energized.

5.3 Identify common valve ports with the same symbol.

6. PORT SYMBOLS

Symbols in diagrams and on the actual component should be in agreement. See Ref. 2, Section 17-6.3.9.

6.1 Arrow symbols are sufficient to indicate the direction of flow through valves containing two ports.

6.1.1 A single arrow (⟶) is sufficient to indicate the direction of preferred flow.

6.1.2 A single arrow (⟶) is sufficient to indicate the direction of free flow.

6.1.3 A double headed arrow (⟵⟶) is sufficient to indicate flow when direction of flow is not significant.

6.2 A and B are symbols identifying the working ports of valves when there are no more than two such ports.

6.2.1 The symbol K identifies the working ports of valves which have three or more such ports. K shall always appear with a suffix digit symbol K1, K2, etc.

6.3 The symbol D identifies a hydraulic *drain* port which usually is connected to an unrestricted line. The symbol D also identifies a pneumatic exhaust port which must not be subjected to restriction or pressure.

6.4 The symbol E identifies the pneumatic *outlet* port that usually supplies a passage to atmosphere. (E on pneumatic valves corresponds to T on hydraulic valves.)

6.4.1 If the valve so marked is used optionally with hydraulics the E port will correspond to T.

6.4.2 E may appear with a suffix symbol. (EA, EB, E1). The suffix symbol takes the same identity as the working port it exhausts.

6.5 The symbol F identifies the controlled *flow* port of a flow control valve.

6.6 The symbol P identifies the *inlet* port of all valves (exception, see 6.1).

6.6.1 P may appear with a suffix symbol (PA, PB, P1) The suffix symbol takes the same identity as the working port it pressurizes.

6.7 The symbol T identifies the *outlet* port usually connected to tank. The symbol T is used only with valves for hydraulic service

6.7.1 If the valve so marked is used optionally with pneumatics, the T port will correspond to E.

6.7.2 T may appear with a suffix symbol (TA, TB, T1) The suffix symbol takes the same identity as the working port which it serves.

6.8 The symbol X:

6.8.1 In pneumatic valves, the symbol X identifies an auxiliary port whose function may be one of the following: (Manufacturer's literature will designate specific use)

6.8.1.1 A pneumatic pilot pressure supply port

6.8.1.2 A pneumatic pilot control port (See option in par. 6.10)

6.8.2 In hydraulic valves, the symbol X identifies an auxiliary port whose function may be one of the following: (Manufacturer's literature will designate specific use)

6.8.2.1 A hydraulic pilot drain port

6.8.2.2 A hydraulic pilot pressure supply port

6.8.2.3 A hydraulic pilot control port

6.8.3 X may appear with a suffix symbol (XA, XB, X1) The suffix symbol takes the same identity as the working port pressurized when the pilot is caused or permitted to actuate.

6.9 The symbol Y:

6.9.1 In pneumatic valves, the symbol Y identifies an auxiliary port whose function may be one of the following: (Manufacturer's literature will designate specific use)

6.9.1.1 A pneumatic pilot pressure supply port

6.9.1.2 A pneumatic pilot control port (See option in par. 6.10)

6.9.2 In hydraulic valves, the symbol Y identifies an auxiliary port whose function may be one of the following: (Manufacturer's

FLUID POWER SYMBOLS AND STANDARDS 151

literature will designate specific use)

6.9.2.1 A hydraulic pilot drain port

6.9.2.2 A hydraulic pilot pressure supply port

6.9.2.3 A hydraulic pilot control port

6.9.3 Y may appear with a suffix symbol (YA, YB, Y1) The suffix takes the same identity as the working port pressurized when the pilot is caused or permitted to actuate.

6.10 The symbol C plus a suffix symbol (CA, CB, C1) identifies a control port into which a pneumatic signal may be introduced to cause valve actuation. The suffix letter takes the same identity as the symbol for the working port which is pressurized when the control signal is given. (See also par. 6.8.1)

6.11 The symbol V identifies a control port which, when vented to a lower reference pressure, causes valve actuation.

6.11.1 V may appear with a suffix symbol (VA, VB, V1) The suffix symbol takes the same identity as the working port which is pressurized when V is vented.

6.12 The number symbols, 1, 2, 3, etc., identify the ports which are not otherwise described. (Manufacturer's literature will designate specific use)

7. EXAMPLES

(Symbols used will be found in Ref. 3)

7.1 Outlet to Atmosphere

7.1.1 Plain Orifice (Unthreaded)

7.1.2 Connectable Orifice (Threaded)

7.1.3 Drain

7.2 Directional Control Valves

7.2.1 Check

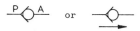

7.2.2 Two-Way

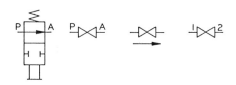

7.2.3 Three-Way

7.2.3.1 Normally Open

7.2.3.2 Normally Closed

7.2.3.3 Diverter

152 VALVE ENGINEERING AND DESIGN DATA

7.2.3.4 Two-Pressure

7.2.3.5 Multipurpose

7.2.4 Four-Way

7.2.4.1 Four Ports

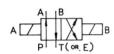

7.2.4.2 Five Ports

 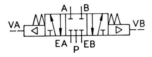

7.2.4.3 Hydraulic Valve with Pneumatic Pilot (EXS & E)

7.2.4.4 Pneumatic Valve with Pilot Supply and Control Supply

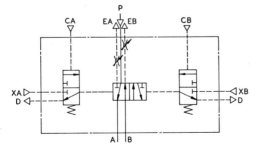

7.2.5 Multi-Directional Multi-Purpose

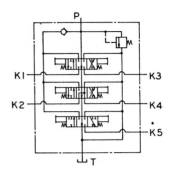

7.3 Flow Control Valves

7.3.1 Fixed Restriction

7.3.2 Adjustable Restriction

7.3.3 Adjustable Restriction with By-pass

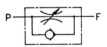

7.3.4 Adjustable, Pressure Compensated with Overload Relief

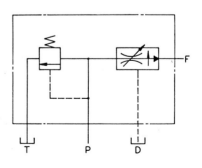

7.4 Pressure Control Valves

7.4.1 Relief

7.4.2 Reducing

7.4.3 Reducing and Relieving (Adjustable)

7.4.4 Relief, Balanced Type

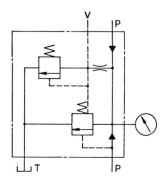

7.4.5 Sequence

7.4.6 Counterbalance

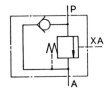

D. AMERICAN NATIONAL STANDARD INTERFACES FOR FOUR-WAY GENERAL PURPOSE INDUSTRIAL PNEUMATIC DIRECTIONAL CONTROL VALVES

(Extracted from American National Standard Interfaces for Four-Way General Purpose Industrial Pneumatic Directional Control Valves (ANSI/B93.33-1974) with the permission of the publisher, the National Fluid Power Association, Inc.)

REFERENCES

1. American National Standard Glossary of Terms for Fluid Power, ANSI/B93.2-1971, and Supplements thereto. (ISO/TC 131/SC 1 (USA-2)3).
2. SI units and recommendations for the use of their multiples and of certain other units, ISO 1000-1973.
3. American National Standard Symbols for Marking Electrical Leads and Ports on Fluid Power Valves, ANSI/B93.9-1969. (ISO/TC 131/SC 5 (USA-3) 10).

INTRODUCTION

In pneumatic fluid power systems, power is transmitted and controlled thru a gas under pressure within an enclosed circuit. A pneumatic directional control valve's primary function is to direct or prevent flow thru selected passages.

Sub-base (or sub-plate) type valves consist of two principle parts—the sub-base or mounting portion and the operating valve portion. All piping is permanently connected to the sub-base. The operating valve portion is bolted to the sub-base; it can easily be removed for service or replacement with simple tools. Air passages between the sub-base and operating valve portion are sealed by resilient sealing means. The mating surfaces of the sub-base and operating valve are called the "interface".

Users of pneumatic valves benefit when valves of the same nominal size from various manufacturers have a common interface—and therefore, can be interchanged when servicing or replacement is required.

1. SCOPE

1.1 To include standard interfaces for a series of six sizes of sub-base 4-way, pneumatic industrial general purpose, directional control valves. The interface configuration will apply to modular and/or individual sub-bases.

154 VALVE ENGINEERING AND DESIGN DATA

1.2 To include the interface fluid power passages, hold down bolt locations and sizes, and provision for an optional electrical connector. (Details of the electrical connector will be set forth in a separate document.)

2. PURPOSE

   2.1 To achieve dimensional interchangeability.

   2.2 To achieve simplification of variety.

3. TERMS AND DEFINITIONS

   For definitions of terms used, see Reference No. 1).

4. UNITS

   4.1 The International System of Units (SI) is used in accordance with Reference No. 2.

   4.2 Rounded conversions to "Customary US" units are given in parentheses following their SI counterpart. The following dimension adjustment tabulation gives an example of how the conversions were achieved for interface type 1:

Established dimension in inches	Dimension millimetre equivalent	Dimension millimetres (rounded off)	Dimension inch equivalent
2	50.8	50	1.968
1¼	31.75	32	1.260
½	12.7	13	.512
⅞	22.225	22	.866
¼	6.35	6	.236

5. PORT IDENTIFICATION

   Use the numbers 1 thru 5 which identify the fluid passages on the interfaces to also identify the associated sub-base port in accordance with Reference No. 3.

6. USE OF OPTIONAL HOLES

   The use of external pilot supply hole, remote pilot signal holes and electrical service location is optional.

7. IDENTIFICATION STATEMENT

   Use the following statement in catalogs and sales literature when electing to comply with this voluntary standard:

   "Interface conforms to American National Standard ANSI/B93.33-1974, Type ____."

E. USA STANDARD DIMENSIONS FOR MOUNTING SURFACES OF SUBPLATE TYPE HYDRAULIC FLUID POWER VALVES

(Extracted from USA Standard Dimensions for Mounting Surfaces of Sub-Plate Type Hydraulic Fluid Power Valves (USAS B93.7-1968) with the permission of the publisher, The National Fluid Power Association, Inc.)

REFERENCES

a. USA Std. B87.1-1965 Decimal Inch
b. USA Std. B93.2-1965 Glossary of Terms for Fluid Power

INTRODUCTION

Fluid Power Systems are those that transmit and control power through the use of a pressurized fluid (liquid or gas) within an enclosed circuit.

Typical components found in such systems are hydraulic valves. This device controls flow direction, pressure, or flow rate of liquids in the enclosed circuit.

For additional descriptions of valves see Ref. b.

1. SCOPE

   1.1 Mounting Surfaces—3,000 psi maximum hydraulic service
      1.1.1 Directional Control Valves
      1.1.2 Two Port Flow Control Valves
      1.1.3 Three Port Flow Control Valves
      1.1.4 Check Valves
      1.1.5 Pilot Operated Check Valves
      1.1.6 Pressure Control Valves—Spool Type
      1.1.7 Pressure Control Valves—Poppet Type
   1.2 Dimensional Criteria
      1.2.1 Minimum surface dimensions
      1.2.2 Sizes and locations of tapped holes for mounting bolts
      1.2.3 Sizes and locations of ports
      1.2.4 Sizes and locations of dowel or rest pins where required
   1.3 General Criteria
      1.3.1 Surface finish and flatness
      1.3.2 Indication of tolerances where pertinent
      1.3.3 Indication of appropriate corner breaks and radii

FLUID POWER SYMBOLS AND STANDARDS 155

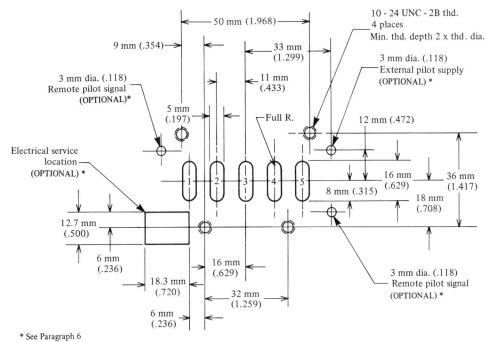

*See Paragraph 6

Fig. 1. Type 1 Interface—Top View.

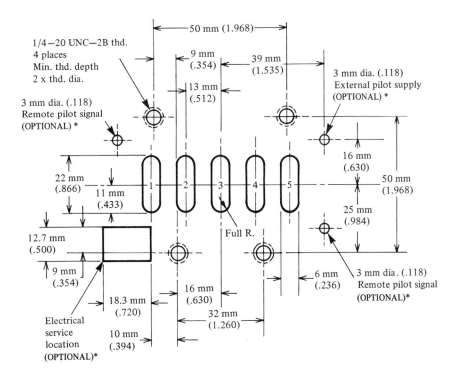

*See Paragraph 6

Fig. 2. Type 2 Interface—Top View.

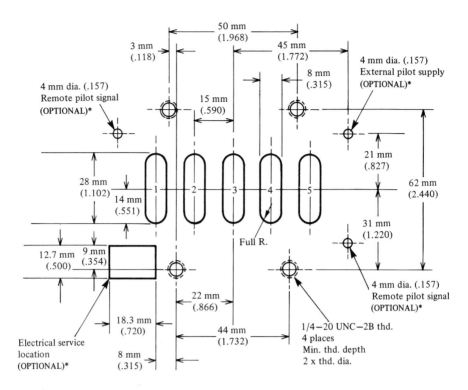

* See Paragraph 6

Fig. 3. Type 4 Interface–Top View.

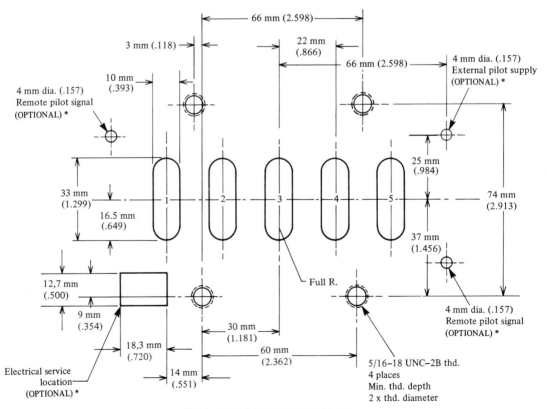

Fig. 4. Type 8 Interface–Top View.

* See Paragraph 6

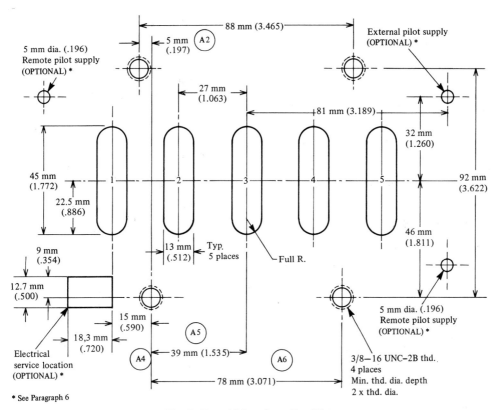

Fig. 5. Type 12 Interface—Top View.

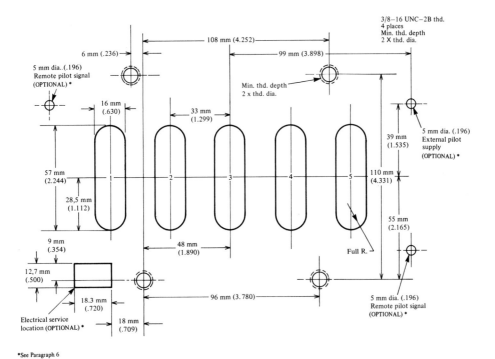

Fig. 6. Type 16 Interface—Top View.

1.3.4 Use decimal inch system (See Reference a.)

2. PURPOSE

The purposes of this recommended standard are as follows:

2.1 Provide a composite reference for new designs
2.2 Simplify interchangeability of sub-plate mounted hydraulic fluid power valves

3. TERMS AND DEFINITIONS

Terms and corresponding definitions applicable to this project are listed in Reference b.

4. GENERAL

Ports are identified by numbers.

Use of an asterisk (*) on a mounting hole or port indicates that this connection is not present in all manufacturers valves and may be eliminated when not specifically required. Individual valve manufacturers should be consulted for specific requirements.

5. MOUNTING SURFACE DRAWING IDENTIFICATION CODING

Drawing identification coding corresponds to the following prefixes:
  D – Directional Control Valves
  2F – Two port Flow Control Valves
  3F – Three Port Flow Control Valves
  C. – Check Valves
  POC – Pilot Operated Check Valves
  P – Pressure Control Valves – Spool Type
  R – Pressure Control Valves – Poppet Type

Drawing identification coding corresponds to the following suffixes:
  02 – ¼ Valves
  03 – ⅜ Valves
  06 – ¾ Valves
  10 – 1¼ Valves

6. TOLERANCES-BREAKS-RADII

The following tolerances apply except when MIN or MAX appear in the drawing notes:

6.1 The tolerance on two place decimals is ±.01
6.2 The tolerance on three place decimals is ±.005
6.3 Break all sharp edges
6.4 Maximum radii .01

FLUID POWER SYMBOLS AND STANDARDS 159

7. CHECK VALVES

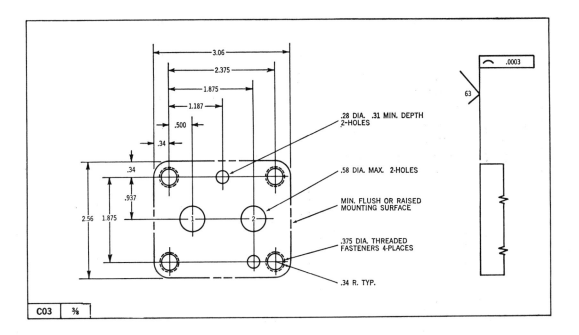

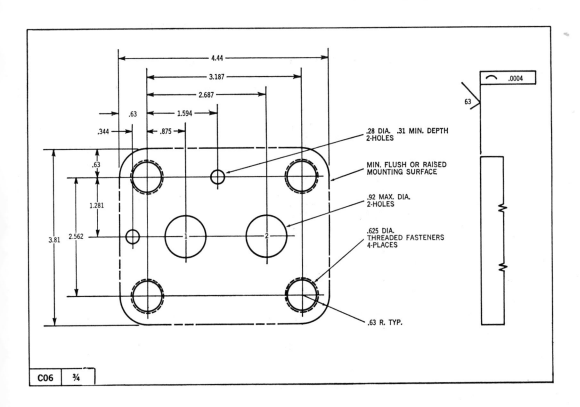

160 VALVE ENGINEERING AND DESIGN DATA

7. CHECK VALVES

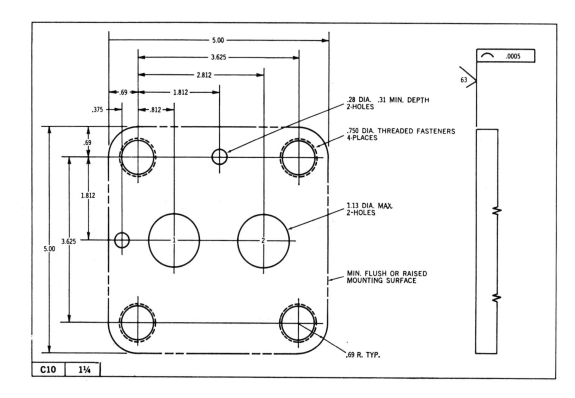

8. PILOT OPERATED CHECK VALVES

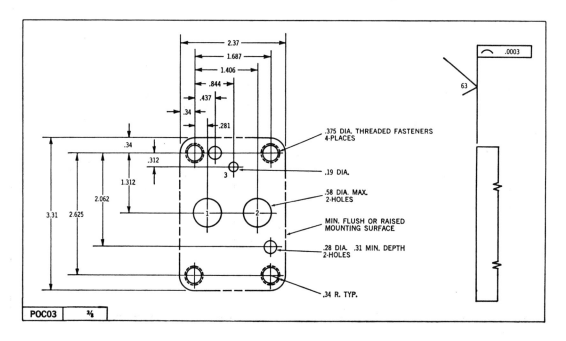

FLUID POWER SYMBOLS AND STANDARDS 161

8. PILOT OPERATED CHECK VALVES

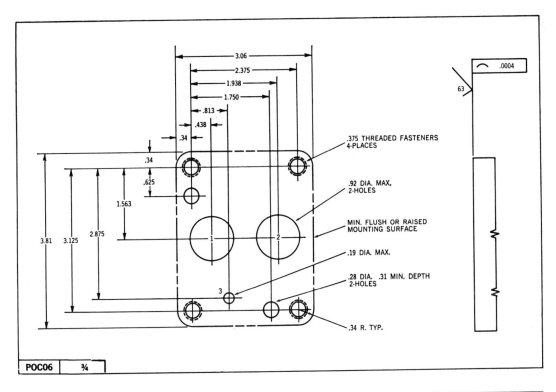

POC06   ¾

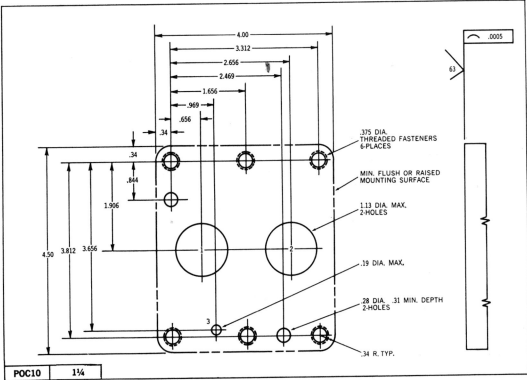

POC10   1¼

9. PRESSURE CONTROL VALVES - SPOOL TYPE

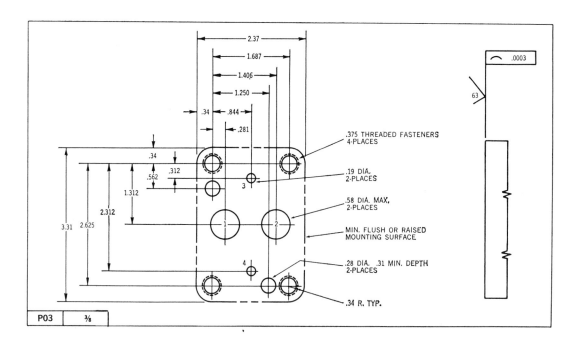

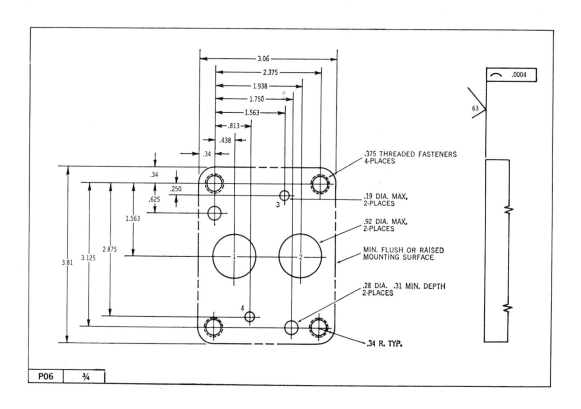

9. PRESSURE CONTROL VALVES - SPOOL TYPE

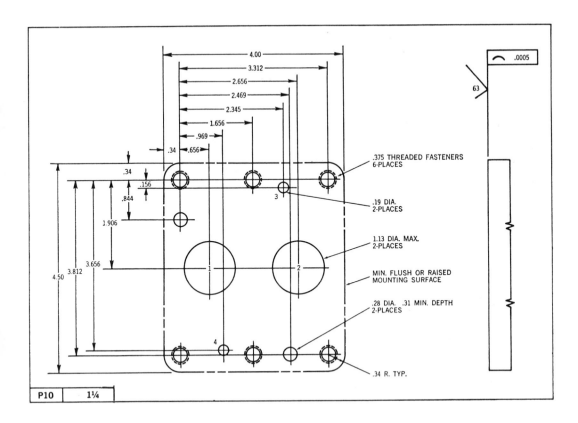

10. PRESSURE CONTROL VALVES - POPPET TYPE

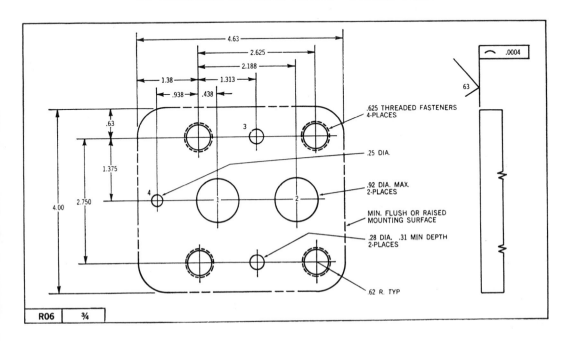

10. PRESSURE CONTROL VALVES - POPPET TYPE

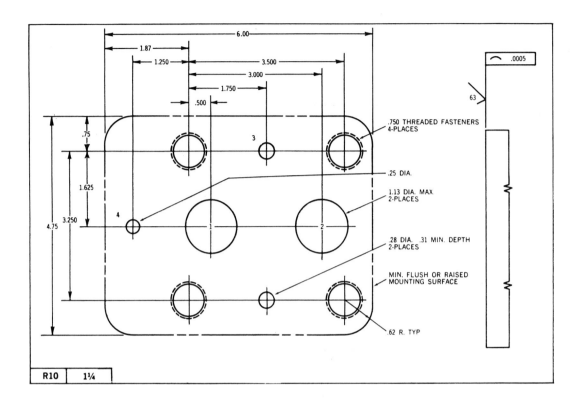

11. DIRECTIONAL CONTROL VALVES

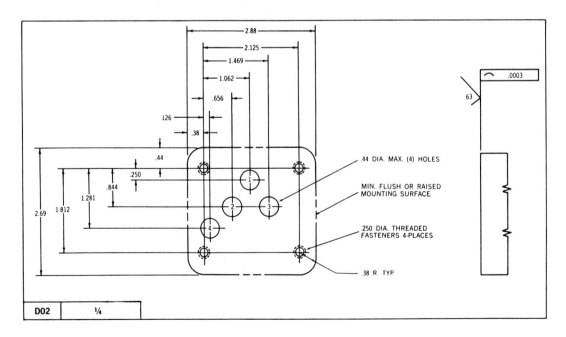

11. DIRECTIONAL CONTROL VALVES

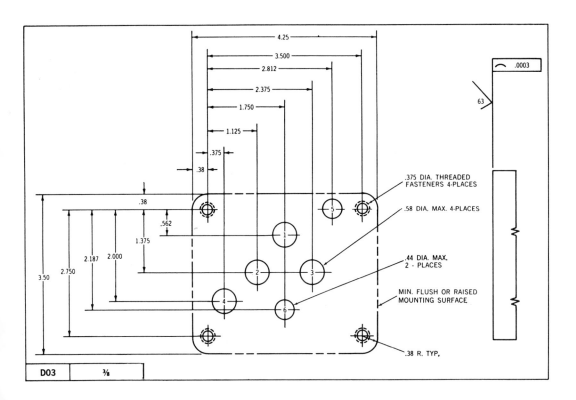

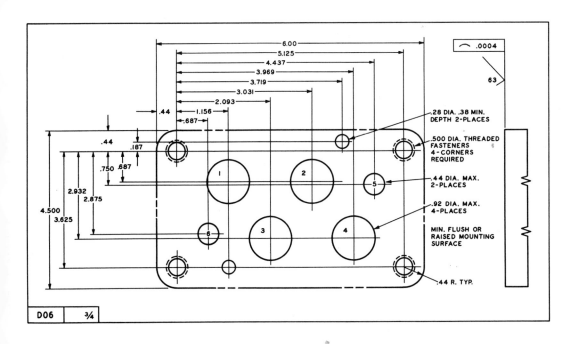

11. DIRECTIONAL CONTROL VALVES

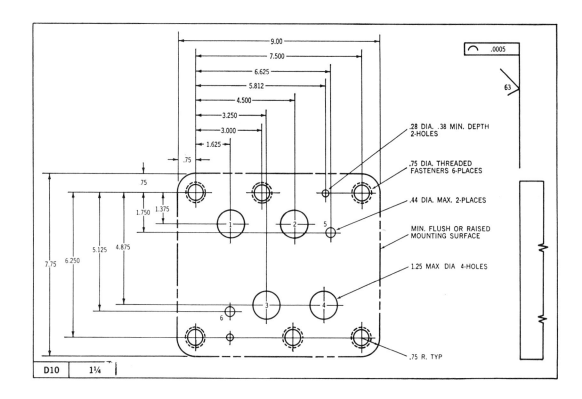

12. TWO PORT FLOW CONTROL VALVES

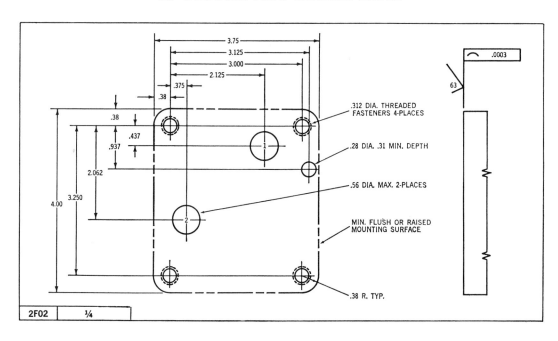

FLUID POWER SYMBOLS AND STANDARDS 167

12. TWO PORT FLOW CONTROL VALVES

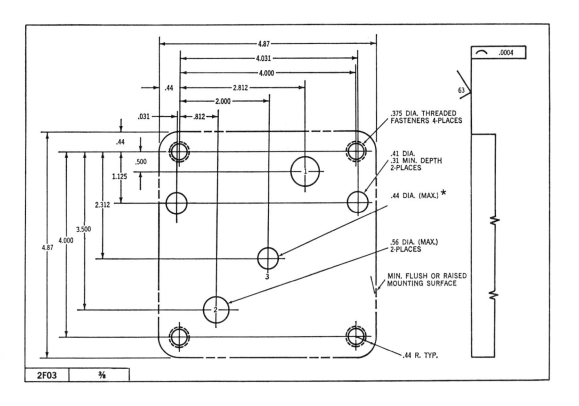

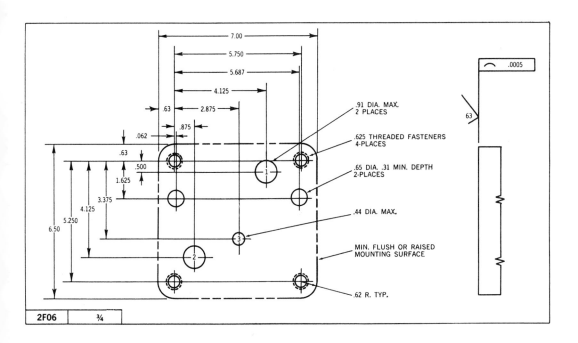

12. TWO PORT FLOW CONTROL VALVES

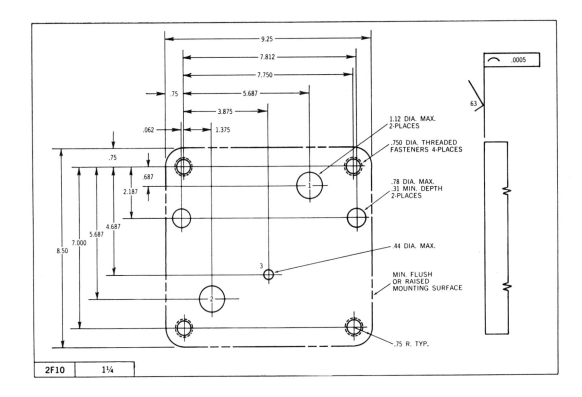

13. THREE PORT FLOW CONTROL VALVES

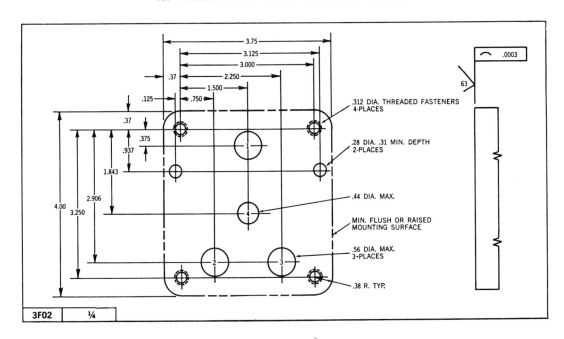

FLUID POWER SYMBOLS AND STANDARDS 169

13. THREE PORT FLOW CONTROL VALVES

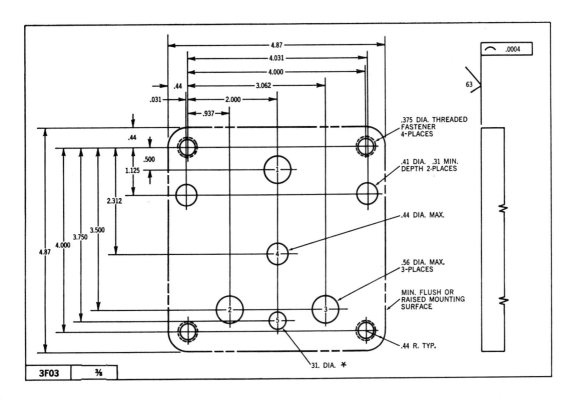

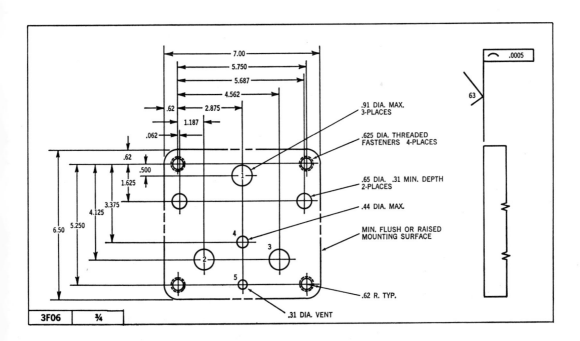

13. THREE PORT FLOW CONTROL VALVES

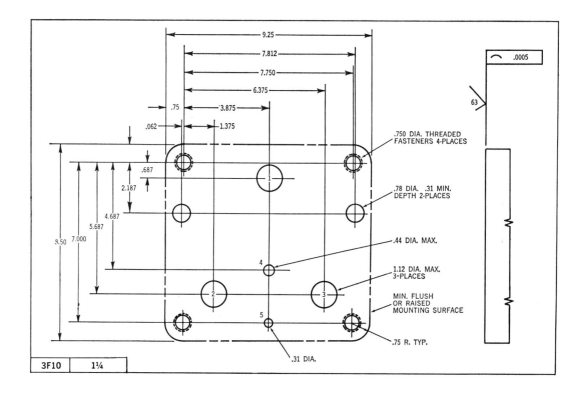

F. MOUNTING SURFACES FOR SUBPLATE TYPE HYDRAULIC DIRECTIONAL CONTROL VALVES—FOR 315 BAR HYDRAULIC SERVICE

(Reproduced by permission of National Fluid Power Association)

REFERENCES

1. American National Standard Glossary of Terms for Fluid Power, ANSI/B93.2-1971, and Supplements thereto. (ISO/TC 131/SC 1 (USA-2) 3)
2. International Standard Rules for the Use of Units of the International System of Units and a Selection of the Decimal Multiples and Sub-Multiples of SI Units, ISO 1000-1973
3. American National Standard Dimensions for Mounting Surfaces of Subplate Type Hydraulic Fluid Power Valves, ANSI/B93.7-1968.
4. American National Standard Decimal Inch, ANSI/B87.1-1965.

MOUNTING SURFACES FOR SUBPLATE TYPE HYDRAULIC DIRECTIONAL CONTROL VALVES—FOR 315 BAR HYDRAULIC SERVICE

INTRODUCTION

In hydraulic fluid power systems, power is transmitted and controlled through a liquid under pressure within an enclosed circuit. Typical components found in such systems are hydraulic valves. These devices control flow direction, pressure or flow rate of liquids in the enclosed circuit.

1. SCOPE

   To include:

   1.1 Subplates—315 bar maximum hydraulic service
      1.1.1 Directional Control Valves
   1.2 Dimensional Criteria
      1.2.1 Minimum surface dimensions

1.2.2 Sizes and locations of tapped holes for mounting bolts
1.2.3 Sizes and locations of ports
1.2.4 Sizes and locations of dowel or rest pins where required

1.3 General Criteria
1.3.1 Surface finish and flatness
1.3.2 Indication of tolerances where pertinent
1.3.3 Indication of appropriate corner breaks and radii
1.3.4 Use decimal inch system per Reference No. 4.

2. PURPOSE

To provide a composite reference for new designs and to simplify interchangeability of subplate mounted, hydraulic, fluid power valves.

3. TERMS AND DEFINITIONS

For definitions of terms used herein, see Ref. No. 1.

4. UNITS

4.1 The "Customary US" units are used herein.

4.2 Approximate conversions to the International System of Units are given per Ref. No. 2. These appear in parentheses following their US counterpart.

5. GENERAL

Ports are identified by numbers as a means of reference only.

6. SUBPLATE IDENTIFICATION CODING

6.1 Subplate identification code prefix is in accordance with Ref. No. 3 as follows: D– Directional Control Valves.

6.2 Subplate identification code suffix is in accordance with Ref. No. 3 as follows:

Table 1. Subplate code suffixes

Code Number	Valve Size
02	1/4"
06	3/4"
10	1 1/4"

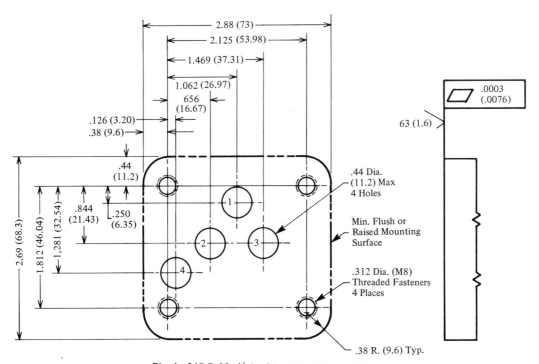

Fig. 1. 315 D 02–1/4 in. Directional Control Valve.

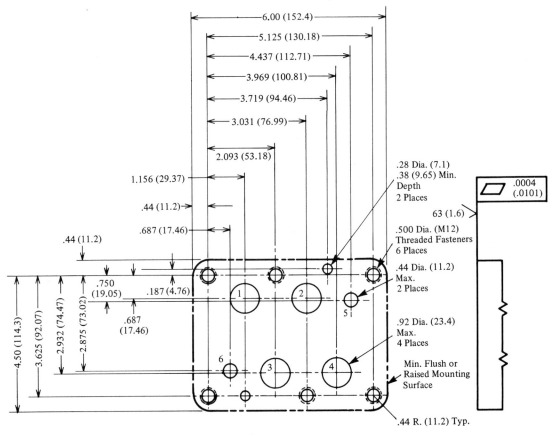

Fig. 2. 315 D 06—³⁄₄ in. Directional Control Valve.

7. TOLERANCES—BREAKS—RADII

7.1 The tolerance on two-place decimals is ±0.010 in.

7.2 The tolerance on three-place decimals is ±0.005 in.

7.3 Break all sharp edges.

7.4 Maximum radii is 0.01 in.

8. IDENTIFICATION STATEMENT

Use the following statement in catalogs and sales literature when electing to comply with this voluntary standard:

8.1 "Mounting surface dimensions conform to NFPA Recommended Standard, NFPA/T3.5.9-1973."

G. METHOD OF DIAGRAMMING FOR MOVING PARTS FLUID CONTROLS

(Reproduced by permission of National Fluid Power Association)

REFERENCES

1. American National Standard Fluid Power Diagrams, ANSI/Y14.17-1966.
2. International Standards Organization Standard Graphical Symbols for Hydraulic and Pneumatic Equipment and Accessories for Fluid Power Transmission, ISO/R 1219-1970. Agrees with ANSI/Y32.10-1967.
3. American National Standard Glossary of Terms for Fluid Power, ANSI/B93.2-1971, and Supplements thereto (ISO/TC 131/SC 1 (USA-2) 3).

BACKGROUND REFERENCES

1. JIC Electrical Standards for Mass Production Equipment—EMP-1-1967.
2. JIC Electrical Standards for General Purpose Machine Tools, EGP-1-1967.
3. General Motors Manufacturing Standards—Basic Electrical Standards for Industrial Equipment, June 1968.
4. General Motors Manufacturing Standards—Fluid Power Standards for Industrial Equipment, April 1967.

FLUID POWER SYMBOLS AND STANDARDS   173

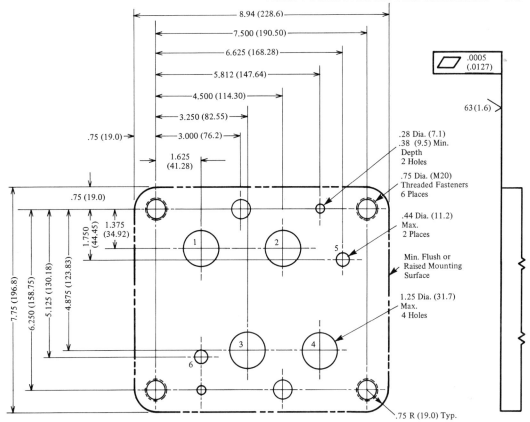

Fig. 3. 315 D 10—1¼ in. Directional Control Valve.

5. Numatics, Inc.—Bulletin 502B, The Numatrol Diagram.
6. Military Standard Graphic Symbols for Logic Diagrams, MIL-STD-806B.
7. American National Standard Graphic Symbols for Logic Diagrams, ANSI/Y32.14-1962.

METHOD OF DIAGRAMMING FOR MOVING PARTS FLUID CONTROLS

PART I—GENERAL

INTRODUCTION

In fluid power systems power is transmitted and controlled thru a fluid (liquid or gas) under pressure within an enclosed circuit.

Moving parts fluid controls achieve control thru the use of devices having moving parts.

With the growth of controls technology, there has arisen within the technology of fluid power a major field of controls endeavor involving the use of air and other fluids as the operating media for logic controls systems.

1. SCOPE

    1.1 To include drafting practices for drawings which depict logic control circuits using moving parts fluid controls, including all power sources, all inputs, all logic functions and all outputs.

    1.2 To include control systems using positive pressure and negative pressure (vacuum).

    1.3 To provide diagrams for logic control circuits. (For power diagrams use Reference No. 1 and Reference No. 2)

2. PURPOSE

    2.1 To develop a uniform means for diagramming moving parts fluid controls.

    2.2 To provide a communication means for industrial and educational purposes.

    2.3 To simplify design, fabrication, analysis and servicing of moving parts fluid controls.

2.4 To promote international understanding and use of moving parts fluid control systems.

3. **TERMS AND DEFINITIONS**

(For definitions of terms not herein defined, see Reference No. 3)

3.1 Circuit, Logic Control. A circuit which gathers and processes information to signal power controls and interfaces.

3.2 Circuit, Power Control. A circuit which directs and regulates fluid power to working devices.

3.3 Control Point. The point where the logic control flow path terminates at the actuator input of a logic control, power control or interface.

3.4 Diagram, Attached Symbols. A diagram in which all functions and connections to component symbols are shown in the symbols.

3.5 Diagram, Detached Symbols. A diagram in which various functions and connections of component symbols are shown by separate symbols located in various places on the diagram.

3.6 Diagram, Detail Logic. A diagram which depicts all logic functions of a circuit including identification and description of logic components, ports and connecting flow paths.

3.7 Diagram, Ladder. A diagram in which inputs are located to the left in a vertical column and outputs in a right vertical column. Interconnecting horizontal flow paths and components give the diagram a ladder appearance.

3.8 Diagram, Logic Control. A diagram which depicts logic control of power controls and interfaces.

3.9 Diagram, Power Control. A diagram which depicts all powered devices and interfaces including identification and description of components and their effect on the system.

3.10 Flow Passage, Controlled. A flow passage whose ability to pass fluid can be changed by the influence of a signal.

3.11 Flow Path. A series of conductors and passages which convey fluid.

3.12 Fluid Memory, Off Return. Fluid memory which receives a momentary signal and produces a change of state which continues to exist after the initiating signal has disappeared, providing an input is present at the supply port of the device. Upon loss of the supply pressure, the device reverts to its initial state.

3.13 Fluid Memory, Retentive. Fluid memory which receives a momentary signal and produces a change of state which continues to exist after the initiating signal has disappeared regardless of the presence or absence of supply pressure to the device. The device returns to its original state only upon receipt of a second reset control signal.

3.14 Fluid Signal. Fluid pressure or flow which can be detected or sensed.

3.15 Fluid Signal, Maintained. A fluid signal which exists indefinitely until caused to disappear by a secondary control action.

3.16 Fluid Signal, Momentary. A fluid signal which exists briefly and then disappears.

3.17 Fluid Signal, Timed. A fluid signal which exists for a definite period of time and then disappears.

3.18 Logic Control Device. Any device employed in a logic control circuit.

3.19 Moving Parts Logic. The technology of achieving logic control by means of fluid devices having moving parts.

3.20 Power Control Device. Any device used in a power control circuit.

3.21 Pressure Switches. This device uses an air signal from the control system to actuate an electric switch, and thereby produces an electrical output. It may be non-adjustable (actuate at a fixed pressure level) or adjustable to actuate at an adjustable pressure level. When the air signal is applied to the device, the switch actuates. When the air signal is removed, the device resets.

3.22 Relay Valve. A logic device which receives control signals and changes flow conditions in one or more controlled flow passages.

3.23 Relay Valve, Free Floating. A relay valve wherein the internal element moves freely without restraint and normally utilizes bias pressure at one control point.

3.24 Relay Valve, One Shot. A relay valve wherein controlled flow passages immediately change conditions when a control point is pressurized by a maintained signal. After a period of time, the controlled flow passages return to their original conditions, even though the control point is pressurized. When the control signal is removed it resets for another operation.

3.25 Relay Valve, Time Delay. A relay valve which creates a time interval between the pressurizing of a control point and a change in the controlled flow passages.

3.26 Relay Valve, Time Delay after Exhausting a Control Point. A relay valve with one control point which receives a maintained signal and causes immediate actuation of controlled flow passages. When the control signal is removed, a time delay occurs before controlled flow passages are reset.

3.27 Relay Valve, Time Delay after Pressurizing a Control Point. A relay valve with one control point which receives a maintained signal and causes a time delay before the controlled flow passages are actuated. The device resets immediately upon exhausting the control point.

3.28 Relay Valve, Time Delay, Detented. A time delay relay valve having "A" and "B" control points arranged to accept and act on momentary signals. A momentary signal into the "A" control point starts actuation and a momentary signal into the "B" control point starts reset.

3.29 Relay Valve, Time Delay, Detented, Delayed Action. A detented time delay relay valve in which a momentary signal into the "A" control point creates a time delay before controlled flow passages are actuated. The device resets immediately upon receipt of a signal in the "B" control point.

3.30 Relay Valve, Time Delay, Detented, Delayed Reset. A detented time delay relay valve in which a momentary signal into the "A" control point produces immediate actuation of the controlled flow passages. A momentary signal to the "B" control point starts the reset action with a time interval before controlled flow passages reset.

3.31 Sensor. Any device which detects a condition in a system and produces a signal.

3.32 Valve, Power Control. A valve which controls fluid power operating working devices.

3.33 Valve, Rotary Selector. A valve which utilizes rotary actuation to connect the inlet to any one of a number of outlets.

3.34 Valves, Pressure Sensing. A device similar to an electrical pressure switch, in which a signal to be sensed enters a control point, and actuates a mechanism which, at the proper pressure level, causes one or more flow passages to change condition. Removal of the signal allows the pressure sensing valve to reset.

4. SYMBOL RULES

4.1 Use symbols to show connections, flow passages and functions of the device represented.

NOTE. Symbols are digital, and do not indicate conditions occurring during transition from one flow condition to another.

4.2 Do not use symbols to indicate construction, or to indicate values, such as pressure, flow rate or other component settings.

4.3 Do not use symbols to indicate the physical location of ports, direction of shifting of spools, poppets or diaphragms, or the position of actuators on actual components.

4.4 Rotate or reverse symbols without altering their meaning.

4.5 Avoid using symbols with words or abbreviations.

NOTE. Symbols capable of crossing language barriers are presented herein.

4.6 Draw symbols so that flow is assumed

176 VALVE ENGINEERING AND DESIGN DATA

to proceed from left to right unless clearly indicated otherwise.

NOTE. Exhaust flow is assumed to proceed from right to left unless clearly indicated otherwise.

4.7 Use a combination of simple symbols to make composite symbols.

4.8 Use composite symbols to represent complex components.

4.9 Use an enclosure to show functional assemblies too complex to be shown by composite symbols.

4.10 Show each symbol with all external ports to be connected for the component to perform its function properly.

NOTE. Do not show exhaust ports, vent ports and unused flow passages which require no external connections.

4.11 Draw each symbol to show the conditions existing when the machine is in the condition specified by the "Initial Conditions" statement in the Sequence of Operations.

4.11.1 Use the Initial Conditions to show the conditions at the start of a normal automatic cycle, with the power and the control air ON.

4.11.2 Where special conditions or circuitry techniques warrant, show the Initial Conditions with the control air and power OFF the machine, and all component symbols shown accordingly.

NOTE. In cases where the particular symbol is incapable of showing the starting condition, the Initial Conditions are of no concern.

5. LINE TECHNIQUES

5.1 Keep line widths approximately equal.

NOTE. Line width does not alter the meaning of the symbols.

5.2 Solid Line _____
Supply line to control system and all lines in the logic control system except output lines.

5.3 Dashed Lines _ _ _ _ _ _ _
Indicate output lines from logic control system to power control points and interfaces.

5.4 Dotted Line ....
Indicates mechanical connection between devices.

5.5 Center Line _ . _ . _ . _ . _
Indicates outline of an enclosure.

NOTE. May also be used around a composite symbol.

5.6 Lines Crossing
Preferred:

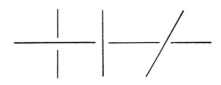

Alternate:

NOTE. Intersection is not necessarily at 90° angle.

5.7 Lines Joining
Preferred:

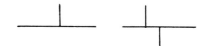

Alternate:

5.8 Lines, Flexible (Continually flexing)

5.9 Sensing Lines
Drawn the same as the line to which they connect.

5.10 Pivot Point
Shown in a mechanical connection.

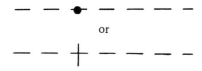

5.11 Line Identification

Identify each line representing a conductor by a number drawn above or to the right of the line.

5.11.1 Use consecutive numbering.

5.11.2 Use each line identification only once in any logic control system.

5.11.3 Use the same identification at all terminals and tie points for each line (conductor).

5.11.4 Use the same identification for all lines (conductors) connected to the same terminal or tie point.

PART II—SYMBOLS—INPUT DEVICES

6. SYMBOLS—MANUAL CONTROLS

6.1 Push Buttons

6.1.1 Terminals (ports) on a device that has an implied exhausting function which is not shown.

6.1.2 Terminals (ports) on a device which has no implied exhausting function. All functions are shown.

6.1.3 Operator and bridge, flush, extended, half-guarded or guarded.

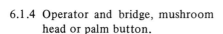

6.1.4 Operator and bridge, mushroom head or palm button.

6.1.5 Spring return action (shown by the presence of the spring symbol).

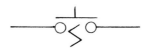

6.1.6 Detented action (shown by the absence of the spring symbol). Solid line shows start condition, dotted line shows actuated condition.

6.1.7 Detented push button, air return.

6.1.8 Latching button, spring return with manually operated mechanical latch.

6.1.8.1 Depress and hold button down.

6.1.8.2 Engage latch.

6.1.8.3 Release button. Latch will hold button depressed.

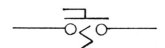

6.1.8.4 To release, pull out latch. Spring will return button to OUT.

NOTE. Latch usually has provision for padlock, so it can be locked in latched position.

6.1.9 Key lock button with lock built into button. Additional note explains any details of key and lock action.

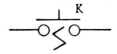

6.1.10 Two-button station, mechanically linked so that when one button is depressed, the second is OUT. Detented in both positions.

178 VALVE ENGINEERING AND DESIGN DATA

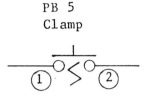

6.1.11 Identify each push button with a PB number, starting with PB1 and running consecutively. There is no significance to the numbers, other than to identify the particular button. Thus, in one circuit the start button might be PB1, while in another circuit it might be PB5.

6.1.11.1 For Spring Return Buttons, put above the symbol the PB number and the function which the button performs in this cycle.

PB 1
Cycle Start

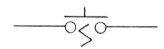

6.1.11.2 For Detented Buttons, put above the symbol the PB number and the function which the button performs in the cycle. At the side put notes telling what happens when the button is pushed in, and when it is pulled out.

PB 6
Emerg. Stop

Pull-Reset
Push-Stop

6.1.11.3 For Port Numbers put under the symbol in small circles, the port numbers on the component which must be connected to provide the function shown.

PB 5
Clamp

6.1.12 See Table 1 for typical push button symbols.

6.2 Selectors and Toggles—Two-Position

6.2.1 Operator and bridge. Show start position by solid arrow. Show actuated position by dotted arrow.

6.2.2 Spring return action. (Shown by the presence of the spring symbol).

6.2.3 Detented action. Show start condition by solid lines. Show actuated conditions by dotted lines.

6.2.4 Typical two-position selector. Indicate spring return to the left position.

6.2.5 Typical two-position selector. Indicate spring return to the right position.

6.2.6 Typical two-position selector. Indicate detented in both positions; indicate left position as starting position.

Table 1. Typical push button symbols.

Action	Symbol	Alternate Symbol
2-way, spring return, normally passing		
2-way, spring return, normally not passing		
3-way, spring return, normally passing		
3-way, spring return, normally not passing		
3-way, detented, passing at start of cycle, push to break		
3-way, detented, not passing at start of cycle, push to make		
3-way, 3 port used as a diverter, spring return, no exhaust in this function, alternate outlet is blocked to reverse flow.		
3-way, 3 port used as a selector, spring return, alternate supply is blocked		
4-way, 4 port 2-position, spring return, alternate output is exhausted		
Dual path 4-way, 5 port spring return, alternate output is exhausted.		

180 VALVE ENGINEERING AND DESIGN DATA

6.2.7 Typical two-position selector. Indicate detented in both positions; indicate right position as starting position.

6.2.8 Identification. Identify each selector with an SV number, starting with SV1 and running consecutively. There is no significance to the numbers other than to identify the particular selector.

6.2.8.1 Above the symbol, put the SV number and the function which the selector performs in the cycle. At each side of the arrows put the action which occurs when the operator is in that position.

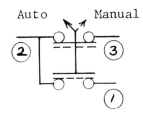

SV 1
Auto-Manual Selector

6.2.8.2 At the bottom of each port, put the port identification which provides the function shown.

6.3 Foot Operated—Two-Position

6.3.1 Treadle operator and bridge, hooded or open.

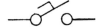

6.3.2 Single treadle spring return action (shown by the presence of the spring symbol).

NOTE. The spring return symbol is necessary to prevent confusion with detented operators. Such a distinction has not been necessary in electrical diagrams because there was not a push button-detented switch until quite recent times.

6.3.2.1 Normally not passing

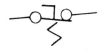

6.3.2.2 Normally passing.

6.3.2.3 Typical four-way action, common inlet

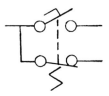

6.3.3 Double treadle, detented in both positions.

6.3.3.1 Not passing at start of cycle

6.3.3.2 Passing at start of cycle

6.3.3.3 Typical four-way action, common inlet

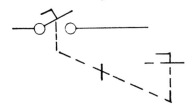

6.3.4 Identify each foot operated valve with an FTV number, starting with FTV1 and running consecutively. There is no significance to the number other than to identify the particular foot operated valve.

FTV 1

6.3.4.1 For Single Treadle Spring Return, put above the symbol the FTV number and the function which the foot operated valve performs in the cycle.

FTV 2
De-clutch Palle

6.3.4.2 For Double Treadle Detented, put the FTV number above the symbol. Next to each treadle put a note describing what happens when that treadle takes command.

FTV 7
Open Door

———Close Door

6.4 Selectors—Three-Position and Multi-Position

  6.4.1 Manually operated (knob, lever, or toggle)

  6.4.2 Foot operated

  6.4.3 Mechanically operated

  6.4.4 Flow passage designations. Represent each flow passage by a horizontal line containing a pair of terminal (port) symbols, followed by a position grid.

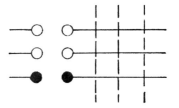

6.4.5 Inlet designations. Indicate a separate inlet passage with a line entering the symbol from the left. An inlet may serve only one flow passage, or it may serve several flow passages. A single line entering the symbol and connecting with several sets of terminal symbols indicates a common connection within the device.

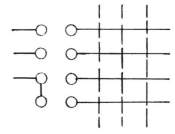

6.4.6 Exhaust designations. Indicate special flow conditions of an exhaust path, if necessary, by a line exiting from the left side of the symbol and terminating in a dashed line. This shows that the particular connection is an exhaust to atmosphere.

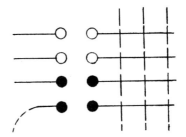

6.4.7 Actuator designations. Show the method of actuating the device by an actuator symbol placed in the top set of terminals. The position of the actuator bridge is not significant, nor is the absence of a spring symbol. The flow condition at the start of the cycle and the method of returning the actuator to the start condition are shown by the position grid.

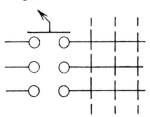

6.4.8 Actuator symbols.

  6.4.8.1 Manual

  6.4.8.2 Foot (single or double treadle)

  6.4.8.3 Mechanical

6.4.9 Position grid. Illustrate the various flow passages offered by the device by a position grid located to the right of the flow path terminals.

6.4.10 Position designations. Represent each position of the operator by a vertical dashed line, with a number above the line.

  6.4.10.1 Use the position number 0 (zero) as the starting position and number all other positions consecutively. All numbers except zero have no significance except to identify the particular position.

6.4.11 Flow passage designations. Indicate by the presence or absence of an "X" at the intersection of the horizontal line and the vertical dashed line the condition of each flow passage for each position of the operator.

"X" means that the flow passage is passing when the operator is in that position.

SV 6
Slide Jog

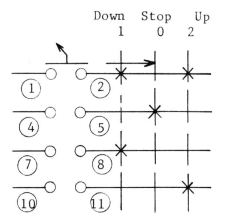

6.4.12 Type of return mechanism. Show spring return to the start position by a horizontal arrow pointing from the position shown back to the start position. (1 to zero)

  6.4.12.1 Show detented action by the lack of a return arrow (2 to zero)

6.4.13 Identification. Identify each selector with a letter combination followed by a number.

The letter combination designates the type of actuator, and is the same letter combination used to identify other devices with this same type of actuation.

  6.4.13.1 Manual—SV plus a number.

  6.4.13.2 Foot—FTV plus a number.

  6.4.13.3 Mechanical—LV plus a number.

  6.4.13.4 The numbers have no significance except to identify the particular device. In assigning these numbers, simply number consecutively

FLUID POWER SYMBOLS AND STANDARDS   183

with other devices using the same type of actuation.

6.4.13.5 Above each symbol put the device number and a note telling the function it performs in the cycle.

6.4.13.6 Above each position of the operator put a note telling what action occurs when the selector is in that position.

6.4.13.7 Under each inlet and outlet terminal symbol put the port identification which provides the function shown.

6.4.14 Rotary selector valves. Use either of the two approved symbols.

6.4.14.1 Identify each rotary selector valve with an RSV number, starting with RSV1 and numbering consecutively. The numbers have no significance other than to identify the device.

6.4.14.2 Under each inlet and outlet terminal put the port identification which provides the function shown.

7. SYMBOLS FOR POSITION SENSORS

7.1 Limit Valves

7.1.1 Mechanical actuator and bridge, spring return.

7.1.2 Mechanical actuator and bridge, detented. Indicate by the double

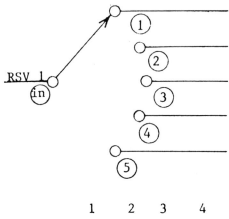

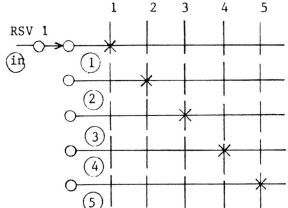

184 VALVE ENGINEERING AND DESIGN DATA

ramp cam that this bridge is detented and is cammed in both directions.

7.1.3 Actuator and bridge, cable actuated, manual reset.

7.1.4 Condition of actuation at start of cycle. Indicate by the position of the actuator and bridge whether or not the actuator is cammed at the start of the cycle.

    7.1.4.1 Actuator NOT actuated at start of cycle. Indicate actuator DOWN as far as it will go.

    7.1.4.2 Actuator ACTUATED at start of cycle. Indicate actuator UP as far as it will go.

7.1.5 Identification. Identify each limit valve by an LV number, starting with LV1 and numbering consecutively. The number has no significance other than to identify the particular device.

    7.1.5.1 Place the LV number and a note telling what machine motion actuates the limit, and under what circumstances above the limit valve symbol.

        LV 7
        Pusher
    Fully to Right

    7.1.5.2 If the limit is actuated on one stroke of the motion, and idles (overrides) on the return stroke, put a brief note explaining this.

    7.1.5.3 Under the symbol, in small circles, put the port identification on the component which must be connected to provide the function shown.

7.1.6 For typical limit valve symbols see Table 2.

7.2 Float-Actuated Valves

    7.2.1 Normally not passing. Goes passing on rising liquid level. Resets on falling level.

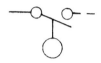

    7.2.2 Normally passing. Goes not passing on rising liquid level. Resets on falling level.

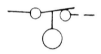

    7.2.3 Identification. Identify each float operated valve by an FV number, starting with FV1, and numbering consecutively. There is no significance to the number other than to identify the particular valve. Above the symbol put a note telling what actuates the valve, and under what conditions.

        FV 5
#1 Tank 500 Gallons
      or more

7.3 Air Jets

    7.3.1 Air jet. Symbol does not indicate whether or not jet is blowing at start of cycle.

Table 2. Typical limit valve symbols.

Action	Symbol
3-way, normally not passing, not actuated at start of cycle	
3-way, normally not passing, actuated (held passing) at start of cycle	
3-way, normally passing, not actuated at start of cycle	
3-way, normally passing, actuated (held not passing) at start of cycle	
4-way, spring return, not actuated at start of cycle	
4-way, spring return, actuated at start of cycle	
Dual 3-way (2-path), spring return, not actuated at start of cycle	
3-way, detented, not passing at start of cycle	
3-way, detented, passing at start of cycle	
3-way, 3 port used as a diverter, spring return, no exhaust in this function, alternate outlet is blocked to reverse flow	
3-way, 3 port used as selector, spring return, alternate supply is blocked.	

186   VALVE ENGINEERING AND DESIGN DATA

7.3.2 Air jet not blocked by target at start of cycle. The target is the device which blocks the jet, and whose position the jet is intended to sense.

7.3.3 Air jet blocked by target at start of cycle.

7.3.4 Identification. Identify each jet by the word "Jet" followed by a number, starting with Jet 1 and numbering consecutively. The number has no significance other than to identify the particular jet.

```
Jet 16
Slide Full Fwd.
```

7.3.4.1 Above each jet symbol, put the jet number and a note telling what machine motion (target) blocks the jet, and under what circumstances.

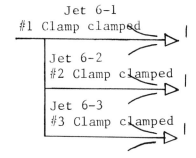

7.3.4.2 In the case of multiple jets, all acting together to actuate one receiving device, all the jets in the group may be given the same jet number, followed by dash numbers, Jet 6-1, Jet 6-2, etc.

8. SYMBOLS FOR PRESSURE SENSORS

8.1 Pressure Sensing Valves

Use the following symbols:

8.1.1 Controlled flow passage which goes passing on rising pressure, and goes not passing on falling pressure. Not pressurized at start of cycle.

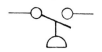

8.1.2 Controlled flow passage which goes not passing on rising pressure and goes not passing on falling pressure. Pressurized at start of cycle.

8.1.3 Controlled flow passage which goes not passing on rising pressure and passing on falling pressure. Not pressurized at start of cycle.

8.1.4 Controlled flow passage which goes not passing on rising pressure and passing on falling pressure. Pressurized at start of cycle.

8.1.5 Pressure sensing valve having more than one controlled flow passage depicted by a composite symbol. (Example shows a 4-way pressure sensing valve.)

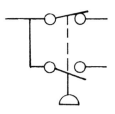

NOTE. Such valves may be used to sense either pressures above atmospheric, or vacuums. In either case, the symbols are the same.

8.1.6 Identification. Identify each pressure sensing valve with a PSV number, starting with PSV1 and numbering consecutively. The numbers have no significance other than to identify the particular pressure sensing valve. Above each symbol, put the PSV number and a note telling what pressure on the machine actuates the pressure sensing valve, and under what circumstances.

PSV 4

Clamp pressure at or above 60 psig

9. SYMBOLS FOR TEMPERATURE SENSORS

9.1 Normally not Passing

Goes passing on increasing temperature. Resets on decreasing temperature.

9.2 Normally Passing

Goes not passing on increasing temperature. Resets on decreasing temperature.

9.3 Identification

9.3.1 Identify each temperature actuated valve by a TAV number, starting with TAV1 and numbering consecutively. The numbers have no significance other than to identify the particular valve.

TAV 1

Cooling water 160°F or higher

9.3.2 Put a note telling what temperature actuates the valve and under what circumstances.

10. SYMBOLS FOR FLOW SENSORS

10.1 Normally Not Passing

Goes not passing when flow reaches required conditions. Resets when flow drops below required conditions.

10.2 Normally Passing

Goes passing when flow is present in required quantity. Resets when flow stops.

10.3 Identification

Identify each flow operated valve with an FLV number, starting with FLV1 and numbering consecutively. The numbers have no significance other than to identify the particular flow operated valve. Put a note telling what flow conditions actuate the valve and under what circumstances.

FLV 3

Cooling water flowing

PART III–POWER CONTROL VALVES AND INTERFACES

11. POWER CONTROL VALVES

11.1 General Principles

11.1.1 Simple and easy to draw.

11.1.2 Compact as possible.

11.1.3 Coded, if possible, by shape to indicate the type of valving action represented, and the type of control signals required to actuate the valve successfully.

NOTE. This is particularly important in view of the international connotations of our work.

11.1.4 Using basic symbol shapes per Reference No. 2.

11.1.5 Capable of representing relay valves of similar action, with suitable changes in the identification.

11.2 Identification of Power Control Valves

11.2.1 Supply a unique identification for each power control valve and each control point on each power control valve.

11.2.2 Use the same identification where it appears on all drawings of electrical control diagrams, pneumatic control diagrams, pneumatic fluid power diagrams or hydraulic fluid power diagrams.

11.2.3 Use the letter "H" as the basic designation for hydraulic power control.

11.2.4 Use the letter "P" as the basic designation for pneumatic power control.

11.2.5 Number each valve in clause 11.2.3 and 11.2.4 consecutively starting with the number 1. The number has no significance other than to identify the particular valve.

EXAMPLE. H1, H2, H3, etc. for hydraulic valves and P1, P2, P3, etc. for pneumatic valves.

11.2.6 Further identify control points on power control valves as follows:

11.2.6.1 Identify the control point on a valve having only one control point solely by the identification of the valve.

EXAMPLE. The control point on a single pilot spring return pneumatic valve P3 is identified as control point P3.

11.2.6.2 Identify by the number of the valve and the letters "A" and "B" the control points on a power control valve which has two control points, one acting to CHANGE the starting condition of the valve, and the other acting to RESTORE the starting condition of the valve.

11.2.6.2.1 "A" actuator indicates the control point which, when it takes command of the valve, causes the machine element to move AWAY from its starting position.

11.2.6.2.2 "B" actuator indicates the control point which, when it takes command of the valve, causes the machine element to move BACK to its starting position.

NOTE. The complete identification of the control point is the identification of the valve followed by the "A" or "B" such as H1A, H1B, etc. for hydraulic valves, P1A, P1B, P2A, P2B, etc. for pneumatic valves.

11.2.7 Use the above noted identification for multiple control points which perform identical functions on a power control valve. In addition, identify by dash numbers added after the usual control point identification, starting with 1, and numbering consecutively. The dash numbers have no significance other than to identify the particular control point.

EXAMPLE. A single pilot spring return valve with multiple control points serving the "A" actuator, any one control point capable of actuating the valve. If the valve identifier is P4, the control points would be identified as P4-1, P4-2, P4-3, etc.

11.2.8 In the case of multiple control points on the "A" actuator of a double pilot detented power control valve, the valve identifier being P6, identify the control points as P6A-1, P6A-2, P6A-3, etc.

NOTE. Clause 11.2.7 anticipates that in the future, developments in pneumatically actuated power control valving will build multiple actuators into the valve. Present standards make no provision for such future developments.

11.2.9 Arrange the power control points in the circuit diagram in a vertical column to the right of all relay valve control points, and use dashed lines leading from the control system to the power control points.

11.2.10 Place a note to the right of each power valve control point telling what machine motion takes place when that control point takes command of its power control valve.

11.2.11 In the case of a control point on a power control valve having control points on both "A" and "B" actuators, place a suitable cross-reference notation on the drawing adjacent to each control point, giving the location on the drawing of all the opposing control points.

11.2.11.1 Put the notation per clause 11.2.11 in the form of a set of brackets to the right of the control point symbol, under the machine motion note, the brackets containing the guide line numbers of the circuit lines containing the opposing control points.

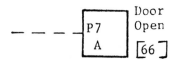

11.3 Symbols for Control Points on Power Control Valves

11.3.1 Control point on a valve having one actuator and spring return or automatic return, the automatic return accomplished entirely within the valve assembly, and requiring no connections from the control system to make it operational.

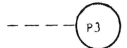

When this control point is pressurized, it takes command of the valve and keeps command as long as it is pressurized. When it is exhausted, the spring (or automatic return) returns the valve to the starting position.

11.3.2 Control points on a double control point detented (or equivalent) power control valve.

When either control point is exhausted, a momentary fluid signal into the opposite control point will cause that control point to take command and shift the valve action. When the fluid signal is exhausted, the detent holds the valve action in the new position until the second control point will cause it to take command, and shift the valve action back to the start position.

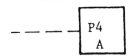

If control pressure is held in either control point, a fluid signal of equal pressure into the opposing control point will not shift the valve until the original control point is exhausted.

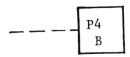

11.3.3 Control points on a double control point 3-position spring centered (or power centered) power control valve.

When both control points are exhausted or at the same pressure, the springs center the valve.

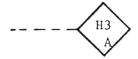

When either control point is pressurized and the opposing control point exhausted, the control point which is pressurized takes command of the valve and shifts it towards the exhausted control point.

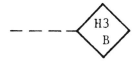

When the opposing control point is pressurized, or the pressurized control point is exhausted again, the springs again center the valve action.

11.3.4 Control points on a double control point spring offset power control valve.

Control point on end opposite the spring.

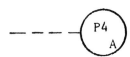

Control point on the spring end.

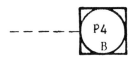

With the spring end control point exhausted, the "A" control point acts like the control point on a single control point spring return valve.

With pressure in the spring end ("B") control point, an equal pressure in the "A" control point will not shift the valve until the "B" control point is exhausted.

If the "B" control point is exhausted, a fluid signal into the "A" control point will shift the valve and it will stay shifted as long as the fluid signal into "A" is maintained.

However, if, while the "A" control point is pressurized and in command of the valve, an equal pressure is introduced into the "B" control point, the two fluid signals will cancel each other, and the spring will return the valve to the starting position EVEN THOUGH THE "A" CONTROL POINT IS STILL PRESSURIZED.

Finally, if the "A" control point is pressurized by a maintained fluid signal, it will take command of the valve and hold the valve shifted against the spring. The "B" control point can then be pressurized and exhausted, and will act like the actuator on a single control point spring return valve.

11.3.5 Cross-hatch all control points which are pressurized at the start of a normal automatic cycle. This cross-hatching is a major aid in understanding the operation of the circuit.

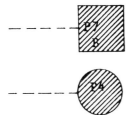

12. SYMBOLS FOR PRESSURE SWITCHES

12.1 Identification of Pressure Actuated Switches

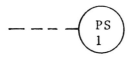

Identify each switch by a PS number, starting with PS1 and numbering consecutively. The number has no significance other than to identify the particular switch.

12.2 Location on the Diagram

12.2.1 Arrange symbols for pressure actuated switches on the circuit diagram in a vertical column to the right of all relay valve control point symbols, and in line with the power control valve control symbols.

12.2.2 Use dashed lines leading from the control system to the pressure actuated switch symbols.

PART IV–ATTACHED METHOD OF DIAGRAMMING (LOGIC DEVICES)

13. GRAPHIC SYMBOLS, ATTACHED METHOD

13.1 Use graphic symbols to express the logic functions of a control circuit.

13.2 Establish symbols that are easy to draw and represent an obvious function.

NOTE. Since all logic functions can be described using standard valve symbols or combinations of valve symbols, use Reference No. 1 to describe each of the basic logic functions.

13.3 Show the descriptive shaped logic symbols with a Truth Table indicating the output state (1 = pressurized, 0 = exhausted) for all possible combinations of input states.

13.4 Use the AND output to assume the 1-state, if and only if, all of the inputs assume the 1-state. The AND is a passive device (See Table 3).

13.5 Use the OR output to assume the 1-state, if any or all, of the inputs assume the 1-state. The OR is a passive device. (See Table 4)

13.6 Use the NOT output to assume the 1-state only when the input assumes the 0-state. The NOT is an active device. (See Table 5)

13.7 Use the Inhibitor output to assume the 1-state when input "B" is in the 1-state and the input "A" is in the 0-state. The inhibitor is a passive device. (See Table 6)

192 VALVE ENGINEERING AND DESIGN DATA

Table 3. Truth table for AND output.

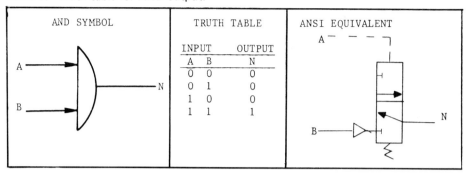

Table 4. Truth table for OR output.

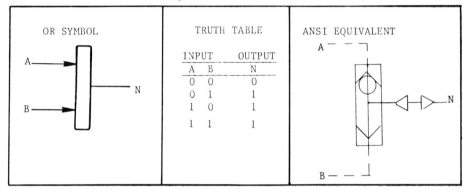

Table 5. Truth table for NOT output.

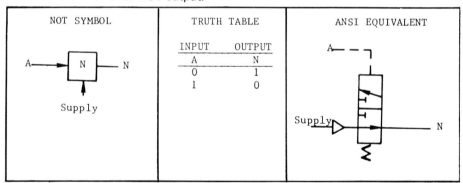

Table 6. Truth table for inhibitor (nonimplication) output.

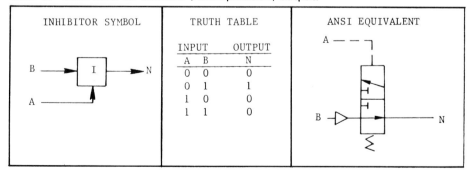

Table 7. Truth table for Fluid Memory device.

FLUID MEMORY SYMBOL	TRUTH TABLE	ANSI EQUIVALENT
A → S. MEM R → N Supply	INPUT OUTPUT A B N 1 0 1 0 1 0	(diagram)

13.8 Use the Fluid Memory as an active device to store a single bit of information. It has two inputs, set (S) and reset (R) and may have one or two outputs. The storage of the single bit of information is maintained *only as long as power is maintained*. (See Table 7)

13.9 Use the Flip-Flop as a sequential device whose output reflects not only the present input condition but the previous input history. (See Table 8). Its storage capability is not affected by power interruptions.

13.11 Use the delay, Timing In output to assume the 1-state after a controlled period of time after the input assumes the 1-state. (See Table 10)

13.12 Use the delay, timing out output (which is in 1-state) to assume the 0-state after a controlled period of time after the input assumes the 0-state. (See Table 11)

13.13 Use the Amplifier as an active device to allow an input signal of low energy to control an output fluid signal of a high energy level.

Table 8. Truth table for Flip-Flop device.

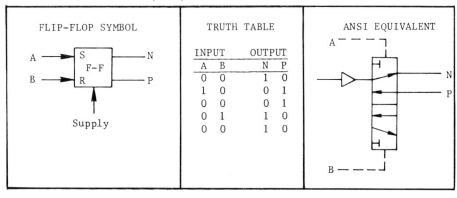

13.10 Use the One-Shot as a passive single shot device to provide a timed single output. When the input assumes the 1-state the output also assumes the 1-state for a predetermined period of time and then returns to the 0-state. (See Table 9.)

13.14 Show the Amplifier, consisting of one or more stages, connected to the AND or NOT symbol to designate a normally 1-state (NOT) or a normally 0-state (AND).

13.15 Use the General Logic Symbol for functions not elsewhere specified.

Table 9. Truth table for one-shot device.

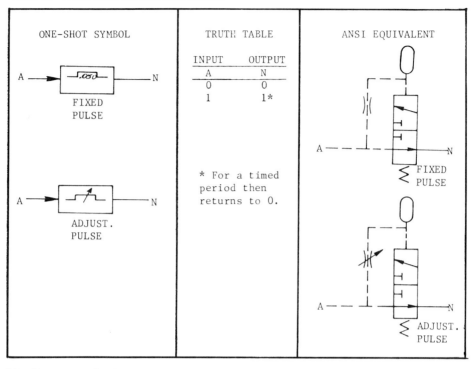

Table 10. Truth table for delay, timing in.

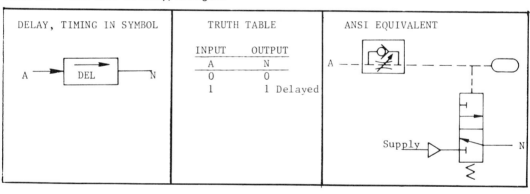

Table 11. Truth table for delay, timing out.

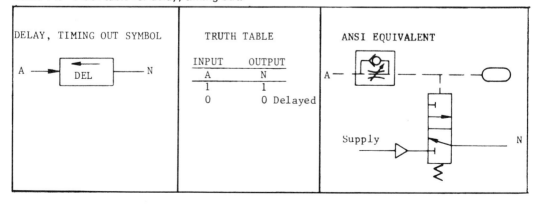

Table 12. Truth table for amplifier device.

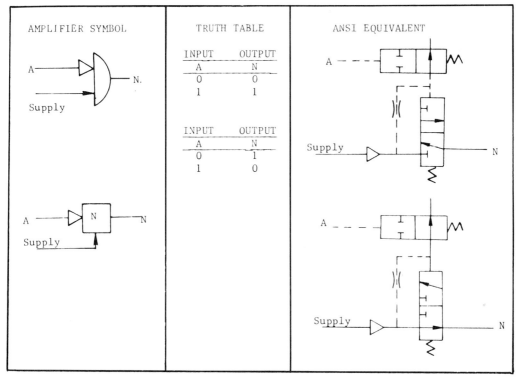

13.15.1
Label this symbol adequately to identify the function performed.

NOTE. It is not intended that this symbol be used for functions which can be logically expressed by a single symbol established in this document.

14. PRACTICE, ATTACHED METHOD

14.1 Multiple Inputs to a Single Function

14.2 Multiple Inputs to Physically Separate Functions with Common Outputs

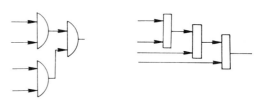

15. DIAGRAM RULES, ATTACHED METHOD

15.1 Arrangement of Symbols

15.1.1 Arrange symbols in the diagram to facilitate the use of direct and straight interconnecting lines.

15.1.2 Prepare lines between symbol inputs and outputs horizontal or vertical with a minimum of line crossing and with spacing to avoid crossing.

15.2 Circuit Arrangement

15.2.1 Arrange the circuit in functional sequence, left to right and top to bottom. Follow this rule rigidly where excessive line crossing impairs the clarity of the diagram.

15.2.2 Show all inputs to the control circuit on the left and all outputs on the right.

NOTE. For clarity of the total system, the operational diagram may be combined and shown connected to the logic diagram.

15.2.3 Use solid lines to represent interconnecting flow paths between logic devices within the control circuit.

15.3 Designations

15.3.1 Use designations of input(s), output(s), flow paths and logic devices to facilitate the checking of circuits. The use of designation is determined by the overall adaptability of the design, assembly, installation and maintenance of the equipment.

15.3.2 Assign all input(s) and output(s) of the control circuit arbitrary designations. These are comprised of letters and combinations of letters and numbers (A, B, C, X, Y, Z, etc. or A1, A2, A3, Z7, A10, etc.) To further identify the inputs from the outputs, the inputs are assigned a letter from the first half of the alphabet with the outputs assigned a letter from the last half of the alphabet (O and I are omitted).

15.3.3 To identify the flow path, use detailed logic diagrams to show each interconnecting flow path (between input(s) or output(s) and logic devices

and between logic devices) and number consecutively.

15.3.4 For logic devices, use detailed logic diagrams to identify each logic device with a numerical designation starting with the number 1 and numbering consecutively.

15.4 Explanatory Notes

Add explanatory notes for clarification of function description where the function is not clear.

NOTE. Where notes are lengthy or must be repeated, they may be shown at a common location with proper reference at the point of application.

15.5 Sequence of Operation

15.5.1 When applicable, list in the order in which they occur, the sequence of operation of the control circuit with reference to input conditions and resulting output conditions.

15.5.2 Number each phase followed by a brief description of the input conditions and the resulting outputs. These sequence statements always supercede a statement concerning the condition of the inputs and outputs as they appear at the start of the sequence.

15.6 Installation

Show data pertinent to installation on the detail logic diagrams. Items to be included are tubing size, operating pressure, degree of filtration required and enclosure (if required).

16. LOGIC DIAGRAMS, ATTACHED METHOD

16.1 Typical Logic Diagrams Showing Logic Function in a Typical Circuit

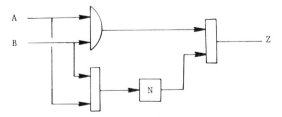

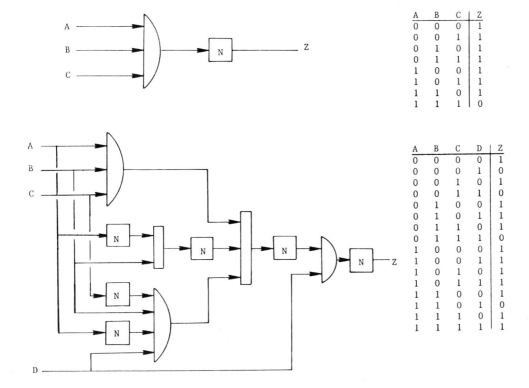

PART V DETACHED METHOD OF DIAGRAMMING (RELAY VALVES AND LOGIC DEVICES)

17. SYMBOLS FOR RELAY VALVES

17.1 Identification of Relay Valves

17.1.1 Identify each relay valve by the letters "RV" followed by a number, starting with the number 1 and numbering consecutively. There is no significance to the numbers other than to identify the particular relay valve.

17.1.2 Identify one relay valve in each circuit as a master relay valve, RVM

17.2 Identification of Special Purpose Relay Valves

Identify either by standard RV numbers, or by special identification which clearly indicates the function when a relay valve is used in conjunction with other devices to provide a special-purpose function (jet position-sensing, pressure-sensing, timing, etc.)

17.3 Special Identification

Use the following rules:

17.3.1 Identify by a JRV number (JRV1, JRV2, etc.) a relay valve which cooperates with a jet in a jet position-sensing function.

17.3.2 Identify by a PSRV number (PSRV1, PSRV2, etc.) a relay valve used as the sensing element in a pressure-sensing function.

17.3.3 Identify a TDRV number (TDRV1, TDRV2, etc.) a relay valve used in a time delay function.

17.3.4 Explain identifications not covered by the above by notes on the drawing.

17.4 Identification of Relay Valve Control Points

Use the following rules for each control point on each relay valve having a unique identification:

17.4.1 Identify the control point on a relay valve having only one control point with the number of that relay only.

EXAMPLE. The control point on a single control point spring return relay valve RV3 carries the identification "RV3".

17.4.2 Identify the control points on a relay valve having two opposing control points as "A" and "B". The complete identification is by the number of the relay valve followed by the "A" or "B".

EXAMPLE. RV4A, RV4B, etc.

NOTES.

1. It is anticipated that new styles of relay valves will make their appearance in the future, having more than one control point in each end of the relay valve. For example, a relay valve similar to a single control point spring return relay valve, except that instead of having just one control point capable of opposing the spring, this relay valve may have several control points on the "A" end, each control point independent of the others and capable of opposing the spring.

2. In the case of a relay valve having multiple control points, all serving the same basic function of actuating the relay valve, these control points would be identified by the usual designation of the control point (e.g., RV6A or RV6B) followed by dash numbers, -1, -2, -3, etc.

3. Thus, a single pilot spring return relay valve with multiple control points opposing the spring would identify these points as RV5-1, RV5-2, RV5-3, etc.

4. In the case of a double pilot relay valve the control points would be identified RV6A-1, RV6A-2, RV6A-3, etc., and RV6B-1, RV6B-2, RV6B-3, etc.

17.5 Symbols for Control Points on Relay Valves

17.5.1 Control point on a relay valve having one actuator and spring return or automatic return. The automatic return is accomplished entirely within the relay valve assembly and requires no connection from the control system.

When this control point is pressurized, it takes command of the relay valve and keeps command as long as it is pressurized. When it is exhausted, the spring (or the automatic return) returns the relay valve to the starting position.

17.5.2 Control points on a double control point detented (or equivalent) relay valve. When either control point is exhausted, a momentary fluid signal into the opposite control point will cause that control point to take command and shift the relay valve.

When that fluid signal is exhausted, the detent holds the relay valve in the new position until the second control point is pressurized.

A momentary fluid signal into the second control point will cause it to assume command and shift the relay valve back to the start position.

If control pressure is held in either control point, a fluid signal of equal pressure into the opposite control point will not shift the relay valve until the original control point is exhausted.

NOTE. These are the symbols for the control points on a true "retentive fluid memory" device, where the relay valve is physically held in position by mechanical or magnetic means.

Thus, once the relay valve has been shifted to a specific position, the control air to the system may be turned on and off any number of times without affecting the position of the relay valve. The only thing which will change its position is a fluid signal into the opposite control point.

17.5.3 Control points on a double control point spring offset relay valve. Control point on the end opposite the spring and control point on the spring end.

When the spring end control point is exhausted, the "A" control point acts exactly like the control point on a single control point spring return relay valve.

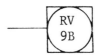

If the "B" control point is exhausted, a fluid signal into the "A" control point will shift the relay valve and it will stay shifted as long as the fluid signal into "A" is maintained. If, however, while the "A" control point is pressurized, and in command of the relay valve, an equal pressure is introduced into the "B" control point, the two fluid signals will cancel each other, and the spring will return the relay valve to the starting position, EVEN THOUGH THE "A" CONTROL POINT IS STILL PRESSURIZED.

Finally, if the "A" control point is pressurized by a maintained fluid signal, it will take command of the relay valve and hold the relay valve shifted against the spring. The "B" control point can then be pressurized and exhausted, and will act like the actuator on a single control point spring return valve.

17.5.4 Control Points on a Free-Floating Relay Valve. These symbols represent the control points on a double-control point relay valve wherein the valving action moves so freely that it has no inherent mechanical restraint, and is normally used with bias pressure in one control point. The "B" control point will normally be the control point connected to the bias pressure.

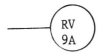

With pressure in the "B" control point, an equal pressure in the "A" control point will not shift the relay valve until the "B" control point is exhausted.

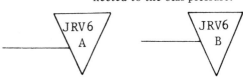

17.6 Symbols for Controlled Flow Passages of Relay Valves

17.6.1 Three-way function, not passing at the start of the cycle.

This may be a controlled flow passage on a detented relay valve, which is not passing at the start of the cycle, or it may be a normally not passing passage on a spring return relay valve with the relay valve in its normal position at the start of the cycle. Exhaust is assumed but not shown.

17.6.2 Three-way function, passing at the start of the cycle. This may be a controlled flow passage on a detented relay valve, or it may be a normally passing passage on a spring return relay valve with the relay valve in its normal condition at the start of the cycle. Exhaust is assumed but not shown.

17.6.3 Three-way function, normally passing but held not passing at the start of the cycle because the relay valve controlled point is pressurized. Used only in the case of spring return or automatic return relay valves. Exhaust is assumed but not shown.

17.6.4 Three-way function, normally not passing but held passing at the start of the cycle because the relay valve control point is pressurized. Used only in the case of spring return or automatic return relay valves. Exhaust is assumed but not shown.

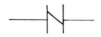

17.6.5 Non-exhausting function. The solid dots indicate that the function is a non-exhausting function.

NOTE. All the three-way symbols under clause 17.6 may be used with the dots to indicate a non-exhausting function.

17.7 Relay Valve Symbol Cross-Reference System

Prepare an identifying and cross-reference system in conjunction with all relay valve control points and controlled flow passage symbols, so that the control points and their associated controlled flow passages may be easily located on the diagram.

17.7.1 Number each horizontal flow path in the diagram, starting with number 1 and numbering consecutively from the top of the diagram downward. Arrange these guide numbers (or line numbers) in a vertical row down the left side of the diagram.

17.7.2 Where the size of the diagram requires it to be separated into several columns on one sheet or continued over from one sheet to another, carry over the guide numbers and continue in sequence on the top of the next column or on the top of the next sheet.

17.7.3 In the case of relay valves having two or more opposing control points, enter a set of brackets to the right of each control point, containing the guide numbers of the lines in the diagram which contain the opposing control points.

FLUID POWER SYMBOLS AND STANDARDS

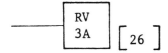

EXAMPLE. A double control point detented relay valve with the "A" control point in line 12 and the "B" control point in line 26.

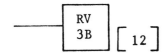

17.7.4 Enter, in parentheses, to the right of each control point, the numbers of the lines where the controlled flow passages of the relay valve are located.

    17.7.4.1 Underline controlled flow passages which a control point causes to go not passing.

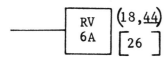

    17.7.4.2 DO NOT underline controlled flow passages which a control point causes to go passing.

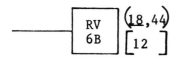

17.8 Relay Valve Controlled Flow Passage Cross-Referencing

    17.8.1 Note above each controlled flow passage symbol the identification of the relay valve control point which, when it takes command of the relay valve, causes the flow condition to change from the condition shown.

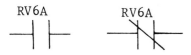

    17.8.2 List only the basic identification of the control points in the case of a relay valve with multiple control points all performing the same function.

    EXAMPLE. A relay valve controlled flow passage where any of the "A" control points, RV6A-1, RV6A-2, or RV6A-3 can change the flow condition. Show only the identification RV6A.

    17.8.3 Where the complexity of the drawing warrants, enter underneath the controlled flow passage symbol the guide numbers of the lines where the control points of the relay valve are located.

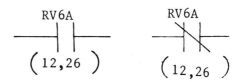

List the locations of the control points which CHANGE the flow condition first.

18. SYMBOLS FOR OFF-RETURN FLUID MEMORY RELAY VALVES

  18.1 Three Versions of the Off-Return Relay Valve

    18.1.1 Single control point. Reset by exhausting the input to the controlled flow passages.

    18.1.2 Double control point. Reset either by exhausting the input to the controlled flow passages, or by applying a momentary control fluid signal to the reset control point.

    18.1.3 Double control point with an external seal-in circuit. Reset by either:

        18.1.3.1 Exhausting the input to the controlled flow passages.

        18.1.3.2 Applying a momentary control fluid

signal to the reset control point.

    18.1.3.3 Interrupting the external seal-in circuit by means of a separate, external control device.

18.2 Symbols for Control Points on Off-Return Relay Valve

Control point on a single control point off-return relay valve.

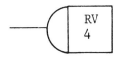

18.3 Symbols for Control Points on a Double Control Point Off-Return

    18.3.1 Control point which causes actuation.

    18.3.2 Control point which causes reset.

18.4 Symbols for Controlled Flow Passages of an Off-Return Relay Valve

    18.4.1 Controlled flow passage which controls the seal-in circuit (internal seal-in circuit).

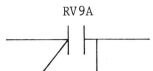

    18.4.2 Controlled flow passage which controls the seal-in circuit (external seal-in circuit) showing a typical interrupter passage in the seal-in circuit (in this case a relay valve RV16).

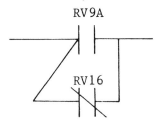

    18.4.3 Controlled flow passage which does not control the seal-in circuit.

18.5 Identification

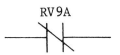

Identify each off-return relay valve in the same manner as other relay valves. Use normal relay valve control point identification and location cross-references.

19. SYMBOLS FOR TIME DELAY RELAY VALVES

19.1 Time Delay Relay Valves—Control Points

For control points on time delay relay valves, use the same symbols as those used for the corresponding action of a relay valve.

    19.1.1 Control point on a single control point time delay relay valve. Use this symbol for both time delay after pressurizing and time delay after exhausting. The symbols for the controlled flow passages indicate whether the action is time delay after pressurizing the control point or time delay after exhausting the control point.

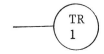

    19.1.2 Control points on a double control point detented time delay relay valve. Use these symbols for both delayed actuation and delayed reset relay valves. The symbols for the controlled flow passages indicate whether the action is that of a delayed actuation or a delayed reset.

19.2 Identification—Time Delay Relay Valves

Identify by a TR number starting with TR1 and numbering consecutively. The number has no significance other than to identify the particular time delay relay valve.

19.2.1 Identify the control point on a single point time delay by the TR number of the time delay relay valve.

19.2.2 Identify the control points on a double control point detented time delay relay valve by the TR number and the letter "A" or "B".

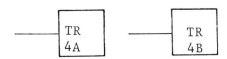

19.2.2.1 Use "A" control point, when it takes command, to start the actuation process.

19.2.2.2 Use "B" control point, when it takes command, to start the reset process.

19.3 Symbols for Time Delay Relay Valves Controlled Flow Passages

Depict each controlled flow passage by a set of terminals and a swinging gate, similar to those used to depict limit valves. Under each swinging gate, use a symbol which indicates the details of the timing action, as follows:

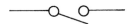

19.3.1 Controlled flow passage on a time delay after pressurization relay valve. Normally not passing. After receipt of a control fluid signal there is a delay before this passage goes passing. It resets immediately when the control fluid signal is removed.

19.3.2 Controlled flow passage on a time delay after pressurization relay valve. Normally passing. After receipt of a control fluid signal there is a delay before this passage goes not passing. It resets immediately when the control fluid signal is removed.

19.3.3 Controlled flow passage on a time delay after exhausting relay valve. Normally not passing. Goes passing immediately upon receipt of a control fluid signal. After the control fluid signal is removed, there is a delay before this passage resets.

19.3.4 Controlled flow passage on a time delay after exhausting relay valve. Normally passing. Goes not passing immediately upon receipt of a control fluid signal. After the control fluid signal is removed, there is a delay before this passage resets.

19.3.5 Controlled flow passage on a double control point detented time delay relay valve with delayed actuation.

Not passing at start of cycle.

Passing at start of cycle.

204 VALVE ENGINEERING AND DESIGN DATA

19.3.6 Controlled flow passages on a double control point detented time delay relay valve with delayed reset.

Not passing at start of cycle.

Passing at start of cycle.

19.4 Identification of Controlled Flow Passages

19.4.1 Above each controlled flow passage symbol put the identification of the time delay relay valve.

TR9

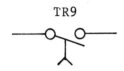

19.4.2 Put the identification of the "A" control point for a double control point detented relay valve.

TR5A

19.5 Relay Symbol Cross-Reference System

Cross-reference time delay relay valve symbols in a manner similar to that used for relay valves, so that control points and their associated passages may be easily located on the diagram.

20. SYMBOLS FOR RESISTANCE DEVICES

20.1 Fixed Resistance

Resists flow in both directions.

20.2 Adjustable Resistance

Resists flow in both directions.

20.3 Fixed Resistance with Free Return Check

20.3.1 Choked flow left to right; free flow right to left.

20.3.2 Choked flow right to left; free flow left to right.

20.4 Adjustable Resistance with Free Return Check

20.4.1 Choked flow left to right; free flow right to left.

20.4.2 Choked flow right to left; free flow left to right.

21. SYMBOLS FOR RELAY VALVES IN A TIMING CIRCUIT

Use relay valves in conjunction with resistance devices to perform timing functions.

21.1 When the resistance device and the relay valve are separate devices, and external connections are made to make them an operational unit, diagram as separate units, with a tube connection between them.

TD 7
Delays start of unclamp

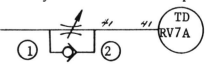

21.2 Identify each resistance device by a TD number, starting with TD1 and numbering consecutively. The numbers have no significance other than to identify the particular component.

Under the TD number put a note telling what action is delayed by this resistance device.

21.3 Depict the relay valve by relay valve symbols.

21.4 Identify the relay valve by standard RV numbers, or by special identification which indicates that it is performing a timing function.

If special identification is used, identify the relay valve by a TDRV number, the number to correspond to the TD number assigned to its resistance device.

21.5 Diagram the relay valve controlled flow passages by relay valve symbols, identified by the RV number or TDRV number, as the case may be.

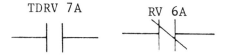

22. SYMBOLS FOR ONE-SHOT RELAY VALVES

NOTE. In certain versions of a one-shot relay valve the device may have two control points. Its action is similar to that of the single control point device, except that when the initiating control fluid signal is removed, the device does not reset. To reset the device the second control point must receive a control fluid signal.

22.1 Control Point Symbols for One-Shot Relay Valves

22.1.1 Single control point—automatic reset when control point is exhausted.

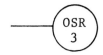

22.1.2 Double control point—requiring a pneumatic reset fluid signal. Identify by "A" the initiating control point, and by "B" the resetting control point.

22.1.3 Use above control point symbols for both fixed pulse length and adjustable pulse length devices.

22.2 Identification

Identify each one-shot relay valve by an OSR number, starting with OSR-1 and numbering consecutively. The numbers have no significance other than to identify the particular one-shot.

22.3 Controlled Flow Passage Symbols for One-Shot Relay Valves

22.3.1 Three-way, normally not passing. Goes passing immediately upon receipt of initiating control fluid signal, returns to not passing after a non-adjustable length of time. Numeral indicates length of fixed pulse in seconds.

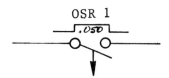

22.3.2 Three-way, normally passing. Goes not passing immediately upon receipt of control fluid signal. Returns to passing after a non-adjustable length of time. Numeral indicates length of fixed pulse in seconds.

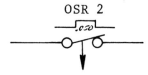

22.3.3 Three-way, normally not passing. Goes passing immediately upon receipt of a control fluid signal. Returns to not passing after an adjustable length of time.

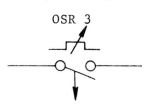

206 VALVE ENGINEERING AND DESIGN DATA

22.3.4 Three-way, normally passing. Goes not passing immediately upon receipt of a control fluid signal. Returns to passing after an adjustable length of time.

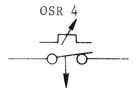

OSR 4

22.3.5 Identification. Identify the one-shot relay valve controlled flow passages by placing the OSR number above the controlled flow passage symbol.

22.4 One-Shot Relay Valve Location Cross-References

Use a cross-reference for control points and controlled flow passages to facilitate location on the diagram, using standard relay valve cross-references.

23. SYMBOLS FOR ELECTRIC-TO-AIR RELAY VALVES

23.1 Identification

Identify each electrical-to-air interface with an ERV (electrical relay valve) number, starting with ERV1 and numbering consecutively. The numbers have no significance other than to identify the particular component.

23.2 Symbols—Actuating Devices

23.2.1 Single Solenoid, Spring or Automatic Return. When the solenoid is energized, one or more controlled flow passages change condition. When the solenoid is de-energized, the controlled flow passages return to the original condition.

ERV 1

23.2.2 Double Solenoid, Detented. When the "A" solenoid is energized momentarily, the controlled flow passages change condition. When the "B" solenoid is energized momentarily, the controlled flow passages return to the original condition.

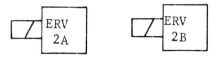

ERV 2A   ERV 2B

23.2.3 Symbols for a Combination of Actuators. When both an air signal AND an electrical signal must be present to cause energization, show the solenoid and the air actuators end to end. When either the air signal OR the electrical signal can cause energization, show the symbols for the solenoid and air actuator side by side.

ERV 3   ERV 4

23.3 Location of Actuator Symbols

23.3.1 Locate the symbols for the ERV actuators in a vertical column on the circuit diagram, in line with the control point symbols for the power control valves.

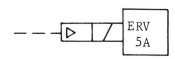

ERV 5A

23.3.2 In the case of combination actuators requiring an air signal from the control system, use a dashed line from the control system to the air actuator.

23.4 Symbols for Controlled Flow Passages

Use standard symbols for relay valve controlled flow passages.

23.5 Identification of Controlled Flow Passages

Identify each controlled flow passage by the ERV number located above the controlled passage symbol. In the case of double solenoid detented relay valves add the letter of the solenoid actuator which causes the controlled flow passage to change from the condition shown.

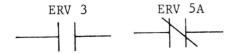

23.6 Symbol Location Cross-Reference

Cross-reference solenoid actuators and controlled flow passages to facilitate location on the diagram, using standard relay valve cross-references.

PART VI—MISCELLANEOUS DEVICES

24. VISUAL INDICATORS

24.1 Spring Return Type, Single Control Point

When control point is pressurized, indicator shows color. When control fluid signal is removed, indicator returns to start conditions.

NOTE. Letter in symbol denotes color.

24.2 Two-Position Detented, Double Control Point

When "A" control point is pressurized by a momentary pulse, indicator changes color.

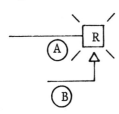

When "B" control point is pulsed, indicator returns to starting condition.

24.3 If necessary, put a note giving details of indicator operation or legend.

 Motors Running

24.4 Identification

Give each visual indicator a VI number, starting with VI-1 and numbering consecutively. The numbers have no significance other than to identify the particular visual indicator.

25. TEST POINT

Use a test point symbol as a special fitting allowing the ready application of a pressure gage to the circuit for troubleshooting purposes.

26. PRESSURE INDICATOR

Use a pressure indicator symbol as a visual indicator installed permanently in the circuit, for the purpose of indicating visually the presence of pressure at that point in the circuit.

27. PRESSURE GAGE

Use a pressure gage symbol as a pressure indicator calibrated in such a way as to indicate not only the presence of pressure, but the exact pressure level.

28. PRESSURE CONTROL VALVE (PRESSURE REGULATOR)

28.1 Adjustable, Relieving Type

28.1.1 Identify each pressure control valve with an "R" number, starting with R1 and numbering consecutively. The number has no significance other than to identify the particular pressure control valve.

208 VALVE ENGINEERING AND DESIGN DATA

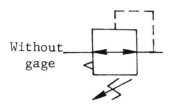

Without gage

28.1.2 Put a note telling to what pressure the pressure control valve is set.

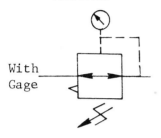

With Gage

29. FLUID CONDITIONERS

29.1 Filters

Identify each filter with an "F" number, starting with F1 and numbering consecutively. The numbers have no significance other than to identify the particular filter.

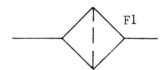

29.2 Lubricators

Identify each lubricator with an "L" number, starting with L1 and numbering consecutively. The numbers have no significance other than to identify the particular lubricator.

29.3 Air Dryers

Identify each air dryer with a "D" number, starting with D1 and numbering consecutively. The numbers have no significance other than to identify the particular air dryer.

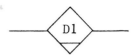

30. CHECK VALVE

30.1 Check Valve, Pilot-Operated

30.1.1 Pilot-operated to open.

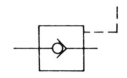

30.1.2 Pilot-operated to close.

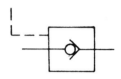

31. SHUTTLE VALVE (TWO-INPUT PASSIVE OR)

Identify each shuttle valve with an "S" number, starting with S1 and numbering consecutively. The number has no significance other than to identify the particular component.

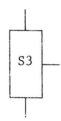

32. IDENTIFICATION STATEMENT

Use the following statement in catalogs and sales literature when electing to comply with this voluntary standard:

32.1 "Method of diagramming conforms to NFPA Recommended Standard, NFPA/T3.28.9-1973"

FLUID POWER SYMBOLS AND STANDARDS 209

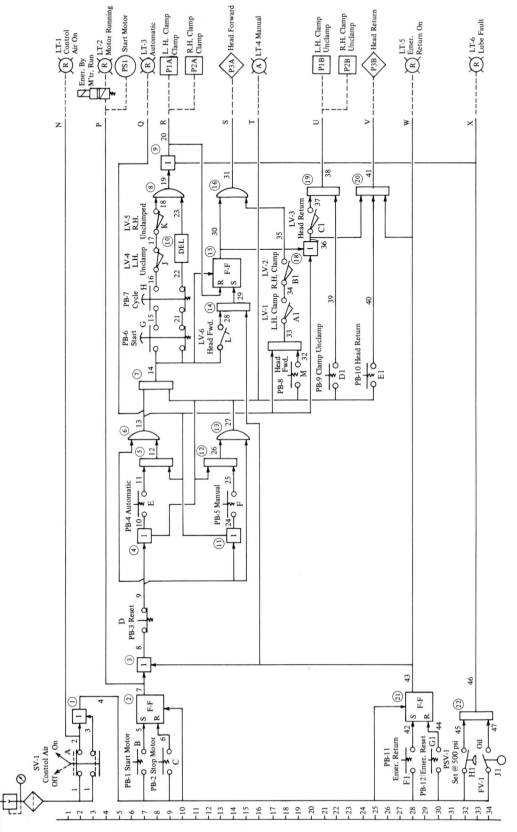

Appendix Y. Sample Diagrams—Attached Method.

210 VALVE ENGINEERING AND DESIGN DATA

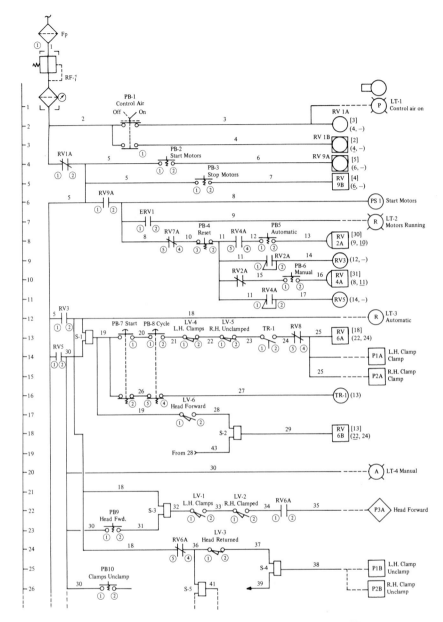

INITIAL CONDITIONS:
1. Control air on.
2. Electrical power on.
3. Hydraulic power off.
4. Left hand clamp unclamped.
5. Right hand clamp unclamped.
6. Head returned.
7. Spindle motor off.
8. Ready to start.

SEQUENCE OF CONTROL OPERATIONS:
AUTOMATIC CYCLE
1. Operator pushes "Start Motors" push button momentarily.
   A. Hydraulic pump motor starts.
   B. Spindle motor starts.
2. Hydraulic pump motor on and spindle motor on energizes ERV1.
   A. No action.
3. Operator pushes "Automatic" push button.
   A. No action – selects automatic mode.
4. Operator pushes two start buttons momentarily and simultaneously.
   A. Left hand clamps clamps.
   B. Right hand clamp clamps.
5. As left hand clamp starts to clamp LV-4 is released.
   A. No action.
6. As right hand clamp starts to clamp LV-5 is released.
   A. No action.
7. Left hand clamp fully clamped makes LV-1 and right hand fully clamped makes LV-2.
   A. Head moves forward.
8. As head moves forward LV-3 is released.
   A. No action.

9. Head fully forward makes LV-6.
   A. Head returns.
10. As head starts to return LV-6 is released.
    A. No action.
11. Head fully returned maked LV-3.
    A. Right hand clamp unclamps.
    B. Left hand clamp unclamps.
12. As right hand clamp starts to unclamp LV-2 is released.
    A. No action.
13. As left hand clamp starts to unclamp LV-1 is released.
    A. No action.
14. Right hand clamp fully unclamped makes LV-5 and left hand clamp unclamped makes LV-4.
    A. No action – cycle complete.

MANUAL CYCLE
1. Operator pushes start "Start Motors" push button momentarily.
   A. Hydraulic pump motor starts.
   B. Spindle motor starts.
2. Hydraulic pump motor on and spindle motor on energizes ERV1.
   A. No action.
3. Operator pushes "Manual" push button.
   A. No action – selects manual mode.

Appendix Z. Sample Diagrams–Detached Method.

VII. Test Methods for Valves

A. PRESSURE DROP TEST FOR FUEL SYSTEM COMPONENTS

In this section some common test methods for fluid components are outlined. The method chosen will depend on cost, available equipment, the intended use of the device, the type of environment it will be exposed to, the operating requirements imposed on the device and the nature of testing required such as military qualification testing, new design research and development, production acceptance testing, sample testing, etc. The test methods presented here are commonly used. However, there may well be other methods of equal importance and suitability for particular cases.

Pressure Drop-Fuels. The recommended method of pressure drop testing of fuel system components is described in ARP 868 published by Society of Automotive Engineers and reproduced here by permission.

In many cases it is necessary to perform the actual measurement with a different fluid from the one the valve will normally carry. For example, a valve intended to carry flammable or highly corrosive fluid will usually be tested with a less dangerous fluid. In such cases the measured pressure drop must be corrected. For cases of turbulent flow where Blasius' equation is valid, the following correction may be used.

$$\frac{\Delta P_1}{\Delta P_2} = \frac{\rho_1}{\rho_2}\left(\frac{\rho_2 \mu_1}{\rho_1 \mu_2}\right)^{1/4}$$

Whether this method is adopted or not, it should be remembered that the important point is that the downstream pressure tap must be located in the region of full recovery well past the location of the vena contracta. Generally this means at least ten diameters downstream from the orifice.

AEROSPACE RECOMMENDED PRACTICE SOCIETY OF AUTOMOTIVE ENGINEERS, Inc. 485 Lexington Ave., New York, N.Y. 10017	ARP 868 Issued 6-25-66 Revised

PRESSURE DROP TEST FOR FUEL SYSTEM COMPONENTS

1. PURPOSE - This recommended practice provides a relatively simple method for establishing static pressure drop through fuel system components in one step, without prior determination of the inlet and the outlet line connection pressure drop normally required.

2. SCOPE - This method of pressure drop test is applicable to any system component using the appropriate fluid under usage consideration.

3. GENERAL REQUIREMENTS

3.1 Fuel - MIL-J-5624 JP-4
    (Note: The use of test fluid is acceptable provided the recorded pressure drop is corrected to fuel in accordance with Type II of Spec. MIL-F-7024.)

3.2 Fuel Flow - 0 to maximum flow, compatible with the design capacity of the component.

4. DETAILED REQUIREMENTS

4.1 Test Set-up - The test set-up is defined schematically in Figure 1 and Figure 2.

4.1.1 Fuel Tank

4.1.1.1 Shape of the tank is optional.

4.1.1.2 The minimum quantity of fuel shall be a volume equivalent to the maximum fuel flow for 2 minutes plus 10 gallons. The tank volume shall be at least 1.1 times the fuel volume.

4.1.1.3 A tank drain shall be provided to permit taking fuel sample, temperature and specific gravity measurement.

4.1.2 Fuel Pump - The pump shall be capable of providing the maximum fuel flow as defined in paragraph 3.2.

4.1.3 Flow Control Valve - A control valve shall be provided to regulate the flow as required.

4.1.4 Flow Meter - A calibrated flow meter shall be used to record fuel flow.

4.1.5 Differential Pressure Gage

4.1.5.1 Two units are required to measure the pressure drop at locations as shown in Figure 1.

4.1.5.2 The pressure gage shall be compatible with the ΔP to be recorded and shall be calibrated for accuracies prior to test.

Copyright 1966 by Society of Automotive Engineers, Inc.   Printed in U.S.A.

ARP 868

4.2 <u>Test Procedure</u>

4.2.1 Establish fuel flow through the system as defined in Figure 1 at incremental flow rate up to the maximum flow specified in paragraph 3.2 and allow each flow condition to stabilize.

4.2.2 At each stabilized flow condition record ΔP_1 and ΔP_2.

4.2.3 Obtain a fuel sample and record fuel temperature and specific gravity.

4.3 <u>Pressure Drop Determination</u> - To determine the net static pressure drop through the component, the procedure is as follows:

4.3.1 Double reading of differential pressure gage No. 1 ($2\Delta P_1$).

4.3.2 Subtract reading of differential pressure gage No. 2 (ΔP_2).

4.3.3 Component static pressure drop: $\Delta P_V = 2\Delta P_1 - \Delta P_2$.

$$2\Delta P_1 = 2\Delta P_L + 2\Delta P_V + 2\Delta P_\ell$$
$$-\Delta P_2 = -2\Delta P_L - \Delta P_V - 2\Delta P_\ell$$
$$2\Delta P_1 - \Delta P_2 = 0 + \Delta P_V + 0$$

PREPARED BY SAE COMMITTEE AE-5, AEROSPACE FUEL, OIL & OXIDIZER SYSTEMS

VALVE ENGINEERING AND DESIGN DATA

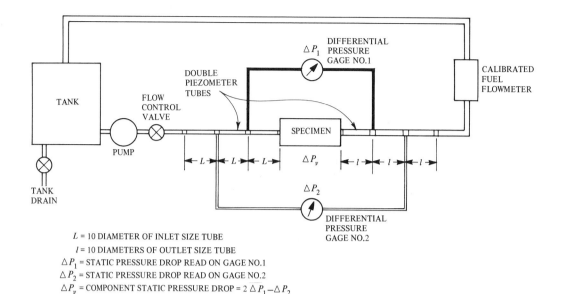

L = 10 DIAMETER OF INLET SIZE TUBE
l = 10 DIAMETERS OF OUTLET SIZE TUBE
ΔP_1 = STATIC PRESSURE DROP READ ON GAGE NO.1
ΔP_2 = STATIC PRESSURE DROP READ ON GAGE NO.2
ΔP_v = COMPONENT STATIC PRESSURE DROP = $2 \Delta P_1 - \Delta P_2$

Fig. 1. Pressure Drop Test Set-Up.

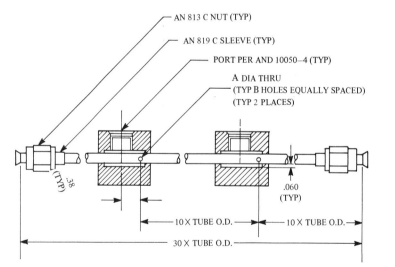

TUBE O.D.	A DIA.	B HOLES
1/4	.040	4
3/8	.060	4
1/2	.060	6
5/8	.098	6
3/4	.098	6
1	.098	6
1-1/4	.098	6
1-1/2	.098	6
1-3/4	.098	6
2	.098	6
2-1/2	.098	6
3	.098	6

NOTE : Bore and holes must be clean and free of burrs.

Fig. 2. Detail—Double Piezometer.

Pressure Drop—Hydraulic Fluids. The recommended method of pressure drop testing of hydraulic system components is given in ARP 24 published by the Society of Automotive Engineers and reproduced here by permission. Note that while these are methods commonly used in the industry they are by no means the only acceptable methods. For futher information consult the standards and literature of various fluid power societies.

AEROSPACE RECOMMENDED PRACTICE

ARP 24B

Issued 2-1-47
Revised 1-31-68

SOCIETY OF AUTOMOTIVE ENGINEERS, Inc. — Two Pennsylvania Plaza, New York, N. Y. 10001

DETERMINATION OF HYDRAULIC PRESSURE DROP

1. PURPOSE

This ARP is intended to serve as an instrument to determine hydraulic pressure drop, utilizing the best known practices for accessories in the hydraulic, fuel, oil, and coolant systems to aerospace vehicles.

2. GENERAL

A fluid flowing through a tube meets a certain amount of resistance due to kinetic and viscous effects. The pressure required to overcome this resistance and to maintain a certain flow rate is known as "pressure drop" or "back pressure." Where the flow of hydraulic fluid is maintained at a certain rate through a horizontal length of tube, discharging to atmosphere, the pressure gage at the rear or upstream end of the tube would show the pressure required to overcome friction and maintain the rate of flow. This pressure is known as the "back pressure" or the drop in pressure due to friction for that length of tube at a certain flow rate. The flow rate is usually expressed in gallons per minute or gpm. The pressure drop is expressed in pounds per square inch or psi at a certain gpm flow; for example, the pressure drop is 50 psi at 3 gpm.

The approximate pressure drop can be determined by applying a fixed flow to a valve inlet port, with pressure gage attached, the outlet port open, and determining the time to fill a container of known capacity. The gage reading would then be the approximate pressure drop at that rate of flow. A piece of tubing, hose, or fitting could be tested for pressure drop in this manner.

All parts of a hydraulic system through which flow is maintained will have a certain pressure drop, including tubing, fittings, valves, etc. When using a hydraulic test stand, the pressure drop through a valve is measured by placing it in a line between the pump and the flow meter, with a pressure gage at each end of the valve. While reading the gages on the valve, the required flow must be stabilized through the valve and flow meter. The difference in readings of the two gages is the approximate pressure drop for the flow shown on the flow meter.

If more accurate results must be obtained, special pressure pick-ups must be used and they must be placed a sufficient distance from any flow disturbance.[1] See typical test set-up shown in Figs. 1 and 2.

Example - The following example is for a 4-way pilot operated solenoid valve. The same set-up and procedure, with modifications, can be used to obtain the pressure drop of any unit.

Test Set-up - Make test set-up as shown in Fig. 1. Use an accurately calibrated pressure gage with sufficient resolution to secure accurate readings. The use of a differential pressure gage (or transducer) eliminates accumulative error between several instruments and gives more accurate results. Use piezometers for pressure pick-up connections; a standard tee may give results considerably in error. Install piezometers at least 4 times the tube inside diameter upstream and at least 10 times tube inside diameter downstream from any other connection, using straight rigid lines.

Fig. 2 shows the set-up to determine the "tare" pressure drop, which is the drop due to fittings and connecting tubing in the circuit. The gross pressure less the tare is the net pressure drop of the specimen or sample. The jumper line must be included in circuit during all pressure drop determinations.

3. EQUIPMENT

The equipment for pressure drop testing should be so constructed that continuous controlled operation can be maintained throughout the test period.

3.1 Tank - The tank should contain some means for controlling temperature variations. The fluid temperatures during pressure drop runs should be held such that the viscosity matches that of the intended service.

Baffles or other devices to remove turbulence and entrained air at the pump section should be provided. Evidence that there is no entrained air should be observed at a transparent tube section in the supply line downstream of both the low pressure tap and the test specimen, or in the transparent flow meter, if such is used.

3.2 Pump - The pump should be mounted as close to the tank as possible to minimize suction losses, and to preclude the possibility of sucking air into the line. The speed of the pump should not be exceptionally high. It is better to use industrial pumps of low rotational speed than aircraft pumps of much higher speeds. Flow variation may be obtained either by throttling the pump outlet and diverting excessive flow or by using a variable flow control directly on the pump.

NOTE: Since low speed pumps give higher pressure ripple, the output must be filtered.

[1] See "ASME Power Test Codes, 1959 Supplement on Instruments & Apparatus, Part 2 - Pressure Measurements" paragraph 20-31.

Copyright 1968 by Society of Automotive Engineers, Inc.

Printed in U.S.A.

ARP 24B

3.3 <u>Flowmeter</u> - Equipment for flow measurement should give accurate and reproducible results; therefore, they must be calibrated periodically by weighing fluid that flows through the instrument in a specific time period or by some other means.[2]

3.4 <u>Pressure Taps</u> - Pressure tap fittings, Fig. 6, or piezometer type tubes, Fig. 7, should be used to measure pressure drop. Tee fittings are considered undesirable. All drillings and flow passages should be smooth with clean intersections. Piezometer type tubes, Fig. 7, are preferred over the pressure tap fittings, Fig. 6.

3.5 <u>Pressure Differential Measurement</u> - The means of measuring pressure differential between piezometer taps may vary with the pressure, fluid, and accuracy required. Accurate gages, strain gage transducers, mercury manometers, and air-test fluid manometers all have application. The best means for measuring pressure differentials is the use of a single gage, where the high and low pressure taps are such as to create a differential reading. This arrangement reduces gage errors to a minimum and also allows the use of a low range pressure gage having graduations which permit more accurate readings. Air fluid columns are usually the most accurate, but the columns become unmanageable at higher pressures. Valves and other means should be provided to allow thorough and positive venting of the manometer connections to eliminate trapped air.

3.6 <u>Test Fluid</u> - Accessories should generally be tested on the fluid with which they will be used. In cases where this is not practical because of safety, facility, availability, etc., one should attempt to select a test fluid which matches the service fluid in viscosity and density (as applicable) as closely as possible to reduce the correction magnitude. Fluid temperature must be measured and recorded at the time pressure is recorded. Viscosity must be periodically checked to determine the datum point.

---

[2]Empirical calibrations of turbine and float flow meters are not considered satisfactory when data accuracy better than 5% is required. Test temperature should be that required to match fluid property (density/viscosity), to which meter is sensitive, to that at which calibrated. Where this cannot be accomplished or is not desirable, corrections must be applied for changed fluid property (density/viscosity).

[3]The slug is the gravitational unit of mass (lb-sec^2/ft). It is defined as the mass which will receive an acceleration of 1 fps when acted upon by an unbalanced force of 1 pound.

[4]Ratio of weight of volume of fluid at the temperature being flowed to that of an equal volume of water at 4 C. Use of data based on water at 60 F introduces an error of approximately 0.1%.

4. THEORY

4.1 <u>Nomenclature</u>

V	= Speed	ft/sec
Q	= Volumetric flow	US gal/min
W	= Weight flow	lb/sec
H	= Pressure head loss	ft
P	= Pressure loss	psi
L	= Length of pipe	ft
D	= Diameter of pipe or passage	in.
A	= Area of passage	in.^{2}
ρ	= Fluid density	Slugs/cu ft[3]
s	= Specific gravity[4] of fluid	(dimensionless)
μ	= Absolute viscosity	lb-sec/ft^2
ν	= Kinematic viscosity	ft^2/sec
g	= Gravitational constant	32.2 ft/sec^2
N_R	= Reynolds number	(dimensionless)
ϕ	= Energy loss coefficient	
f	= Friction loss coefficient	Darcy (dimensionless)
S	= Dimensionless factor	
T	= Dimensionless factor	

4.2 <u>Fundamental Information</u> - Several factors affect energy losses. Such factors are: type of flow, viscosity of fluid, surface roughness, sudden flow passage area and direction changes through fittings and tube bends, etc.

This energy loss is dissipated in the form of heat and, for a good first estimate, when the flow is turbulent, can be taken as being proportional to the fluid velocity squared.

It is convenient to express this in terms of velocity head:

$$H_{turb} = \phi \frac{V^2}{2g} \qquad (1)$$

where: ϕ is a coefficient of energy loss whose magnitude varies considerably for different types of flow impeders and, in general, is a function of the nature of flow.

The flow velocity is usually of little practical interest and is most conveniently interpreted in terms of flow rate, using either volumetric or weight measure. Also measuring pressure in psi, and for a circular passageway of diameter D, equation (1) can be written:

For volumetric measure:

$$P_{turb} = \phi \frac{0.00112 \, sQ^2}{D^4} \qquad (2)$$

For weight measure:

$$P_{turb} = \phi \frac{0.05809 \, W^2}{sD^4} \qquad (2A)$$

ARP 24B

Equations (2) and (2A) can be used for non-circular passageways, provided D is an equivalent diameter computed from the equation:

$$D' = 1.128 \sqrt{A} \quad (3)$$

where: A is the area of the passageway.

The nature of flow is dependent upon the ratio of inertial to viscous forces in the stream of fluid. This ratio is known as Reynolds number, N_R, and is a dimensionless factor given by:

$$N_R = \frac{\rho VD}{\mu} \quad (4)$$

or

$$N_R = \frac{VD}{\nu} \quad (5)$$

For laminar flow, where the viscous forces are predominant, the ratio N_R is small. In a turbulent flow the inertial forces predominate and therefore the value of N_R is large.

Substituting rate of flow in place of V in equation (4):
For volumetric measure:

$$N_R = \frac{0.0660 \, sQ}{\mu D} \quad (6)$$

For weight measure:

$$N_R = \frac{0.4750 \, W}{\mu D} \quad (6A)$$

Most of the flow encountered in aircraft installations is turbulent. In general, for straight tubing, laminar flow is predominant for values of N_R below about 1400, and becomes fully turbulent above about 3600 (see Fig. 3). When the flow is disturbed by the presence of bends and fittings a turbulent condition is found to prevail down to $N_R = 1000$ or less.

The value of ϕ, equations (2) and (2A), must be determined experimentally for hydraulic fittings and tube bends, etc., over a range of Reynolds numbers relevant to system requirements.

For the special case of straight smooth tubing, much classical work has been done. Pressure head loss may be computed from the following equation, derived from the Darcy - Weisbach law:

$$H = f \frac{LV^2}{2Dg} \quad (7)$$

in which the energy loss coefficient ϕ, equation (1), is replaced by a coefficient of friction "f".

i.e.

$$\phi = f \frac{L}{D} \quad (8)$$

From the theory of laminar flow it can be shown that:

$$f_{lam} = \frac{64}{N_R} \quad (9)$$

The coefficient "f" for turbulent flow, from experimental work by Blasius, can be expressed by:

$$f_{turb} = \frac{0.316}{N_R^{0.25}} \quad (10)$$

By substituting flow rate and pressure in place of V and H in equation (7), then combining with equations (9) and (10), the following expressions can be found:

For laminar-flow, and using volumetric measure (gpm):

$$P_{lam} = \frac{13.07 \, \mu LQ}{D^4} \quad (11)$$

For laminar-flow, and using weight measure (lb/sec):

$$P_{lam} = \frac{94.04 \, \mu LW}{sD^4} \quad (11A)$$

For turbulent flow, and using volumetric measure (gpm):

$$P_{turb} = \frac{0.008422 \, \mu^{0.25} \, s^{0.75} \, Q^{1.75} \, L}{D^{4.75}} \quad (12)$$

For turbulent flow, and using weight measure (lb/sec):

$$P_{turb} = \frac{0.2662 \, \mu^{0.25} \, W^{1.75} \, L}{sD^{4.75}} \quad (12A)$$

It will be noted that English units of measure are specified in the above equations. If metric units or a combination of metric and English units are used (metric units are commonly used for fluid viscosity), then care should be taken to properly compensate for these changes. See Table I for table of systems of units and Table II for viscosity conversion factors. Since viscosities and density varies with applied pressure, pressure drops should be measured at hydraulic pressure levels encountered in the design environment. Caution should be exercised in use of the equations contained in this document to assure that the parameters are in the correct units. For example, in equation (4), D must be converted from inches to feet to make the equation dimensionless.

5. CHART METHOD OF SOLUTION FOR STRAIGHT SMOOTH PIPES

In order to simplify calculations, particularly in the turbulent flow region, the classical equations for flow in straight tubes can be written in terms of dimensionless factors S and T:

$$S = \frac{1}{\mu}\left(\frac{\Delta P}{L} \cdot D^3 \rho\right)^{1/2} = N_R \left(\frac{f}{2}\right)^{1/2} \quad (13)$$

$$T = \frac{1}{\mu}\left(\frac{\Delta P}{L} \cdot Q^3 \cdot \rho^4\right)^{1/5} = \frac{N_R}{4}\left(8\pi^3 f\right)^{1/5} \quad (14)$$

ARP 24B

From the right hand sides of equations (13) and (14), S and T are plotted against "f" in Figs. 4 and 5.

The left hand sides of equations (13) and (14) are expressed in terms of five convenient elements of flow relationship:

$$(1)\,\mu;\ (2)\,\rho;\ (3)\,D;\ (4)\,\frac{\Delta P}{L};\ (5)\,Q$$

Also Fig. 3 shows N_R plotted against "f", as derived from classical work.

By using the left hand sides of equations (13) and (14) in conjunction with Figs. 3, 4 and 5, either S, T, N_R or "f" can be found (depending which is most convenient for a given problem) in terms of the four known elements. Having computed the values of this factor, the factor containing the 5th, or unknown, element can be read from the relevant curve. From this, the unknown element of flow or pressure drop, etc., is then calculated.

TABLE I

SYSTEMS OF UNITS

QUANTITY	SYMBOL	SYSTEM	
		ABSOLUTE	GRAVITATIONAL (Engineering)
Mass	M	M (lb)	FTi^2/L [slug (lb sec^2 ft)]
Force	F	ML/Ti^2 [Poundal (lb ft/sec^2)]	F (lb)
Density	ρ	M/L^3 (lb/ft^3)	FTi^2/L^4 [slug/ft^3 (lb sec^2/ft^4)]
Absolute Viscosity	ρ	M/LTi (lb/ft sec)	FTi/L^2 (lb sec/ft^2)
Kinematic Viscosity	$\nu = \mu/\rho$	L^2/Ti (ft^2/sec)	L^2/Ti (ft^2 sec)

PREPARED BY

THE FLUIDS PANEL OF SUBCOMMITTE A-6C,

FLUID POWER DISTRIBUTION ELEMENTS,

OF COMMITTEE A-6, AEROSPACE FLUID POWER & CONTROL TECHNOLOGIES

ARP 24B

TABLE II

VISCOSITY CONVERSION FACTORS*

Multiply by appropriate entry to obtain ↳	Centipoise	Poise	g_F sec cm^{-2}	lb_F sec in^{-2}	lb_F sec ft^{-2}	lb_F hr in^{-2}
Centipoise	1	1×10^{-2}	1.0197×10^{-5}	1.4504×10^{-7}	2.0886×10^{-5}	4.0289×10^{-11}
Poise	$1. \times 10^2$	1	1.0197×10^{-3}	1.4504×10^{-5}	2.0886×10^{-3}	4.0289×10^{-9}
g_F sec cm^{-2}	9.8067×10^4	9.8067×10^2	1	1.4224×10^{-2}	2.0482	3.9510×10^{-6}
lb_F sec in^{-2}	6.8947×10^6	6.8947×10^4	7.0505×10^1	1	1.4400×10^2	2.7778×10^{-4}
lb_F sec ft^{-2}	4.7880×10^4	4.7880×10^2	4.8823×10^{-1}	6.9445×10^{-3}	1	1.9290×10^{-6}
lb_F hr in^{-2}	2.4821×10^{10}	2.4821×10^8	2.5310×10^5	3.6000×10^3	5.1841×10^5	1
lb_F hr ft^{-2}	1.7237×10^8	1.7237×10^6	1.7577×10^3	2.5001×10^1	3.6001×10^3	6.9446×10^{-3}
g_M sec^{-1} cm^{-1}	1×10^2	1	1.0197×10^{-3}	1.4504×10^{-5}	2.0886×10^{-3}	4.0289×10^{-9}
lb_M sec^{-1} in^{-1}	1.7858×10^4	1.7858×10^2	1.8210×10^{-1}	2.5901×10^{-3}	3.7298×10^{-1}	7.1948×10^{-7}
lb_M sec^{-1} ft^{-1}	1.4882×10^3	1.4882×10^1	1.5175×10^{-2}	2.1585×10^{-4}	3.1083×10^{-2}	5.9958×10^{-8}
lb_M hr^{-1} in^{-1}	4.9605	4.9605×10^{-2}	5.0582×10^{-5}	7.1947×10^{-7}	1.0361×10^{-4}	1.9985×10^{-10}
lb_M hr^{-1} ft^{-1}	4.1338×10^{-1}	4.1338×10^{-3}	4.2152×10^{-6}	5.9957×10^{-8}	8.6339×10^{-6}	1.6655×10^{-11}

Multiply by appropriate entry to obtain ↳	lb_F hr ft^{-2}	lb_M sec^{-1} in^{-1}	lb_M hr^{-1} ft^{-1}	slug sec^{-1} in^{-1}	slug hr^{-1} ft^{-1}	g_M sec^{-1} cm^{-1}
Centipoise	5.8016×10^{-9}	5.5998×10^{-5}	2.4191	1.7405×10^{-6}	7.5188×10^{-2}	1×10^{-2}
Poise	5.8016×10^{-7}	5.5998×10^{-3}	2.4191×10^2	1.7405×10^{-4}	7.5188	1
g_F sec cm^{-2}	5.6895×10^{-4}	5.4916	2.3723×10^5	1.7068×10^{-1}	7.3733×10^3	9.8067×10^2
lb_F sec in^{-2}	4.0000×10^{-2}	3.8609×10^2	1.6679×10^7	1.2000×10^1	5.1840×10^5	6.8947×10^4
lb_F sec ft^{-2}	2.7778×10^{-4}	2.6812	1.1583×10^5	8.3335×10^{-2}	3.6000×10^3	4.7880×10^2
lb_F hr in^{-2}	1.4400×10^2	1.3899×10^6	6.0044×10^{10}	4.3199×10^4	1.8662×10^9	2.4821×10^8
lb_F hr ft^{-2}	1	9.6524×10^3	4.1698×10^8	3.0000×10^2	1.2960×10^7	1.7237×10^6
g_M sec^{-1} cm^{-1}	5.8016×10^{-7}	5.5998×10^{-3}	2.4191×10^2	1.7405×10^{-4}	7.5188	1
lb_M sec^{-1} in^{-1}	1.0360×10^{-4}	1	4.3200×10^4	3.1081×10^{-2}	1.3427×10^3	1.7858×10^2
lb_M sec^{-1} ft^{-1}	8.6339×10^{-6}	8.3333×10^{-2}	3.6000×10^3	2.5902×10^{-3}	1.1189×10^2	1.4882×10^1
lb_M hr^{-1} in^{-1}	2.8779×10^{-8}	2.7778×10^{-4}	1.2000×10^1	8.6337×10^{-6}	3.7297×10^{-1}	4.9605×10^{-2}
lb_M hr^{-1} ft^{-1}	2.3983×10^{-9}	2.3148×10^{-5}	1	7.1946×10^{-7}	3.1081×10^{-2}	4.1336×10^{-3}

*From WADC TR 58-638 Vol I Part I

220 VALVE ENGINEERING AND DESIGN DATA

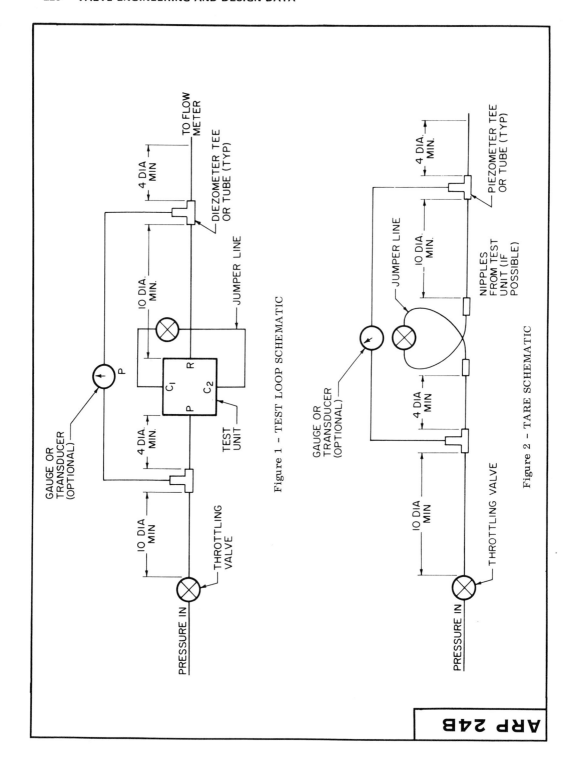

Figure 1 - TEST LOOP SCHEMATIC

Figure 2 - TARE SCHEMATIC

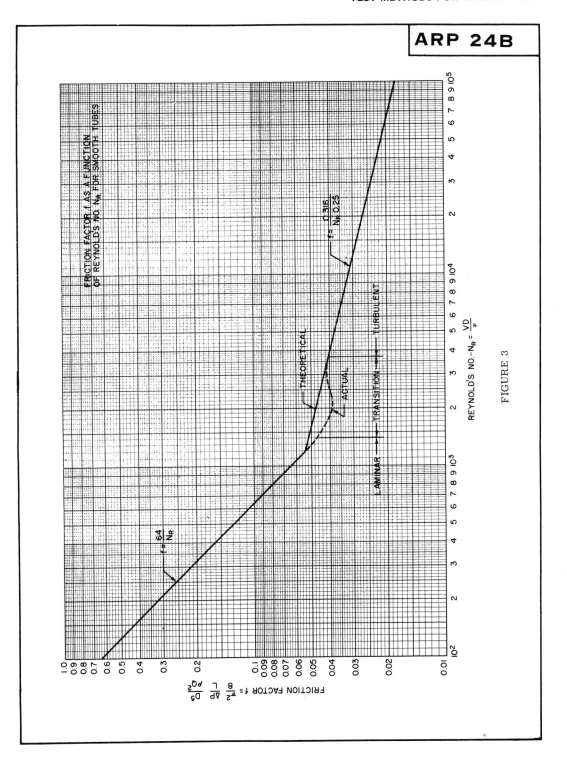

FIGURE 3

222 VALVE ENGINEERING AND DESIGN DATA

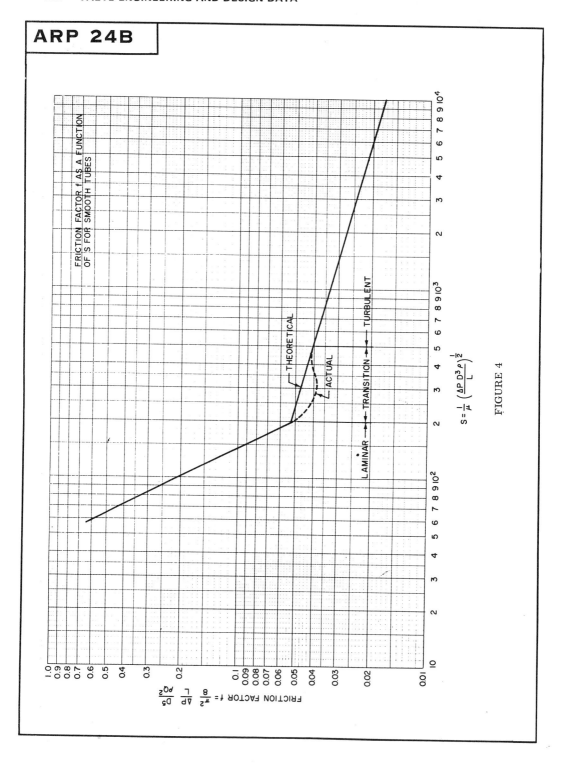

FIGURE 4

ARP 24B

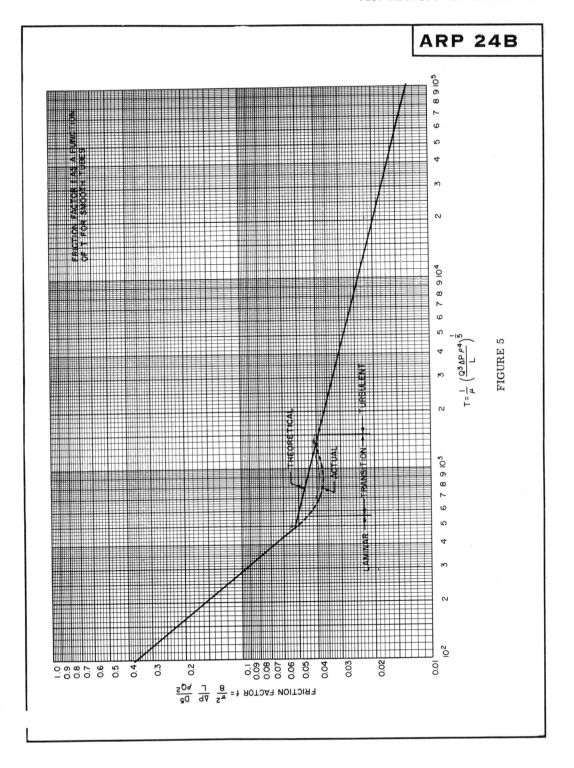

FIGURE 5

224 VALVE ENGINEERING AND DESIGN DATA

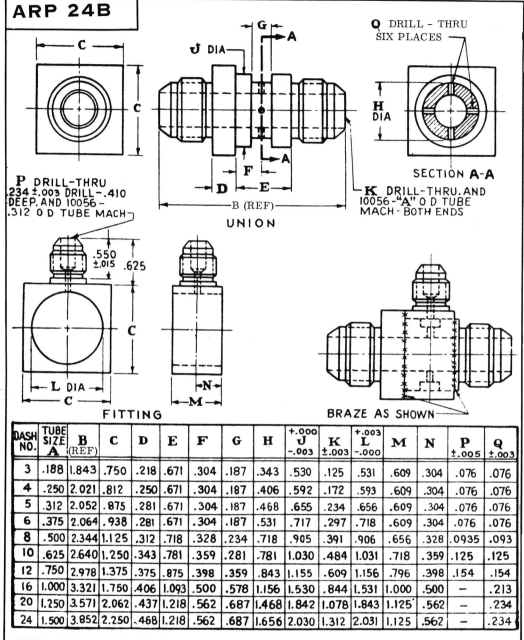

ARP 24B

DASH NO.	TUBE SIZE A	B (REF)	C	D	E	F	G	H	J +.000 -.003	K ±.003	L +.003 -.000	M	N	P ±.005	Q ±.003
3	.188	1.843	.750	.218	.671	.304	.187	.343	.530	.125	.531	.609	.304	.076	.076
4	.250	2.021	.812	.250	.671	.304	.187	.406	.592	.172	.593	.609	.304	.076	.076
5	.312	2.052	.875	.281	.671	.304	.187	.468	.655	.234	.656	.609	.304	.076	.076
6	.375	2.064	.938	.281	.671	.304	.187	.531	.717	.297	.718	.609	.304	.076	.076
8	.500	2.344	1.125	.312	.718	.328	.234	.718	.905	.391	.906	.656	.328	.0935	.093
10	.625	2.640	1.250	.343	.781	.359	.281	.781	1.030	.484	1.031	.718	.359	.125	.125
12	.750	2.978	1.375	.375	.875	.398	.359	.843	1.155	.609	1.156	.796	.398	.154	.154
16	1.000	3.321	1.750	.406	1.093	.500	.578	1.156	1.530	.844	1.531	1.000	.500	—	.213
20	1.250	3.571	2.062	.437	1.218	.562	.687	1.468	1.842	1.078	1.843	1.125	.562	—	.234
24	1.500	3.852	2.250	.468	1.218	.562	.687	1.656	2.030	1.312	2.031	1.125	.562	—	.234

NOTES: Tolerance ±0.010 unless noted.
"C" dimension may be increased as necessary to allow use of AND 10050 boss for the pressure pick-off.

FIGURE 6
PRESSURE TAP FITTING

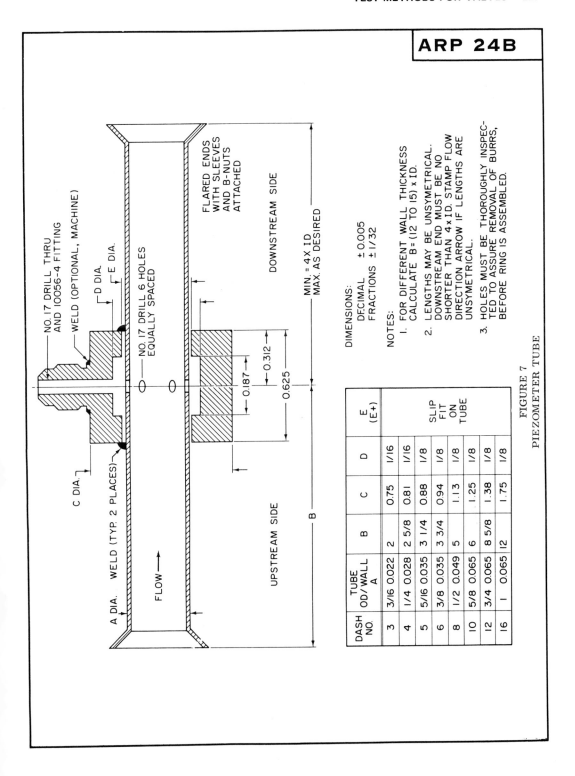

FIGURE 7
PIEZOMETER TUBE

B. CAPACITY TESTS

For valves to be used for flow control, it is necessary to have a measurement of the flow coefficient C_v. Also useful is knowledge of the piping factor F_p and pressure drop limit X_t. Widely accepted methods for experimental determination of these quantities for liquid and gas flow are given in the following papers.

Control Valve Capacity Test Procedure for Compressible Fluids. (Reproduced by permission of Instrument Society of America from ISA S39.4, 1974)

Control Valve Capacity Test Procedure for Incompressible Fluids. (Reproduced by permission of Instrument Society of America from ISA S38.2, 1972)

S39.4

Control Valve Capacity Test Procedure
For Compressible Fluids

1 SCOPE

This Standard is based on the developed mathematical equations of ISA Standard S39.3, Valve Sizing Equations for Compressible Fluids, and is confined to determining data for the Valve Sizing Coefficient C_v, the Piping Geometry Factor F_p, and the pressure-drop limitation x_T.

This procedure shall be applicable only to compressible fluids in turbulent flow.

This Standard is intended for valves used in pressure or flow control of process fluids and is not intended to apply to fluid power components as defined in the National Fluid Power Association (NFPA) Proposed Recommended Standard T.3.5.7.

2 TEST SYSTEM

2.1 A basic flow test system as shown in Figure 1 includes:

    a. Test Specimen
    b. Test Manifold
    c. Throttling Valves
    d. Flow Measuring Device
    e. Pressure Taps
    f. Temperature Sensor

2.2 Test Specimen

The test specimen is any valve or combination of valve and pipe reducers and expanders for which C_v, x_T and related data are required.

2.3 Test Manifold

The upstream and downstream piping adjacent to the test specimen shall conform to the nominal size of the test specimen connection and to the length requirements of Table 1.

The piping on either side of the test specimen shall be schedule 40 pipe for valves through 10 in. size having a pressure rating up to 600 lb ANSI. Pipe having 0.375 in. wall may be used for 12 through 24 in. sizes. An effort should be made to match the inside diameter at the inlet and outlet diameter of the test specimen with the inside diameter of the adjacent piping for valves outside the above limits.

The inside surface shall be reasonably free of flaking rust or mill scale and without irregularities which could cause excessive turbulence.

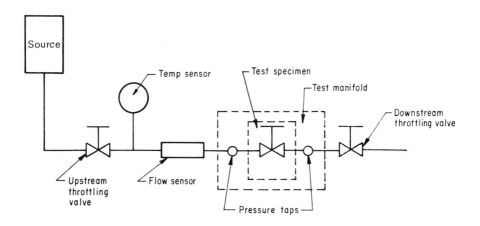

Figure 1. Basic Flow Test System

VALVE ENGINEERING AND DESIGN DATA

Instrument Society of America

TABLE 1 MANIFOLD PIPING REQUIREMENTS
STANDARD TEST MANIFOLD

A	B	C	D	Standard Test Manifold Configuration
From Table I	1/2 to 2-1/2 Nom. Pipe Diam.	4 to 6 Nom Pipe Diam.	From Table I	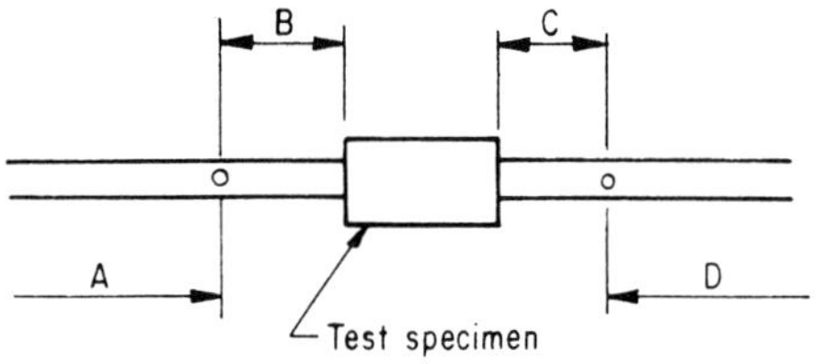

Distances A and D from upstream and downstream sources of flow disturbance to the test manifold pressure taps are given in the remainder of this table.

Information concerning the design of straightening vanes can be found in "ASME Power Test Codes Supplement" PTC 19.5; 4-1959 - Chapter 4 part 5, Section 5, Section 44.

The piping lengths of TABLE I are based upon "ASME Power Test Codes Supplement" PTC 19.5; 4-1959 - Figure 16 for orifice plates with contraction ratio (B) of 0.7.

Control Valve Capacity Test Procedure
For Compressible Fluids

S39.4

TABLE I - MANIFOLD PIPING REQUIREMENTS (CONTINUED)

LENGTHS "A" AND "D" ARE LENGTHS OF STRAIGHT PIPE UPSTREAM AND DOWNSTREAM RESPECTIVELY TO SOURCES OF FLOW DISTURBANCE
Length "A" minimums given in table.
Length "D" shall be 1 or more nominal pipe diameters in all cases.

SOURCE OF DISTURBANCE	PIPING ARRANGEMENTS	LENGTH "A" IN NOMINAL PIPE DIAMETERS	TYPICAL SYSTEM CONFIGURATION
General	System Incorporating a Straightening Vane	8	
Reducers and Expanders	General	12	
Control Valve Upstream		15	
Ells or Tube Turns.	Fittings in Different Planes	24	
Long Radius Bends		15	

230 VALVE ENGINEERING AND DESIGN DATA

Instrument Society of America

TABLE I - MANIFOLD PIPING REQUIREMENTS (CONTINUED)

LENGTHS "A" AND "D" ARE LENGTHS OF STRAIGHT PIPE UPSTREAM AND DOWNSTREAM RESPECTIVELY TO SOURCES OF FLOW DISTURBANCE

Length "A" minimums given in table.
Length "D" shall be 1 or more nominal pipe diameters in all cases.

SOURCE OF DISTURBANCE	PIPING ARRANGEMENTS	LENGTH "A" IN NOMINAL PIPE DIAMETERS	TYPICAL SYSTEM CONFIGURATION
Ells or Tube Turns	Fittings in Different Plane	30	
Long Radius Bends		22	
Drum or Tank Supply. Separator In Line. Angular Approach Using "Y"'s, Ells, or Tees.	Fittings In The Same Plane	15	

S39.4

Control Valve Capacity Test Procedure
For Compressible Fluids

TABLE I - MANIFOLD PIPING REQUIREMENTS (CONTINUED)

LENGTHS "A" AND "D" ARE LENGTHS OF STRAIGHT PIPE UPSTREAM AND DOWNSTREAM RESPECTIVELY TO SOURCES OF FLOW DISTURBANCE

Length "A" minimums given in table.
Length "D" shall be 1 or more nominal pipe diameters in all cases.

SOURCE OF DISTURBANCE	PIPING ARRANGEMENTS	LENGTH "A" IN NOMINAL PIPE DIAMETERS	TYPICAL SYSTEM CONFIGURATION
Ells, Tube Turns	Fittings In The Same Plane	18	
Long Radius Bends	Fittings In The Same Plane	18	
Tees or "Y"'s Used As Spool Pieces	Fittings In The Same Plane	10	

2.4 Throttling Valves

The upstream and downstream throttling valves are used to control the ΔP across the test specimen and also to maintain a specific downstream pressure. There are no restrictions as to size or style of these valves except where choked flow testing is anticipated.

2.5 Flow Measurement

The flow measuring instrument may be any device which meets the specified accuracy.

The resolution and repeatability shall be within 0.5 percent for the purpose of testing for x_T by the method in 4.2.3.

The flowmeter may be located upstream or downstream of the test specimen.

The purpose of the flow measuring device, of whatever principle, is to determine true time average flow rate. Flow must be determined with an accuracy of ±2 percent of the measured value. The flow measuring device shall be calibrated as frequently as necessary to maintain specified accuracy.

2.6 Pressure Taps

Pressure taps shall be provided for measurement of ΔP across the test specimen and static pressure, and taps shall conform to Figure 2. In no case may any fitting protrude inside the pipe. The nominal diameter of the manifold piping in which the tap is to be placed shall be used in calculating the distance from the tap to the test specimen connection.

The upstream and downstream pipe taps shall be horizontal or above to reduce the possibility of dirt collection in the pressure taps.

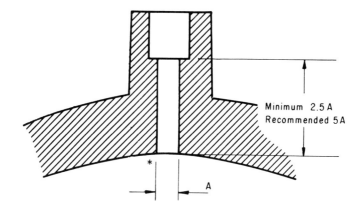

SIZE OF PIPE	A NOT EXCEEDING	A NOT LESS THAN
less than 2 in.	1/4 in. (6mm)	1/8 in. (3mm)
2 to 3 in.	3/8 in. (9mm)	1/8 in. (3mm)
4 to 8 in.	1/2 in. (13mm)	1/8 in. (3mm)
10 in. and greater	3/4 in. (19mm)	1/8 in. (3mm)

*Edge of hole must be clean and sharp or slightly rounded, free from burrs, wire edges or other irregularities.

Any suitable method of making the physical connection is acceptable if recommendations are adhered to.

Figure 2. Recommended Pressure Connection

S39.2

Control Valve Capacity Test Procedure
For Incompressible Fluids

1 SCOPE

ISA-S39.2 is based on the developed mathematical equations of ISA-S39.1, Valve Sizing Equations for Incompressible Fluids, and is confined to determining data for the Valve Sizing Coefficient C_V, the Liquid Pressure Recovery Factor F_L, the Piping Geometry Factor F_P, the Reynolds Number Factor F_R and the Liquid Critical Pressure Ratio Factor F_F.

This procedure shall be applicable only to incompressible Newtonian fluids in turbulent flow.

ISA-S39.2 is intended for valves used in pressure or flow control of process liquids and is not intended to apply to fluid power components as defined in the National Fluid Power Association (NFPA) Proposed Recommended Standard T.3.5.7.

2 TEST SYSTEM

2.1 A basic flow test system as shown in Figure 1 includes:

1. Test Specimen
2. Test Manifold
3. Throttling Valves
4. Flow Measuring Device
5. Pressure Taps
6. Temperature Sensor

2.2 Test Specimen

The test specimen is any valve or combination of valve and pipe reducers and expanders for which C_V and related data are required.

2.3 Test Manifold

The upstream and downstream piping adjacent to the test specimen shall conform to the nominal size of the test specimen connection and to the length requirements of Table 1.

The piping on either side of the test specimen shall be schedule 40 pipe for valves through 10 in. size having a pressure rating up to 600 lb ANSI. Pipe having 0.375 in. wall may be used for 12 in. through 24 in. sizes. An effort should be made to match the inlet and outlet diameter of the test specimen with the inside diameter of the adjacent piping for valves outside the above limits.

The inside surface shall be reasonably free of flaking rust or mill scale and without irregularities which could cause excessive turbulence.

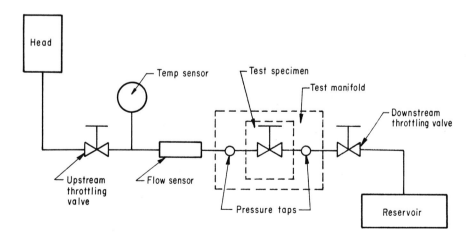

Figure 1. Basic Flow Test System

TABLE 1 MANIFOLD PIPING REQUIREMENTS
STANDARD TEST MANIFOLD

A	B	C	D	Standard Test Manifold Configuration
From Table I	1/2 to 2-1/2 Nom. Pipe Diam.	4 to 6 Nom Pipe Diam.	From Table I	

Distances A and D from upstream and downstream sources of flow disturbance to the test manifold pressure taps are given in the remainder of this table.

Information concerning the design of straightening vanes can be found in "ASME Power Test Codes Supplement" PTC 19.5; 4-1959 - Chapter 4 part 5 - Section 5.

S39.2   Control Valve Capacity Test Procedure For Incompressible Fluids

TABLE I - MANIFOLD PIPING REQUIREMENTS (CONTINUED)

LENGTHS "A" AND "D" ARE LENGTHS OF STRAIGHT PIPE UPSTREAM AND DOWNSTREAM RESPECTIVELY TO SOURCES OF FLOW DISTURBANCE
Length "A" minimums given in table.
Length "D" shall be 1 or more nominal pipe diameters in all cases.

SOURCE OF DISTURBANCE	PIPING ARRANGEMENTS	LENGTH "A" IN NOMINAL PIPE DIAMETERS	TYPICAL SYSTEM CONFIGURATION
General	System Incorporating a Straightening Vane	8	
Reducers and Expanders	General	12	
Control Valve Upstream		15	
Ells or Tube Turns.	Fittings in Different Planes	24	
Long Radius Bends		15	

TABLE I - MANIFOLD PIPING REQUIREMENTS (CONTINUED)

LENGTHS "A" AND "D" ARE LENGTHS OF STRAIGHT PIPE UPSTREAM AND DOWNSTREAM RESPECTIVELY TO SOURCES OF FLOW DISTURBANCE
Length "A" minimums given in table.
Length "D" shall be 1 or more nominal pipe diameters in all cases.

SOURCE OF DISTURBANCE	PIPING ARRANGEMENTS	LENGTH "A" IN NOMINAL PIPE DIAMETERS	TYPICAL SYSTEM CONFIGURATION
Ells or Tube Turns	Fittings in Different Plane	30	
Long Radius Bends		22	
Drum or Tank Supply. Separator In Line. Angular Approach Using "Y"'s, Ells, or Tees.	Fittings In The Same Plane	13	

S39.2   Control Valve Capacity Test Procedure For Incompressible Fluids

TABLE I - MANIFOLD PIPING REQUIREMENTS (CONTINUED)

LENGTHS "A" AND "D" ARE LENGTHS OF STRAIGHT PIPE UPSTREAM AND DOWNSTREAM RESPECTIVELY TO SOURCES OF FLOW DISTURBANCE

Length "A" minimums given in table.
Length "D" shall be 1 or more nominal pipe diameters in all cases.

SOURCE OF DISTURBANCE	PIPING ARRANGEMENTS	LENGTH "A" IN NOMINAL PIPE DIAMETERS	TYPICAL SYSTEM CONFIGURATION
Ells, Tube Turns	Fittings In The Same Plane	18	
Long Radius Bends	Fittings In The Same Plane	18	
Tees or "Y"'s Used As Spool Pieces	Fittings In The Same Plane	10	

Instrument Society of America

2.4 Throttling Valves

The upstream and downstream throttling valves are used to control the ΔP across the test specimen and also to maintain a specific downstream pressure. There are no restrictions as to size or style of these valves except where choked flow testing is anticipated. Choked flow at the upstream valve is not permissible.

2.5 Flow Measurement

The flow measuring instrument may take the form of a rotameter, orifice plate, magnetic flowmeter, turbine meter or other device which meets the specified accuracy. A weight-time method may be used.

The flowmeter may be located upstream or downstream of the test specimen. The flowmeter shall be positioned in the flow test loop where leakage and entrained air will not compromise accuracy.

The purpose of the flow measuring device, of whatever principle, is to determine true time average volumetric flow rate. Flow must be determined with an accuracy of ±2 percent of the measured value. The flow measuring device shall be calibrated at least once a year by comparison with a standard of flow measurement.

2.6 Pressure Taps

Pressure taps shall be provided for measurement of ΔP across the test specimen and static pressure, and taps shall conform to Figure 2. In no case may any fitting protrude inside the pipe. The

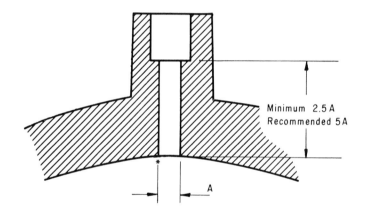

SIZE OF PIPE	A NOT EXCEEDING	A NOT LESS THAN
less than 2 in.	1/4 in. (6mm)	1/8 in. (3mm)
2 to 3 in.	3/8 in. (9mm)	1/8 in. (3mm)
4 to 8 in.	1/2 in. (13mm)	1/8 in. (3mm)
10 in. and greater	3/4 in. (19mm)	1/8 in. (3mm)

*Edge of hole must be clean and sharp or slightly rounded, free from burrs, wire edges or other irregularities.

Any suitable method of making the physical connection is acceptable if recommendations are adhered to.

Figure 2. Recommended Pressure Connection

VIII Valve Noise Calculation

An old problem associated with fluid flow that is assuming new and rather large proportions is the generation of noise in valves and other systems. In many places new OSHA regulations are forcing the insulation, modification or replacement of existing installations. In large installations the cost of isolating noisy valves may become very large. Further, as the allowed noise levels are reduced, the problem becomes magnified. One obvious approach to the noise problem is to design valves to be as silent as possible. The following material is presented as a guide to the valve designer. For further information, the reader is referred to various texts on accoustics.

METHODS FOR CALCULATING VALVE NOISE

(Reproduced by permission of Honeywell Inc. Process Control Div.) Fort Washington, Pennsylvania)

Note: Data in Tables 1 and 3, and curve No. 4 may vary with style of valve and from one manufacturer to another. Consult manufacturers data.

METHODS FOR CALCULATING VALVE NOISE

Noise level requirements have been established, by law, to protect the health of industrial workers. The Occupational Safety and Health Act (OSHA) as well as the modified Walsh-Healey Public Contracts Act include provisions for protection against the effects of exposure to sound exceeding specific levels for a specific duration as shown in the accompanying chart.

It is the legal responsibility of plant operating management and, ultimately, equipment suppliers, to provide and operate plant equipment at sound levels that do not endanger the health or safety of employees. Apart from legal requirements, it is considered good business practice to maintain comfortable sound levels in plant operating areas since noise tends to irritate and distract workers, adversely affecting their attitudes and efficiency.

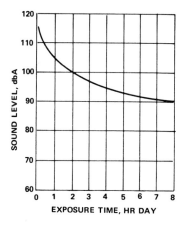

DEFINITION OF TERMS

A definition of terms commonly used in describing and calculating sound and noise levels will be helpful in understanding the causes, effects and remedies for excessive noise related to control valve usage.

Sound
A form of vibration that propagates through an elastic material such as air by alternately compressing and releasing molecules at a frequency characteristic of the medium.

Hz Sound Frequency
The number of sound waves that pass through a position per second; the number of times that the sound pressure varies through a complete cycle (compression and release) per second.

Ps Sound Pressure
The sound wave has a pressure that fluctuates (cycles) above and below atmospheric pressure to produce the auditory sensation of sound. Sound pressure is the root mean square (rms) value of the pressure as it varies from a high value at compression to a low value at expansion.

db Decibel
A dimensionless number expressing the ratio of two numerical values on a logarithmic scale. In acoustical terms, the decibel generally relates sound pressure level or a sound power level to a selected reference level.

SPL Sound Pressure Level
Expressed in decibels, SPL of a sound is 20 times the logarithm to the base 10 of the ratio of the pressure of this sound to a reference pressure which is usually taken as 0.0002 microbar.

$$db = 20 \, Log_{10} \frac{measured \ level}{reference \ level}$$

dbA
Sound pressure level measured in decibels on the "A" weighted frequency scale of a sound level meter. The "A" weighted network closely simulates the human ear sensitivity which has a peak frequency response at approximately 1000 Hz.

Attenuation
A weakening or reduction of sound pressure level.

Other terms
C_v—Nominal valve capacity coefficient
P_1—Upstream pressure, psia
P_2—Downstream pressure, psia (actual)
P_v—Vapor pressure of fluid, psia
F_L—Valve pressure recovery ratio, dimensionless
Z_1—Point of incipient cavitation
Z_2—Point where maximum sound pressure level exists in liquid flow
ΔP—Pressure differential, $P_1 - P_2$, psi
$\Delta P/P_1$—Pressure ratio (pressure differential divided by upstream pressure)

VALVE NOISE

Although many factors contribute to excessively high noise levels in processing plants, one of the most important of these is related to the operation of valves in piping systems carrying air, gases, steam and liquids. Since noise is attributable to energy conversion, it will increase and require greater control as the demand for energy, in terms of steam power, natural gas and chemicals increases.

Understanding and thoroughly analyzing the causes of valve associated noise is the logical first step in achieving a reduction in noise to acceptable levels. With a proper understanding of the causes of noise it is frequently possible, in the design of new installations, to avoid acoustical problems.

SOUND PRESSURE LEVEL OF COMMON NOISES

SPL (db)	Source of Sound
155	Nearby Siren
140	Jet aircraft (threshold of pain)
130	Hydraulic press
120	Loud automobile horn
110	Planers; routers
100	Subway; propeller plane
90	Symphony orchestra
80	Heavy traffic
70	Noisy office
60	Conversational speech
50	Private business office
40	Library
30	Recording studio
20	Electric clock (10 feet)
10	Rustle of grass
0	Threshold of hearing

CAUSES OF VALVE NOISE

Three conditions are generally considered as the major causes of valve noise: mechanical vibration, cavitation and turbulent, or aerodynamic, noise.

Mechanical vibration is caused by random pressure fluctuations within the valve body and by fluid impingement upon movable or flexible parts of the valve. It causes the particular component to vibrate at its natural frequency. The chief source of noise due to mechanical vibration is the lateral movement of the valve plug relative to the guide surfaces. Noise from this type of vibration sounds like a metallic rattling.

When a valve component resonates at its natural frequency the noise is recognizable by its characteristic single-pitch, high frequency, tone.

Control valve noise due to mechanical vibration is less prevalent than other types of noise due to improved valve design.

Cavitation requires two stages. The first is the transformation of some of the liquid into the vapor state, forming bubbles. The second is the implosion or collapse of these bubbles as they return to the liquid state.

In a control valve, the fluid stream is accelerated as it flows through a restricted area of the orifice or valve trim and reaches maximum velocity at the vena contracta. If this velocity is sufficient, the pressure at the vena contracta may be reduced to the vapor pressure of the liquid. The first stage of cavitation occurs at this point. Downstream, the fluid decelerates resulting in increased pressure above the vapor pressure causing the bubbles to collapse or implode. Cavitation can cause valve damage and vibration problems and is the major cause of high noise levels in liquid flow.

Conditions tending to cause this type of noise can be accurately predicted. Both the noise and the valve damage that can result from cavitation can be avoided by selecting appropriate limits for the service conditions.

Aerodynamic noise is the major cause of valve noise. This noise is a result of turbulent flow of steam, air and other gases. Turbulence can result from obstructions in the flow path, rapid expansion or deceleration of the high velocity gas exiting from the valve, or sharp turns or bends in the system piping. Mach 1.0 velocity (sonic flow) always creates high noise levels, however, high noise levels can also be generated as low as mach 0.4 and at low pressure drops where large mass flows occur.

The velocity of the fluid stream and the extent of the turbulent area control the noise level of the disturbance. Factors to be considered in controlling this type of noise are velocity of the flowing medium, flow rate, upstream pressure, valve pressure drop, piping configuration, and fluid physical properties.

CALCULATION OF VALVE NOISE

Different methods of calculation are required for valve noise caused by liquids and gases:

Liquids—In studies of valve noise generated by liquids, Honeywell found that by varying the pressure ratio $\Delta P/P_1$, and measuring the noise level generated (refer to Fig. 1), a narrow band (labeled L) had no substantial noise emission. This was followed by a larger band (T) which was characterized by a moderate rise of sound pressure level. From a certain point, Z_1 the sound pressure level increased sharply to reach a maximum value Z_2. Z_1 is the point where cavitation begins and Z_2 is the point where the maximum sound pressure level exists. Over a certain range, the radiated acoustic power will then remain nearly constant, and will finally be reduced again if the pressure differential is increased further. Explanation for this phenomenon is that the outlet pressure P_2 approaches the vapor pressure P_v, i.e. the differential value of $P_2 - P_v$ approaches zero. Under this condition, continuous evaporation with a comparatively stable vapor phase will prevail, rather than flashing evaporation with subsequent sudden condensation. This, then, represents a lessening of the degree of cavitation and thus a decrease of the sound pressure level.

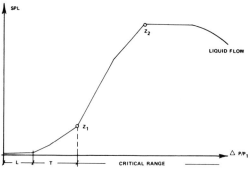

SOUND PRESSURE LEVEL – PRESSURE RATIO CHARACTERISTIC

HYDRODYNAMIC NOISE

The equation for calculating noise level caused by liquids is:

$SPL = SPL_1 + SPL_2 - SPL_3$

where: SPL=Sound Pressure Level (dbA)
SPL_1=Determined from Curve #1
SPL_2=Determined from Curve #2
SPL_3=Determined from Table #2

Data needed for Curve #1
1. Flow rate (Q), gpm
2. Pressure Differential (ΔP), psi
3. Inlet Pressure (P_1), psia
4. Liquid Vapor Pressure (P_v), at operating temperature, psia
5. Calculated C_v and Valve C_v to determine Z_1 and Z_2 (%) from Table #1

Data needed for Curve #2
1. Inlet Pressure (P_1), psia
2. Z_1 from Table #1

Data needed for Table #2
1. Pipe size, inches
2. Pipe schedule

Procedure:

Step 1—Determine SPL_1

A. Calculate $\dfrac{\Delta P}{P_1 - P_v} \times 100$

242 VALVE ENGINEERING AND DESIGN DATA

(1) if value of $\frac{\Delta P}{P_1-P_v} \times 100$ is less than or equal to Z_2 (table #1), subtract Z_1 (from table #1) from this value and read dbA value from Curve #1.

(2) if value of $\frac{\Delta P}{P_1-P_v} \times 100$ is greater than Z_2 but less than $Z_2 + 15$, use Z_2 in place of $\frac{\Delta P}{P_1-P_v}$ and subtract Z_1 from this value. Read dbA value from Curve #1.

(3) if value of $\frac{\Delta P}{P_1-P_v} \times 100$ is greater than $Z_2 + 15$, calculate dbA per step (2) above and then deduct 0.4 dbA for each 1% above $Z_2 + 15$.

Step 2. Determine SPL_2
  A. From Curve #2, determine dbA correction using inlet pressure and Z_1 (from Table #1).

Step 3. Determine SPL_3
  A. Select dbA correction for pipe size and schedule from Table #2.

Example #1

To determine maximum sound pressure level from available data.

A. *Known:*
Medium = water
t = 267°F
P_1 = 120 psia
P_2 = 80 psia
ΔP = 40 psi
P_v = 40 psia
Q = 600 gpm
C_v calc = 95
C_v valve = 120 (3" 9200)
Z_1 = 30%
Z_2 = 59%
Pipe Size = 3 inch
Pipe Schedule = 40

B. *Solution*
1. Determine SPL_1

(a) $\frac{\Delta P}{P_1-P_v} \times 100 = \frac{40}{120-40} \times 100 = 50\%$ (which is less than Z_2)

(b) From Curve # 1, $\frac{\Delta P}{P_1-P_v} - Z_1 = 50 - 30 = 20\%$

at intersection of 20% and flow of 600 gpm, read 77 dbA

2. Determine SPL_2
  (a) From Curve #2, at intersection of P_1 = 120 psia and Z_1 = 30%, read +4 dbA.

3. Determine SPL_3
  (a) From Table #2, 3 inch schedule 40 pipe has SPL correction of 1

4. SPL = $SPL_1 + SPL_2 - SPL_3 = 77 + 4 - 1 = 80$ dbA

Example #2

To determine maximum sound pressure level from available data.

A. *Known:*
Medium = water
t = 200°F.
P_1 = 250 psia
P_2 = 75 psia
ΔP = 175 psi
P_v = 11.5 psia
Q = 940 gpm
C_v calc. = 70
C_v valve = 90 (2½" 9200)
Z_1 = 30%
Z_2 = 59%
Pipe Size = 2½"
Pipe Schedule = 40

B. *Solution*
1. Determine SPL_1

(a) $\frac{\Delta P}{P_1-P_v} \times 100 = \frac{175}{250-11.5} \times 100 = 73\%$
(which is greater than Z_2)

(b) From Curve #1, using Z_2 in place of $\frac{\Delta P}{P_1-P_v}$ and subtracting Z_1, read (59−30) 29% vs. 940 gpm to obtain 83 dbA

2. Determine SPL_2
  (a) From Curve #2, at intersection of P_1 = 250 psia and Z_1 = 30%, read +11 dbA.

3. Determine SPL_3
  (a) From Table #2, 2½" schedule 40 pipe has SPL correction of 1
4. SPL = $SPL_1 + SPL_2 - SPL_3 = 83 + 11 - 1 = 93$ dbA

Example #3

To determine maximum sound pressure level from available data.

A. *Known:*
Medium = water
t = 250°F.
P_1 = 150 psia
P_2 = 50 psia
ΔP = 100 psi
P_v = 30 psia
Q = 400 gpm
C_v calc = 40
C_v valve = 53 (2" 9200)
Z_1 = 32%
Z_2 = 61%
Pipe Size = 2"
Pipe Schedule = 40

B. *Solution*
1. Determine SPL

(a) $\frac{\Delta P}{P_1-P_v} \times 100 = \frac{100}{150-30} \times 100 = \frac{100}{120} \times 100 = 83\%$
(which is greater than $Z_2 + 15$)

(b) From Curve #1, using Z_2 in place of $\frac{\Delta P}{P_1-P_v}$ and subtracting Z_1 read (61−32) 29% vs. 400 gpm to obtain 82 dbA. Since $\frac{\Delta P}{P_1-P_v} \times 100 > Z_2 + 15$ by 7 (83−76), deduct 2.8 dbA (0.4 × 7) from 82 dbA so that SPL_1 = 79.2 dbA.

2. Determine SPL_2
  (a) From Curve #2, at intersection P_1 = 150 psia and Z_1 = 32%, read 6 dbA.

3. Determine SPL_3
  (a) From Table #2, 2" schedule 40 pipe has SPL correction of 1.

4. SPL = $SPL_1 + SPL_2 - SPL_3 = 79.2 + 6 - 1 = 84.2$ dbA

AERODYNAMIC NOISE

Gases—The magnitude of aerodynamic noise caused by the turbulent flow of gases can be predicted. The following equation can be used to determine the anticipated sound pressure level, SPL.

$$SPL = Sq + Sp + Sv + Sg - Sr$$

where: SPL = Sound Pressure Level, dbA
Sq = Component for flow rate—determined from Curve #3
Sp = Component for inlet pressure P_1 as a function of valve type—determined from Curve #4
Sv = Correction factor resulting from pressure ratio P_1/P_2—obtain from Table #6
Sg = Correction factor for specific gravity—obtain from Table #5
Sr = Correction factor resulting from damping effect due to piping—obtain from Table #4

Example #1

To determine maximum sound pressure level from available data:

A. *Known*
Medium = natural gas
t = 200°F.
P_1 = 155 psia
ΔP = 40 psi
P_2 = 115 psia
Q = 1,500,000 scfh
SG = .6
C_v calc. = 295 (6"-9200)
Pipe Size = 6 inch
Pipe Schedule = 40

B. *Solution*
1. Determine Sq from Curve #3
  (a) Find F_L from Table 3 = .98
      $C_v \times F_L = 295 \times .98 = 289$
  (b) From Curve #3, at $C_v \times F_L = 289$, read Sq = 46 dbA

2. Determine Sp from Curve #4
  (a) At intersection of P_1 = 155 psia and Series 9200 Curve, read Sp = 58 dbA

3. Determine Sv from Table #6
   (a) Calculate $\frac{P_1}{P_2} = \frac{155}{115} = 1.35$
   (b) From Table #6, estimate Sv = 0.4
4. Determine Sg from Table #5
   (a) Sg = −1
5. Determine Sr from Table #4
   (a) 6 inch schedule 40 pipe has Sr of 20 dbA
6. SPL = Sq + Sp + Sv + Sg − Sr = 46 + 58 + 0.4 − 1 − 20 = 83.4 dbA

Example #2

To determine maximum sound pressure level from available data:

A. *Known*
   Medium—superheated steam
   t = 638°F C_v calc = 34.4 (2" 9100)
   P_1 = 115 psia   Pipe Size = 2 inch
   ΔP = 80 psi   Pipe Schedule = 40
   P_2 = 35 psia
   Q = 6,000 lb./hr.

B. *Solution*
1. Determine Sq from Curve #3
   (a) Find F_L from Table 3 = .98
       $C_v \times F_L$ = 34.4 × .98 = 33.7
   (b) Find Curve #3 at $C_v \times F_L$ = 33.7, read Sq = 29 dbA
2. Determine Sp from Curve #4
   (a) At intersection of P_1 = 115 and Series 9100 Curve, read Sp = 55.5 dbA
3. Determine Sv from Table #6
   (a) Calculate $\frac{P_1}{P_2} = \frac{115}{35} = 3.3$
   (b) From Table #6, estimate Sv = 29 dbA
4. Determine Sg from Table #5
   (a) Sg = −3 dbA
5. Determine Sr from Table #4
   (a) 2 inch schedule 40 pipe has Sr of 15.5 dbA
6. SPL = Sq + Sp + Sv + Sg − Sr = 29 + 55.5 + 29 − 3 − 15.5 = 95 dbA

TABLE 1

Valve Type	Flow to	Variation of Z_1, Z_2 with Ratio of Calc. C_v Vs Valve C_v.*							
		100%		75%		50%		25%	
		Z_1 %	Z_2 %	Z_1 %	Z_2 %	Z_1 %	Z_2 %	Z_1 %	Z_2 %
1120	open	20	50	23	49	28	49	33	51
1220	open	22	56	26	55	30	55	36	57
1400	open	16	55	20	60	24	65	30	72
9100	open	19	47	32	61	43	70	52	75
9200	open	19	47	32	61	43	70	52	75

*$\frac{\text{Calculated } C_v}{\text{Valve } C_v} \times 100 = -\%$. Extrapolate for %'s between 100, 75, 50 and 25%.

TABLE 2

Pipe Size	1	1½	2	2½	3	4	6	8	10	12	14	16
Pipe Schedule												
40	1	1	1	1	1	3	3	3	3	3	3	3
80	3	3	3	3	3	5	5	5	6	6	6	6
160	5	5	5	5	5	6	6	8	8	8	8	8

SPL_1 Damping effect of piping, dbA

TABLE 3

Valve Type	Plug Type	Flow to	F_L (at 100% of C_v)
1120	%c, LC	open	.95
1220	%c, LC	open	.95
1400	%C, LC, %VP	open	.95
9100	%VP, LVP	open	.98
9200	%VP, LVP	open	.98

F_L Factor

TABLE 4

Pipe Size	1	1½	2	3	4	6	8	10	12
Pipe Schedule									
40	14.5	15	15.5	18	18.5	20	21	22	22.5
80	17	17.5	18	20.5	21	23	24.5	25.5	26.5
160	19	20	21.5	23	24.5	27	28.5	30.3	31.5

Sr Damping effect of piping, dbA

TABLE 5

Gas	Sg (dbA)
Acetylene	−1
Air	0
Ammonia	−2
Carbon dioxide	+1
Carbon monoxide	0
Ethane	−1
Ethylene	−1
Helium	−6.5
Hydrogen	−10
Methane	−1
Natural gas	−1
Nitrogen	0
Oxygen	+0.5
Propane	+1
Saturated steam	−2
Superheated steam	−3

Sg Specific Gravity Correction (for most gases, this effect is negligible)

TABLE 6

P_1/P_2	F_L 1.0	.95	.9	.8	.7	.6
1.1	−16.0	−11.0	−5.5	−1.0	8.0	10.1
1.2	−11.0	−6.0	+1.5	+8.25	12.2	13.6
1.3	−6.0	0.0	6.65	12.4	14.35	15.35
1.4	−.25	+6.45	10.75	14.6	15.75	16.55
1.5	+5.1	10.1	13.6	16.0	16.90	17.35
1.6	10.1	13.55	15.7	17.05	17.75	18.05
1.7	14.75	16.6	17.3	17.95	18.4	18.55
1.8	18.4	18.45	18.65	18.85	18.9	19.0
1.9	20.95	20.2	19.75	19.6	19.45	19.35
2.0	23.3	21.5	20.6	20.25	19.9	19.65
2.5	28.3	26.05	24.0	22.85	21.5	20.85
3.0	29.6	27.8	25.65	24.1	22.43	21.55
3.5	30.45	28.7	26.45	24.7	23.1	22.0
4.0	31.15	29.22	26.98	25.15	23.6	22.35
5.0	31.97	29.95	27.65	25.85	24.15	22.85
6.0	32.55	30.45	28.1	26.35	24.45	23.1
7.0	32.9	30.8	28.4	26.7	24.7	23.3
8.0	33.2	31.1	28.68	27.0	24.9	23.5
9.0	33.5	31.3	28.9	27.25	25.1	23.65
10.0	33.72	31.5	29.12	27.45	25.25	23.8
15.0	34.48	32.1	29.9	28.12	25.87	24.38
20.0	34.9	32.45	30.48	28.5	26.3	24.8

Sv in dbA (P_1/P_2 Pressure Ratio)

Note: Data in Tables 1, 2, 3, 4, and 6 are subject to change as additional test data are obtained.

244 VALVE ENGINEERING AND DESIGN DATA

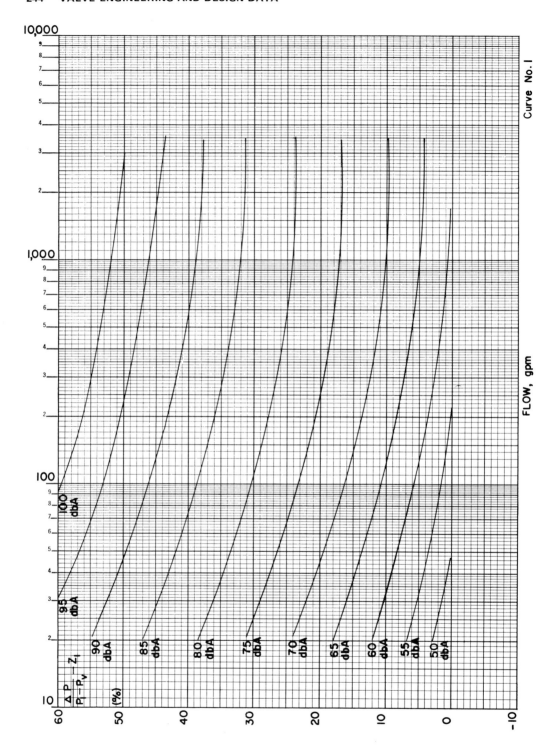

Curve No. 1

VALVE NOISE CALCULATION 245

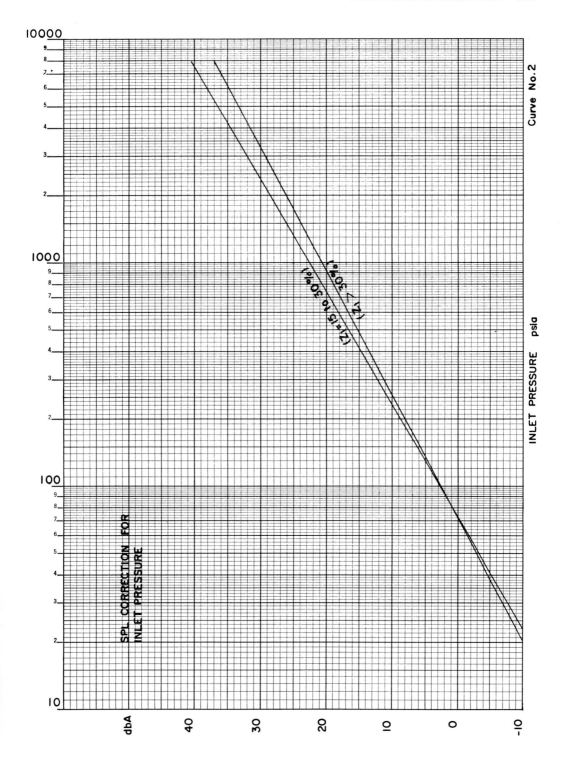

246 VALVE ENGINEERING AND DESIGN DATA

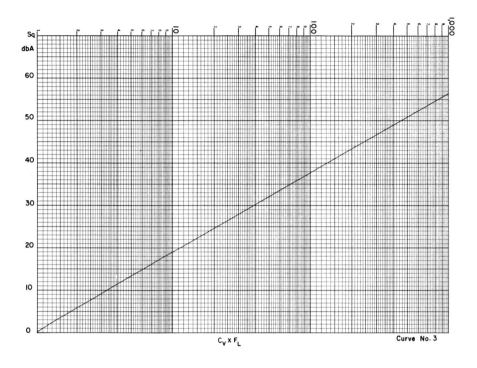

Curve No. 3

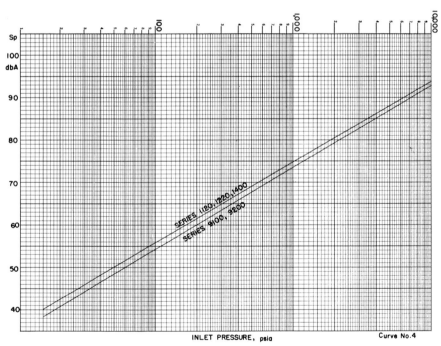

Curve No. 4

VALVE NOISE CALCULATION 247

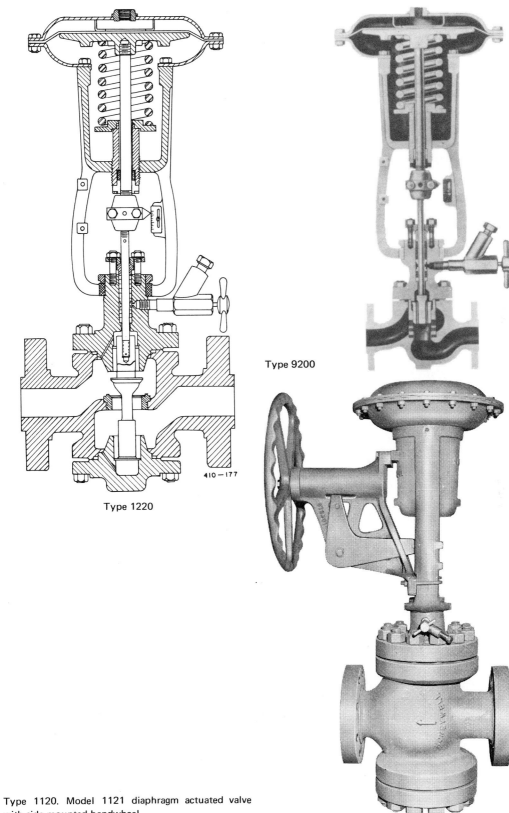

Type 1220

Type 9200

Type 1120. Model 1121 diaphragm actuated valve with side mounted handwheel.

248 VALVE ENGINEERING AND DESIGN DATA

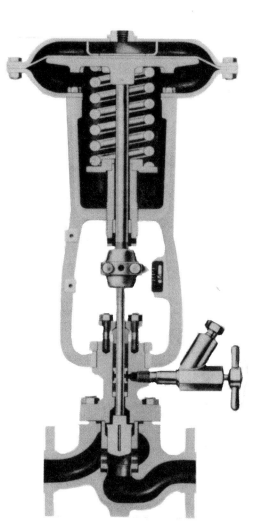

Type 9100. Model 9101 single seated.

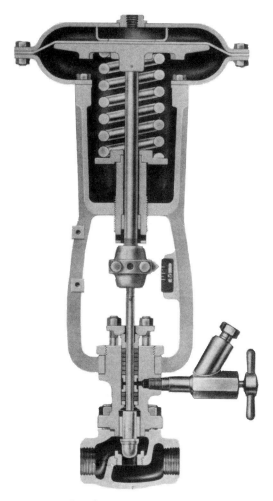

Type 1400. Fig. 1. Model 1401 diaphragm actuated.

APPENDIX A
Metric-English Conversion Factors*

*(Reproduced by Permission of Design News Magazine)

Basic Unit of Measurement-Length (Meter)

The following charts give conversion factors for the basic unit of length and the units that are derived from this basic unit.

LENGTH

U.S. TO METRIC	METRIC TO U.S.
1 inch = 25.40 millimeters	1 millimeter = 0.03937 inch
1 inch = 2.540 centimeters	1 centimeter = 0.3937 inch
1 foot = 30.480 centimeters	1 meter = 39.37 inches
1 foot = 0.3048 meter	1 meter = 3.2808 feet
1 yard = 91.440 centimeters	1 meter = 1.0936 yards
1 yard = 0.9144 meter	1 kilometer = 0.62137 mile
1 mile = 1.609 kilometers	

To convert from	to	Multiply by
angstrom	meter (m)	$1.000\,000 \times 10^{-10}$
astronomical unit	meter (m)	$1.495\,98 \times 10^{11}$
caliber	meter (m)	$2.540\,000 \times 10^{-4}$
fathom	meter (m)	$1.828\,800$
fermi (*fermtometer*)	meter (m)	$1.000\,000 \times 10^{-15}$
foot	meter (m)	$3.048\,000 \times 10^{-1}$
foot (*U.S. survey*)	meter (m)	$1200/3937$
foot (*U.S. survey*)	meter (m)	$3.048\,006 \times 10^{-1}$
inch	meter (m)	$2.540\,000 \times 10^{-2}$
league (*International nautical*)	meter (m)	$5.556\,000 \times 10^{3}$
league (*statute*)	meter (m)	$4.828\,032 \times 10^{3}$
league (*U.K. nautical*)	meter (m)	$5.559\,552 \times 10^{3}$
light year	meter (m)	$9.460\,55 \times 10^{15}$
micron	meter (m)	$1.000\,000 \times 10^{-6}$
mil	meter (m)	$2.540\,000 \times 10^{-5}$
mile (*international nautical*)	meter (m)	$1.852\,000 \times 10^{3}$
mile (*U.K. nautical*)	meter (m)	$1.853\,184 \times 10^{3}$
mile (*U.S. nautical*)	meter (m)	$1.852\,000 \times 10^{3}$
mile (*U.S. statute*)	meter (m)	$1.609\,344 \times 10^{3}$
parsec	meter (m)	$3.083\,74 \times 10^{16}$
pica (*printer's*)	meter (m)	$4.217\,518 \times 10^{-3}$
point (*printer's*)	meter (m)	$3.514\,598 \times 10^{-4}$
rod	meter (m)	$5.029\,200$
statute mile (*U.S.*)	meter (m)	$1.609\,344 \times 10^{3}$
yard	meter (m)	$9.144\,000 \times 10^{-1}$

AREA

U.S. TO METRIC	METRIC TO U.S.
1 sq. inch = 645.16 sq. millimeters	1 sq. millimeter = 0.00155 sq. inch
1 sq. inch = 6.4516 sq. centimeters	1 sq. centimeter = 0.1550 sq. inch
1 sq. foot = 929.03 sq. centimeters	1 sq. meter = 10.7640 sq. feet
1 sq. foot = 0.0929 sq. meter	1 sq. meter = 1.196 sq. yards
1 sq. yard = 0.836 sq. meter	1 sq. hectometer = 2.471 acres
1 acre = 0.4047 sq. hectometer	1 hectare = 2.471 acres
1 acre = 0.4047 hectare	1 sq. kilometer = 0.386 sq. mile
1 sq. mile = 2.59 sq. kilometers	

AREA (Continued)

To convert from	to	Multiply by
acre	meter² (m²)	$4.046\ 856 \times 10^3$
barn	meter² (m²)	$1.000\ 000 \times 10^{-28}$
circular mil	meter² (m²)	$5.067\ 075 \times 10^{-10}$
foot²	meter² (m²)	$9.290\ 304 \times 10^{-2}$
inch²	meter² (m²)	$6.451\ 600 \times 10^{-4}$
mile² (*U.S. statute*)	meter² (m²)	$2.589\ 988 \times 10^6$
section	meter² (m²)	$2.589\ 988 \times 10^6$
township	meter² (m²)	$9.323\ 957 \times 10^7$
yard²	meter² (m²)	$8.361\ 274 \times 10^{-1}$

VOLUME (Capacity)

U.S. TO METRIC	METRIC TO U.S.
1 fluid ounce = 2.957 centiliters = 29.57 cm³	1 centiliter = 10 cm³ = 0.338 fluid ounce
1 pint (liq.) = 4.732 deciliters = 473.2 cm³	1 deciliter = 100 cm³ = 0.0528 pint (liq.)
1 quart (liq.) = 0.9463 liter = 0.9463 dm³	1 liter = 1 dm³ = 1.0567 quarts (liq.)
1 gallon (liq.) = 3.7853 liters = 3.7853 dm³	1 liter = 1 dm³ = 0.26417 gallon (liq.)

To convert from	to	Multiply by
acre-foot	meter³ (m³)	$1.233\ 482 \times 10^3$
barrel (*oil, 42 gal*)	meter³ (m³)	$1.589\ 873 \times 10^{-1}$
board foot	meter³ (m³)	$2.359\ 737 \times 10^{-3}$
bushel (*U.S.*)	meter³ (m³)	$3.523\ 907 \times 10^{-2}$
cup	meter³ (m³)	$2.365\ 882 \times 10^{-4}$
fluid ounce (*U.S.*)	meter³ (m³)	$2.957\ 353 \times 10^{-5}$
foot³	meter³ (m³)	$2.831\ 685 \times 10^{-2}$
gallon (*Canadian liquid*)	meter³ (m³)	$4.546\ 122 \times 10^{-3}$
gallon (*U.K. liquid*)	meter³ (m³)	$4.546\ 087 \times 10^{-3}$
gallon (*U.S. dry*)	meter³ (m³)	$4.404\ 884 \times 10^{-3}$
gallon (*U.S. liquid*)	meter³ (m³)	$3.785\ 412 \times 10^{-3}$
gill (*U.K.*)	meter³ (m³)	$1.420\ 652 \times 10^{-4}$
gill (*U.S.*)	meter³ (m³)	$1.182\ 941 \times 10^{-4}$
inch³	meter³ (m³)	$1.638\ 706 \times 10^{-5}$
liter	meter³ (m³)	$1.000\ 000 \times 10^{-3}$
ounce (*U.K. fluid*)	meter³ (m³)	$2.841\ 305 \times 10^{-5}$
ounce (*U.S. fluid*)	meter³ (m³)	$2.957\ 353 \times 10^{-5}$
peck (*U.S.*)	meter³ (m³)	$8.809\ 768 \times 10^{-3}$
pint (*U.S. dry*)	meter³ (m³)	$5.506\ 105 \times 10^{-4}$
pint (*U.S. liquid*)	meter³ (m³)	$4.731\ 765 \times 10^{-4}$
quart (*U.S. dry*)	meter³ (m³)	$1.101\ 221 \times 10^{-3}$
quart (*U.S. liquid*)	meter³ (m³)	$9.463\ 529 \times 10^{-4}$
stere	meter³ (m³)	$1.000\ 000$
tablespoon	meter³ (m³)	$1.478\ 676 \times 10^{-5}$
teaspoon	meter³ (m³)	$4.928\ 922 \times 10^{-6}$
ton (*register*)	meter³ (m³)	$2.831\ 685$
yard³	meter³ (m³)	$7.645\ 549 \times 10^{-1}$

The following Nomogram provides a quick solution to problems involving inches and millimeters.

Example:
Convert 8 inches to millimeters

Solution:
Locate 8 inches on the Nomogram and read the answer of 203 mm.

Basic Unit of Measurement—Time (Second)

The following charts give conversion factors for the basic unit of time and the units that are derived from this basic unit.

TIME

To convert from	to	Multiply by
day (*mean solar*)	second (s)	$8.640\ 000 \times 10^4$
day (*sidereal*)	second (s)	$8.616\ 409 \times 10^4$
hour (*mean solar*)	second (s)	$3.600\ 000 \times 10^3$
hour (*sidereal*)	second (s)	$3.590\ 170 \times 10^3$
minute (*mean solar*)	second (s)	$6.000\ 000 \times 10$
minute (*sidereal*)	second (s)	$5.983\ 617 \times 10$
month (*mean calendar*)	second (s)	$2.628\ 000 \times 10^6$
second (*sidereal*)	second (s)	$9.972\ 696 \times 10^{-1}$
year (*calendar*)	second (s)	$3.153\ 600 \times 10^7$
year (*sidereal*)	second (s)	$3.155\ 815 \times 10^7$
year (*tropical*)	second (s)	$3.155\ 693 \times 10^7$

ACCELERATION

To convert from	to	Multiply by
foot/second²	meter/second² (m/s²)	$3.048\ 000 \times 10^{-1}$
free fall, standard	meter/second² (m/s²)	$9.806\ 650$
gal (*galileo*)	meter/second² (m/s²)	$1.000\ 000 \times 10^{-2}$
inch/second²	meter/second² (m/s²)	$2.540\ 000 \times 10^{-2}$

VELOCITY (INCLUDES SPEED)

To convert from	to	Multiply by
foot/hour	meter/second (m/s)	$8.466\ 667 \times 10^{-5}$
foot/minute	meter/second (m/s)	$5.080\ 000 \times 10^{-3}$
foot/second	meter/second (m/s)	$3.048\ 000 \times 10^{-1}$
inch/second	meter/second (m/s)	$2.540\ 000 \times 10^{-2}$
kilometer/hour	meter/second (m/s)	$2.777\ 778 \times 10^{-1}$
knot (*international*)	meter/second (m/s)	$5.144\ 444 \times 10^{-1}$
mile/hour (*U.S. statute*)	meter/second (m/s)	$4.470\ 400 \times 10^{-1}$
mile/minute (*U.S. statute*)	meter/second (m/s)	$2.682\ 240 \times 10$
mile/second (*U.S. statute*)	meter/second (m/s)	$1.609\ 344 \times 10^3$
mile/hour (*U.S. statute*)	kilometer/hour	$1.609\ 344$

The following Nomogram provides a quick solution to problems involving the conversion from frequency to time units.

Example:

A component vibrates at 60 cps for 2 days. Determine the total number of cycles on the component.

Solution:

Construct a line from 60 on the F scale (cps) to 2 on the T scale (days). Where this line intersects the total number of cycles scale, read the answer of 10,000,000.

METRIC-ENGLISH CONVERSION FACTORS 253

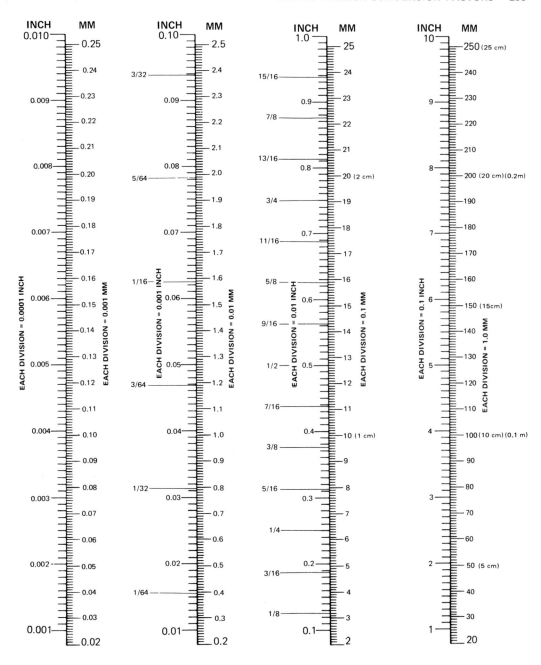

254 VALVE ENGINEERING AND DESIGN DATA

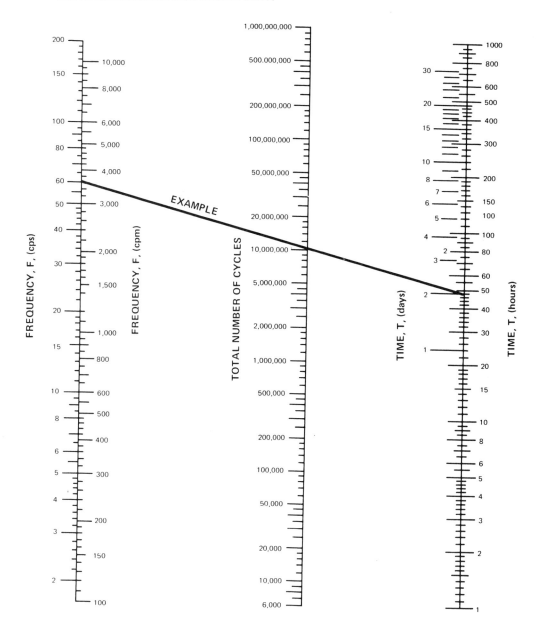

Basic Unit of Measurement - Mass (Kilogram)

The following charts give conversion factors for the basic unit of mass and the units that are derived from this basic unit.

MASS (Weight)

U.S. TO METRIC		METRIC TO U.S.	
1 ounce (dry)	= 28.35 grams	1 gram	= 0.03527 ounce
1 pound	= 0.4536 kilogram	1 kilogram	= 2.2046 pounds
1 short ton (2000 lb.)	= 907.2 kilograms	1 metric ton	= 2 204.6 pounds
1 short ton (2000 lb.)	= 0.9072 metric ton	1 metric ton	= 1.102 tons (short)

MASS

To convert from	to	Multiply by
carat *(metric)*	kilogram (kg)	$2.000\ 000 \times 10^{-4}$
grain	kilogram (kg)	$6.479\ 891 \times 10^{-5}$
gram	kilogram (kg)	$1.000\ 000 \times 10^{-3}$
hundredweight *(long)*	kilogram (kg)	$5.080\ 235 \times 10$
hundredweight *(short)*	kilogram (kg)	$4.535\ 924 \times 10$
kilogram-force-second^2/meter *(mass)*	kilogram (kg)	$9.806\ 650$
kilogram-mass	kilogram (kg)	$1.000\ 000$
ounce-mass *(avoirdupois)*	kilogram (kg)	$2.834\ 952 \times 10^{-2}$
ounce-mass *(troy or apothecary)*	kilogram (kg)	$3.110\ 348 \times 10^{-2}$
pennyweight	kilogram (kg)	$1.555\ 174 \times 10^{-3}$
pound-mass *(lbm avoirdupois)*	kilogram (kg)	$4.535\ 924 \times 10^{-1}$
pound-mass *(troy or apothecary)*	kilogram (kg)	$3.732\ 417 \times 10^{-1}$
slug	kilogram (kg)	$1.459\ 390 \times 10$
ton *(assay)*	kilogram (kg)	$2.916\ 667 \times 10^{-2}$
ton *(long, 2240 lbm)*	kilogram (kg)	$1.016\ 047 \times 10^{3}$
ton *(metric)*	kilogram (kg)	$1.000\ 000 \times 10^{3}$
ton *(short, 2000 lbm)*	kilogram (kg)	$9.071\ 847 \times 10^{2}$
tonne	kilogram (kg)	$1.000\ 000 \times 10^{3}$

ENERGY OR WORK

To convert from	to	Multiply by
British thermal unit *(International table)*	joule (J)	$1.055\ 056 \times 10^{3}$
British thermal unit *(mean)*	joule (J)	$1.055\ 87 \times 10^{3}$
British thermal unit *(thermochemical)*	joule (J)	$1.054\ 350 \times 10^{3}$
British thermal unit *(39 F)*	joule (J)	$1.059\ 67 \times 10^{3}$
British thermal unit *(60 F)*	joule (J)	$1.054\ 68 \times 10^{3}$
calorie *(International Table)*	joule (J)	$4.186\ 800$
calorie *(mean)*	joule (J)	$4.190\ 02$
calorie *(thermochemical)*	joule (J)	$4.184\ 000$
calorie *(15 C)*	joule (J)	$4.185\ 80$
calorie *(20 C)*	joule (J)	$4.181\ 90$
calorie *(kg, International Table)*	joule (J)	$4.186\ 800 \times 10^{3}$
calorie *(kg, mean)*	joule (J)	$4.190\ 02 \times 10^{3}$
calorie *(kg, thermochemical)*	joule (J)	$4.184\ 000 \times 10^{3}$
electron volt	joule (J)	$1.602\ 10 \times 10^{-19}$
erg	joule (J)	$1.000\ 000 \times 10^{-7}$
foot-pound-force	joule (J)	$1.355\ 818$
foot-poundal	joule (J)	$4.214\ 011 \times 10^{-2}$
joule *(International of 1948)*	joule (J)	$1.000\ 165$
kilocalorie *(International Table)*	joule (J)	$4.186\ 800 \times 10^{3}$
kilocalorie *(mean)*	joule (J)	$4.190\ 02 \times 10^{3}$
kilocalorie *(thermochemical)*	joule (J)	$4.184\ 000 \times 10^{3}$
kilowatt-hour	joule (J)	$3.600\ 000 \times 10^{6}$

To convert from	to	Multiply by
kilowatt-hour *(international of 1948)*	joule (J)	3.60059×10^6
ton *(nuclear equivalent of TNT)*	joule (J)	4.20×10^9
watt-hour	joule (J)	3.600000×10^3
watt-second	joule (J)	1.000000

FORCE

To convert from	to	Multiply by
dyne	newton (N)	1.000000×10^{-5}
kilogram-force	newton (N)	9.806650
kilopound-force	newton (N)	9.806650
kip	newton (N)	4.448222×10^3
ounce-force *(avoirdupois)*	newton (N)	2.780139×10^{-1}
pound-force *(lbf avoirdupois)*	newton (N)	4.448222
pound-force *(lbf avoirdupois)*	kilogram-force	4.535924×10^{-1}
poundal	newton (N)	1.382550×10^{-1}

POWER

To convert from	to	Multiply by
Btu *(International Table)* /hour	watt (W)	2.930711×10^{-1}
Btu *(thermochemical)* /second	watt (W)	1.054350×10^3
Btu *(thermochemical)* /minute	watt (W)	1.757250×10
Btu *(thermochemical)* /hour	watt (W)	2.928751×10^{-1}
calorie *(thermochemical)* /second	watt (W)	4.184000
calorie *(thermochemical)* /minute	watt (W)	6.973333×10^{-2}
erg/second	watt (W)	1.000000×10^{-7}
foot-pound-force/hour	watt (W)	3.766161×10^{-4}
foot-pound-force/minute	watt (W)	2.259697×10^{-2}
foot-pound-force/second	watt (W)	1.355818
horsepower *(550 ft. lbf/s)*	watt (W)	7.456999×10^2
horsepower *(boiler)*	watt (W)	9.80950×10^3
horsepower *(electric)*	watt (W)	7.460000×10^2
horsepower *(metric)*	watt (W)	7.35499×10^2
horsepower *(water)*	watt (W)	7.46043×10^2
horsepower *(U.K.)*	watt (W)	7.4570×10^2
kilocalorie *(thermochemical)* /minute	watt (W)	6.973333×10
kilocalorie *(thermochemical)* /second	watt (W)	4.184000×10^3
watt *(international of 1948)*	watt (W)	1.000165

PRESSURE OR STRESS (FORCE/AREA)

To convert from	to	Multiply by
atmosphere *(normal=760 torr)*	newton/meter² (N/m²)	1.013250×10^5
atmosphere *(technical=1 kgf/cm²)*	newton/meter² (N/m²)	9.806650×10^4
bar	newton/meter² (N/m²)	1.000000×10^5
centimeter of mercury *(0 C)*	newton/meter² (N/m²)	1.33322×10^3
centimeter of water *(4 C)*	newton/meter² (N/m²)	9.80638×10
decibar	newton/meter² (N/m²)	1.000000×10^4
dyne/centimeter²	newton/meter² (N/m²)	1.000000×10^{-1}
foot of water *(39.2F)*	newton/meter² (N/m²)	2.98898×10^3
gram-force/centimeter²	newton/meter² (N/m²)	9.806650×10
inch of mercury *(32 F)*	newton/meter² (N/m²)	3.386389×10^3
inch of mercury *(60 F)*	newton/meter² (N/m²)	3.37685×10^3
inch of water *(39.2 F)*	newton/meter² (N/m²)	2.49082×10^2
inch of water *(60 F)*	newton/meter² (N/m²)	2.4884×10^2
kilogram-force/centimeter²	newton/meter² (N/m²)	9.806650×10^4
kilogram-force/meter²	newton/meter² (N/m²)	9.806650
kilogram-force/millimeter²	newton/meter² (N/m²)	9.806650×10^6
kip/inch²	newton/meter² (N/m²)	6.894757×10^6
millibar	newton/meter² (N/m²)	1.000000×10^2
millimeter of mercury *(0 C)*	newton/meter² (N/m²)	1.333224×10^2
newton/meter²	pascal (pa)	1.000000

METRIC-ENGLISH CONVERSION FACTORS

pascal	newton/meter² (N/m²)	1.000 000
poundal/foot²	newton/meter² (N/m²)	1.488 164
pound-force/foot²	newton/meter² (N/m²)	4.788 026 × 10
pound-force/inch² *(psi)*	newton/meter² (N/m²)	6.894 757 × 10³
pound-force/inch² *(psi)*	kilogram-force/mm²	7.030 696 × 10⁻⁴
psi	newton/meter² (N/m²)	6.894 757 × 10³
torr *(mm Hg, 0 C)*	newton/meter² (N/m²)	1.333 22 × 10²

The following Nomogram provides a quick solution to problems involving pounds and grams.

Example: Convert 500 grams to pounds.

Solution: Locate 500 grams on the Nomogram and read the answer of 1.1 lb.

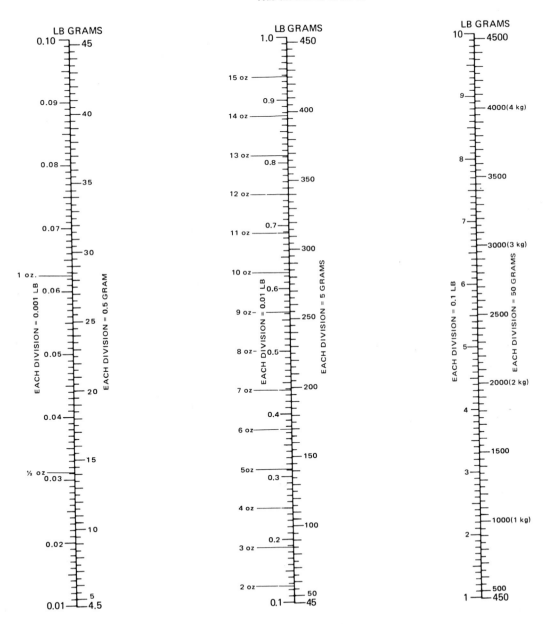

Basic Unit of Measurement—Temperature (Kelvin)

The following chart gives conversion factors for the basic unit of temperature.

TEMPERATURE

To convert from	to	Multiply by
degree Celsius	kelvin (K)	$t_K = t_C + 273.15$
degree Fahrenheit	kelvin (K)	$t_K = (t_F + 459.67)/1.8$
degree Rankine	kelvin (K)	$t_K = t_R/1.8$
degree Fahrenheit	degree Celsius	$t_C = (t_F - 32)/1.8$
kelvin	degree Celsius	$t_C = t_K - 273.15$

The following Nomograms provide a quick solution to problems involving the conversion from one temperature scale to another.

Example:
Given 32 deg F, determine its kelvin, celsius and rankine value.

Solution:
On chart 2, draw a straight line parallel to the bottom of the chart and read the corresponding values: kelvin = 273.15 deg, celsius = 0 deg and rankine = 491.69 deg.

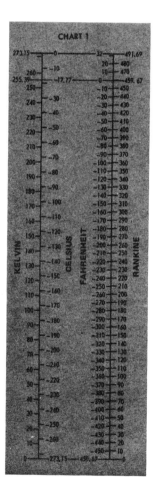

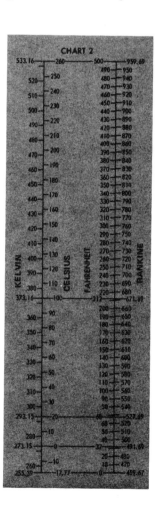

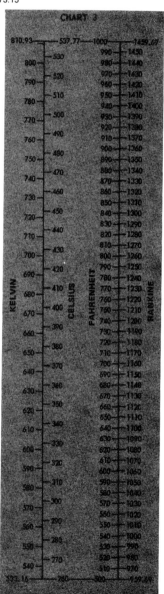

METRIC-ENGLISH CONVERSION FACTORS 259

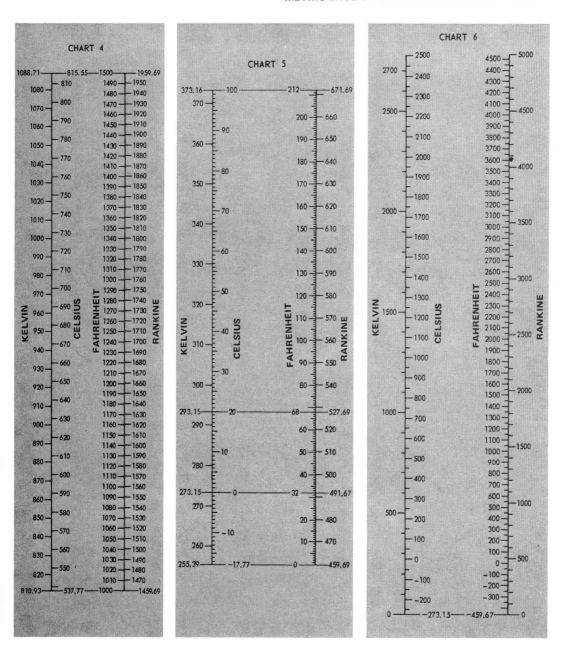

Basic Unit of Measurement—Electric Current (Ampere)

The following chart gives conversion factors for the basic unit of electric current and the units that are derived from this basic unit.

ELECTRICITY AND MAGNETISM

To convert from	to	Multiply by
abampere	ampere (A)	$1.000\,000 \times 10$
abcoulomb	coulomb (C)	$1.000\,000 \times 10$
abfarad	farad (F)	$1.000\,000 \times 10^9$
abhenry	henry (H)	$1.000\,000 \times 10^{-9}$
abmho	mho	$1.000\,000 \times 10^9$
abohm	ohm (Ω)	$1.000\,000 \times 10^{-9}$
abvolt	volt (V)	$1.000\,000 \times 10^{-8}$
ampere (*international of 1948*)	ampere (A)	$9.998\,35 \times 10^{-1}$
ampere-hour	coulomb (C)	$3.600\,000 \times 10^3$
coulomb (*international of 1948*)	coulomb (C)	$9.998\,35 \times 10^{-1}$
EMU of capacitance	farad (F)	$1.000\,000 \times 10^9$
EMU of current	ampere (A)	$1.000\,000 \times 10$
EMU of electric potential	volt (V)	$1.000\,000 \times 10^{-8}$
EMU of inductance	henry (H)	$1.000\,000 \times 10^{-9}$
EMU of resistance	ohm (Ω)	$1.000\,000 \times 10^{-9}$
ESU of capacitance	farad (F)	$1.112\,6 \times 10^{-12}$
ESU of current	ampere (A)	$3.335\,6 \times 10^{-10}$
ESU of electric potential	volt (V)	$2.997\,9 \times 10^2$
ESU of inductance	henry (H)	$8.987\,6 \times 10^{11}$
ESU of resistance	ohm (Ω)	$8.987\,6 \times 10^{11}$
farad (*international of 1948*)	farad (F)	$9.995\,05 \times 10^{-1}$
faraday (*based on carbon-12*)	coulomb (C)	$9.648\,70 \times 10^4$
faraday (*chemical*)	coulomb (C)	$9.649\,57 \times 10^4$
faraday (*physical*)	coulomb (C)	$9.652\,19 \times 10^4$
gamma	tesla (T)	$1.000\,000 \times 10^{-9}$
gauss	tesla (T)	$1.000\,000 \times 10^{-4}$
gilbert	ampere-turn	$7.957\,747 \times 10^{-1}$
henry (*international of 1948*)	henry (H)	$1.000\,495$
maxwell	weber (Wb)	$1.000\,000 \times 10^{-8}$
oersted	ampere/meter (A/m)	$7.957\,747 \times 10$
ohm (*international of 1948*)	ohm (Ω)	$1.000\,495$
ohm-centimeter	ohm-meter ($\Omega \cdot$ m)	$1.000\,000 \times 10^{-2}$
statampere	ampere (A)	$3.335\,640 \times 10^{-10}$
statcoulomb	coulomb (C)	$3.335\,640 \times 10^{-10}$
statfarad	farad (F)	$1.112\,650 \times 10^{-12}$
stathenry	henry (H)	$8.987\,554 \times 10^{11}$
statmho	mho	$1.112\,650 \times 10^{-12}$
statohm	ohm (Ω)	$8.987\,554 \times 10^{11}$
statvolt	volt (V)	$2.997\,925 \times 10^2$
unit pole	weber (Wb)	$1.256\,637 \times 10^{-7}$
volt (*international of 1948*)	volt (V)	$1.000\,330$

Basic Unit of Measurement - Luminous Intensity (Candela)

The following chart gives conversion factors for the basic unit of luminous intensity and units that are derived from this basic unit.

LIGHT

To convert from	to	Multiply by
footcandle	lumen/meter² (lm/m²)	1.076 391 × 10
footcandle	lux (lx)	1.076 391 × 10
footlambert	candela/meter² (cd/m²)	3.426 3
lux	lumen/meter² (lm/m²)	1.000 000

Almost anyone can select an incandescent lamp for a standard application. The lamp manufacturer gives nominal operating voltage, current and the light output you can expect. In addition, the lamps will be designed for some nominal life. Problems begin if you wish to use a lamp in a nonstandard application. For instance, what happens if you raise the operating voltage—what is your expected operating life then? How much more light can you expect from the lamp? Answers to these questions can be found by using the following nomogram.

Example: Given a small vacuum lamp, extend its life by a factor of 5.

Solution: The first step is to tabulate the manufacturer's data for the lamp in question. Write down the rated voltage, current and lamp life. Connect a line from 5 on the lamp life scale through the V life point on the E axis and extend it to the lamp operating voltage scale. Where this line intersects the lamp operating voltage scale, read the value of 0.89. This tells us that multiplying the lamp's design voltage by the factor 0.89 will extend the lamp's life by a factor of 5.

While the life may be extended, what does this do to the light output or luminous flux? Starting at the point just found, 0.89, connect a line to the V luminous flux point on the E axis and extend it to the luminous flux scale. Where this line intersects the luminous flux scale, read the value of 0.67. This means that under our new operating conditions, the light output will amount to only 67% of the normal or rated value. In the same manner, values for the other lamp parameters can be found.

Note that there are two points given for each factor along the E axis. G is for a gas-filled lamp while V is for a vacuum lamp. As a general rule, lamps having a rated current of less than 0.4 amp are vacuum, larger lamps are usually gas-filled.

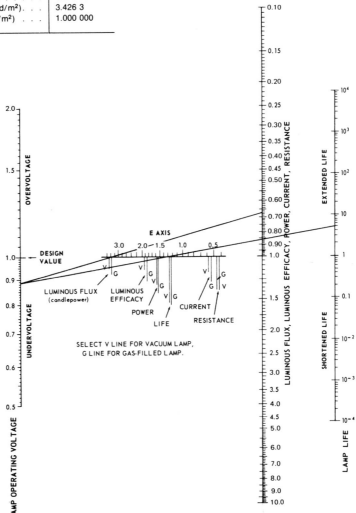

Miscellaneous Units of Measurement

The following charts give conversion factors for the miscellaneous metric units.

BENDING MOMENT OR TORQUE

To convert from	to	Multiply by
dyne-centimeter	newton-meter (N·m)	$1.000\ 000 \times 10^{-7}$
kilogram-force-meter	newton-meter (N·m)	$9.806\ 650$
ounce-force-inch	newton-meter (N·m)	$7.061\ 552 \times 10^{-3}$
pound-force-inch	newton-meter (N·m)	$1.129\ 848 \times 10^{-1}$
pound-force-foot	newton-meter (N·m)	$1.355\ 818$

BENDING MOMENT OR TORQUE/LENGTH

To convert from	to	Multiply by
pound-force-foot/inch	newton-meter/meter (N·m/m)	$5.337\ 866 \times 10$
pound-force-inch/inch	newton-meter/meter (N·m/n)	$4.448\ 222$

ENERGY/AREA TIME

To convert from	to	Multiply by
Btu (*thermochemical*)/foot²-second	watt/meter² (W/m²)	$1.134\ 983 \times 10^4$
Btu (*thermochemical*)/foot²-minute	watt/meter² (W/m²)	$1.891\ 489 \times 10^2$
Btu (*thermochemical*)/foot²-hour	watt/meter² (W/m²)	$3.152\ 481$
Btu (*thermochemical*)/inch²-second	watt/meter² (W/m²)	$1.634\ 246 \times 10^6$
calorie (*thermochemical*)/centimeter²-minute	watt/meter² (W/m²)	$6.973\ 333 \times 10^2$
erg-centimeter²-second	watt/meter² (W/m²)	$1.000\ 000 \times 10^{-3}$
watt/centimeter²	watt/meter² (W/m²)	$1.000\ 000 \times 10^4$

FORCE/LENGTH

To convert from	to	Multiply by
pound-force/inch	newton/meter (N/m)	$1.751\ 268 \times 10^2$
pound-force/foot	newton/meter (N/m)	$1.459\ 390 \times 10$

METRIC-ENGLISH CONVERSION FACTORS

HEAT

To convert from	to	Multiply by
Btu (*thermochemical*)·in./s·ft.²·deg F (k_1 thermal conductivity)	watt/meter-kelvin (W/m·K)	$5.188\ 732 \times 10^2$
Btu (*International Table*)·in./s·ft.²·deg F (*k*, thermal conductivity)	watt/meter-kelvin (Wm/m·K)	$5.192\ 204 \times 10^2$
Btu (*thermochemical*)·in./h·ft.²·deg F (*k*, thermal conductivity)	watt/meter-kelvin (W/m·K)	$1.441\ 314 \times 10^{-1}$
Btu (*International Table*)·in./h·ft.²·deg F (*k*, thermal conductivity)	watt/meter-kelvin (W/m·K)	$1.442\ 279 \times 10^{-1}$
Btu (*International Table*)/ft.²	joule/meter² (J/m²)	$1.135\ 653 \times 10^4$
Btu (*thermochemical*)/ft.²	joule/meter² (J/m²)	$1.134\ 893 \times 10^4$
Btu (*International Table*)/h·ft.²·deg F (*C*, thermal conductance)	watt/meter²-kelvin (W/m²·K)	$5.678\ 263$
Btu (*thermochemical*)/h·ft.²·deg F (*C*, thermal conductance)	watt/meter²-kelvin (W/m²·K)	$5.674\ 466$
Btu (*International Table*)/pound-mass	joule/kilogram (J/kg)	$2.326\ 000 \times 10^3$
Btu (*thermochemical*)/pound-mass	joule/kilogram (J/kg)	$2.324\ 444 \times 10^3$
Btu (*International Table*)/lbm·deg F (*c*, heat capacity)	joule/kilogram-kelvin (J/kg·K)	$4.186\ 800 \times 10^3$
Btu (*thermochemical*)/lbm·deg F (*c*, heat capacity)	joule/kilogram-kelvin (J/kg·K)	$4.184\ 000 \times 10^3$
Btu (*International Table*)/s·ft.²·deg F	watt/meter²-kelvin (W/m²·K)	$2.044\ 175 \times 10^4$
Btu (*thermochemical*)/s·ft.²·deg F	watt/meter²-kelvin (W/m²·K)	$2.042\ 808 \times 10^4$
cal (*thermochemical*)/cm²	joule/meter² (J/m²)	$4.184\ 000 \times 10^4$
cal (*thermochemical*)/cm²·s	watt/meter² (W/m²)	$4.184\ 000 \times 10^4$
cal (*thermochemical*)/cm·s·deg C	watt/meter-kelvin (W/m·K)	$4.184\ 00 \times 10^2$
cal (*International Table*)/g	joule/kilogram (J/kg)	$4.186\ 800 \times 10^3$
cal (*International Table*)/g·deg C	joule/kilogram-kelvin (J/kg·K)	$4.186\ 800 \times 10^3$
cal (*thermochemical*)/g	joule/kilogram (J/kg)	$4.184\ 000 \times 10^3$
cal (*thermochemical*)/g·deg C	joule/kilogram-kelvin (J/kg·K)	$4.184\ 000 \times 10^3$
clo	kelvin-meter²/watt (K·m²/W)	$2.003\ 712 \times 10^{-1}$
deg F·h·ft.²/Btu (*thermochemical*) (R_1 thermal resistance)	kelvin-meter²/watt (K·m²/W)	$1.762\ 280 \times 10^{-1}$
deg F·h·ft.²/Btu (*International Table*) (*R*, thermal resistance)	kelvin-meter²/watt (K·m²/W)	$1.761\ 102 \times 10^{-1}$
ft.²/h (thermal diffusivity)	meter²/second (m²/s)	$2.580\ 640 \times 10^{-5}$

MASS/VOLUME (INCLUDES DENSITY AND MASS CAPACITY)

To convert from	to	Multiply by
grain (lbm avoirdupois/7000)/gallon (*U.S. liquid*)	kilogram/meter³ (kg/m³)	$1.711\ 806 \times 10^{-2}$
gram/centimeter³	kilogram/meter³ (kg/m³)	$1.000\ 000 \times 10^3$
ounce (*avoirdupois*)/gallon (*U.K. liquid*)	kilogram/meter³ (kg/m³)	$6.236\ 027$
ounce (*avoirdupois*)/gallon (*U.S. liquid*)	kilogram/meter³ (kg/m³)	$7.489\ 152$
ounce (*avoirdupois*) (mass)/inch³	kilogram/meter³ (kg/m³)	$1.729\ 994 \times 10^3$
pound-mass/foot³	kilogram/meter³ (kg/m³)	$1.601\ 846 \times 10$
pound-mass/inch³	kilogram/meter³ (kg/m³)	$2.767\ 990 \times 10^4$
pound-mass/gallon (*U.K. liquid*)	kilogram/meter³ (kg/m³)	$9.977\ 644 \times 10$
pound-mass/gallon (*U.S. liquid*)	kilogram/meter³ (kg/m³)	$1.198\ 264 \times 10^2$
slug/foot³	kilogram/meter³ (kg/m³)	$5.153\ 788 \times 10^2$
ton (*long, mass*)/yard³	kilogram/meter³ (kg/m³)	$1.328\ 939 \times 10^3$

MASS/AREA

To convert from	to	Multiply by
ounce-mass/yard².	kilogram/meter² (kg/m²).	$3.390\ 575 \times 10^{-2}$
pound-mass/foot².	kilogram/meter² (kg/m²).	$4.882\ 428$

MASS/TIME (INCLUDES FLOW)

To convert from	to	Multiply by
perm (0 C).	kilogram/newton-second (kg/N·s).	$5.721\ 35 \times 10^{-11}$
perm (23 C).	kilogram/newton-second (kg/N·s).	$5.745\ 25 \times 10^{-11}$
perm-inch (0 C).	kilogram-meter/newton-second (kg·m/N·s)	$1.453\ 22 \times 10^{-12}$
perm-inch (23 C).	kilogram-meter/newton-second (kg·m/N·s)	$1.459\ 29 \times 10^{-12}$
pound-mass/second.	kilogram/second (kg/s)	$4.535\ 924 \times 10^{-1}$
pound-mass/minute.	kilogram/second (kg/s)	$7.559\ 873 \times 10^{-3}$
ton (*short, mass*)/hour.	kilogram/second (kg/s)	$2.519\ 958 \times 10^{-1}$

VISCOSITY

To convert from	to	Multiply by
centipoise.	newton-second/meter² (N·s/m²)	$1.000\ 000 \times 10^{-3}$
centistoke	meter²/second (m²/s)	$1.000\ 000 \times 10^{-6}$
foot²/second	meter²/second (m²/s)	$9.290\ 304 \times 10^{-2}$
poise	newton-second/meter² (N·s/m²)	$1.000\ 000 \times 10^{-1}$
poundal-second/foot²	newton-second/meter² (N·s/m²)	$1.488\ 164$
pound-mass/foot-second.	newton-second/meter² (N·s/m²)	$1.488\ 164$
pound-force-second/foot²	newton-second/meter² (N·s/m²)	$4.788\ 026 \times 10$
rhe	meter²/newton-second (m²/N·s)	$1.000\ 000 \times 10$
slug/foot-second	newton-second/meter² (N·s/m²)	$4.788\ 026 \times 10$
stoke	meter²/second (m²/s)	$1.000\ 000 \times 10^{-4}$

VOLUME/TIME (INCLUDES FLOW)

To convert from	to	Multiply by
foot³/minute	meter³/second (m³/s)	$4.719\ 474 \times 10^{-4}$
foot³/second	meter³/second (m³/s)	$2.831\ 685 \times 10^{-2}$
inch³/minute	meter³/second (m³/s)	$2.731\ 177 \times 10^{-7}$
yard³/minute	meter³/second (m³/s)	$1.274\ 258 \times 10^{-2}$
gallon (*U.S. liquid*)/day	meter³/second (m³/s)	$4.381\ 264 \times 10^{-8}$
gallon (*U.S. liquid*)/minute	meter³/second (m³/s)	$6.309\ 020 \times 10^{-5}$

Miscellaneous Metric Nomograms

The following nomograms provide a quick solution to metric conversion problems.

Example:
Convert 30 psi to kg/sq cm.

Solution:
Locate 30 on the psi scale and read the answer of 2.1 kg/sq cm.

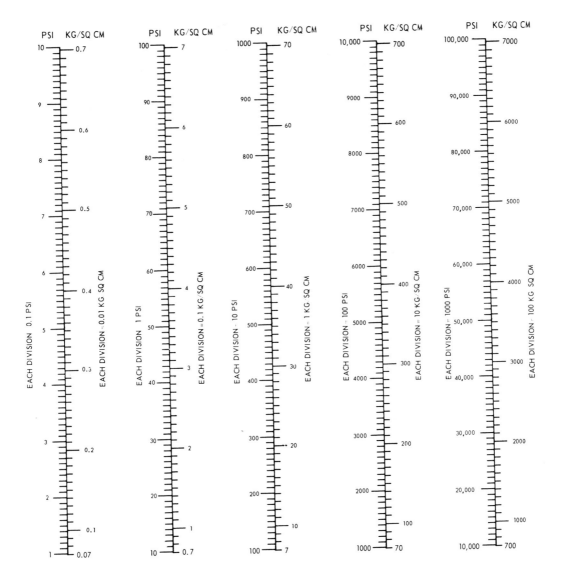

VALVE ENGINEERING AND DESIGN DATA

Example:
Convert 600 cu in to cc.

Solution:
Locate 600 on the cu in scale and read the answer of 9800 cc.

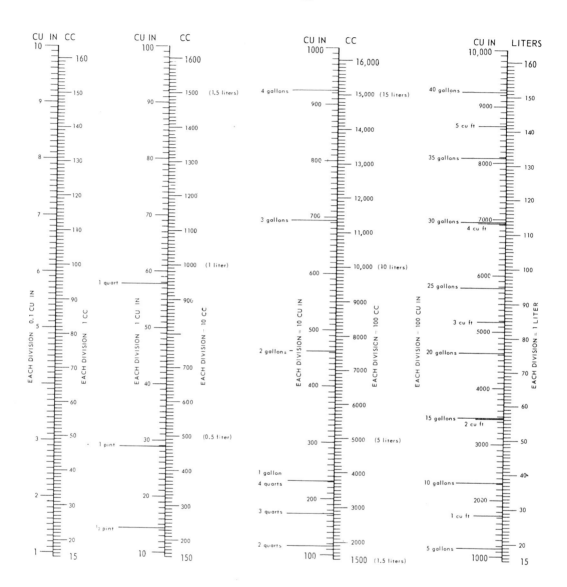

METRIC-ENGLISH CONVERSION FACTORS 267

Example:
Convert 20 mph to knots, ft/min, ft/sec, km/hr, m/min and m/sec.

Solution:
Locate 20 mph on the mph scale and construct a horizontal line parallel to the base. Where this line intersects the various scales, read the answers of: 17.5 knots, 1750 ft/min, 30 ft/sec, 32 km/hr, 540 m/min and 9.0 m/sec.

The same procedure is used on the bottom nomogram.

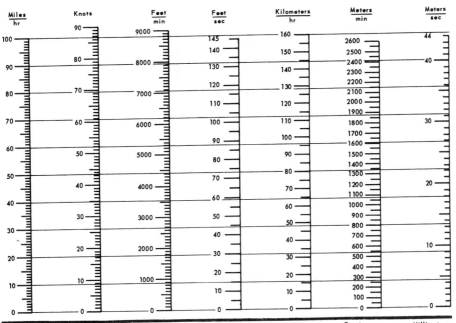

Selected SI Units for Fluid Power Usage

Quantity	Symbol*	Customary U.S. Unit	Practical SI Units (preferred unit underlined)			Conversion from U.S. to SI	Notes
angular velocity (also see rotational frequency)	ω	radian per second (rad/s)	rad/s				
area	A or S	square inch (in^2)	$\underline{m^2}$	cm^2	mm^2	$in^2 \times 6.452 = cm^2$	
bulk modulus (liquids)	K	pounds per square inch (psi)	$\underline{bar}$	N/m^2		$\dfrac{psi}{14.50} = bar$	1
capacity (displacement)	V	cubic inches per revolution (cipr)	ℓ/r	mℓ/r		$cipr \times 16.39 = mℓ/r$	
coefficient of thermal expansion (cubic)	α	per °F (1/°F)	1/K				2
dynamic viscosity	μ	centipoise (cP)	$\underline{cP}$	P	Pa s		3
efficiency	η	percent	$\underline{percent}$				
force	F	pound (f) lb(f)	$\underline{N}$	kN		$lb(f) \times 4.448 = N$	9
frequency	f	cycles per second (cps)	Hz	kHz		1 cps = 1 Hz	
kinematic viscosity	ν	Saybolt Universal Seconds (SUS)	cSt	m^2/s		Per ANSI/Z11.129-1972 (ASTM/D2161-1971)	4
length	l	inch (in)	m	$\underline{mm}$	μm	$in \times 25.40 = mm$	9
linear velocity	v	feet per second (ft/s)	m/s			$ft/s \times 0.3048 = m/s$	
mass	m	pound (m) lb(m)	Mg	$\underline{kg}$	g	$lb(m) \times 0.4536 = kg$	5
mass density	ρ	pound (m) per cubic foot lb(m)/ft^3	$\underline{kg/m^3}$	kg/dm^3	kg/ℓ	$lb(m)/ft^3 \times 16.02 = kg/m^3$	
mass flow	M	pound (m)/second lb(m)/s	$\underline{kg/s}$	g/s		$\dfrac{lb(m) \times 0.4536}{s} = kg/s$	9
power	P	horsepower (HP)	W	$\underline{kW}$		$HP \times .7457 = kW$	
pressure (above atmospheric)	p	pounds per square inch (psi)	$\underline{bar}$	mbar Pa	kPa	$\dfrac{psi}{14.50} = bar$	6, 9
pressure (below atmospheric)	p	inches of mercury, abs (in Hg)	bar, abs	Pa	kPa	in Hg, abs (@60°F) × 0.03377= bar, abs	6

* Symbol for Fluid Power Usage

(Conversion factors and SI units for Fluid Power reproduced by Permission of National Fluid Power Association.)

METRIC-ENGLISH CONVERSION FACTORS

Quantity	Symbol*	Customary U.S. Unit	Practical SI Units (preferred unit underlined)	Conversion from U.S. to SI	Notes
quantity of heat	Q_c	British Thermal Unit (BTU)	$\underline{J}$ kJ MJ	BTU × 1055 = J	
rotational frequency (shaft speed)	n	revolutions per minute (RPM)	r/s $\underline{r/min}$		
specific heat capacity	c	British Thermal Units per pound mass degree Fahrenheit BTU/lb(m)°F	J/(kg K)	BTU/lb(m)°F × 4187 = J/kg°K	
stress (materials)	σ	pounds per square inch (psi)	$\underline{daN/mm^2}$ MPa	$\dfrac{psi}{1450.0}$ = daN/mm²	9
surface roughness		microinch (μin)	μm	Per ISO/R 1302-1971.	
temperature (customary)	θ	degree Fahrenheit (°F)	$\underline{°C}$ grade N	$\dfrac{°F - 32}{1.80}$ = °C	9
temperature (interval)		degree Fahrenheit (°F)	$\underline{°C}$	$\dfrac{°F - 32}{1.80}$ = °C	9
temperature (thermodynamic)	T	Rankine (°R)	$\underline{K}$	$\dfrac{°R}{1.80}$ = K	9
time	t	second (s)	s $\underline{min}$ h		
torque (moment of force)	T	pounds (f)-inch lb(f)-in	$\underline{N\,m}$ kN m mN m	lb(f)-in × 0.1130 = N m	
volume	V	gallon (US gal)	$\underline{m^3}$ dm³ $\underline{(\ell)}$ cm³ (mℓ)	US gal × 3.785 = ℓ	7
volumetric flow (gases)	Q(ANR)	standard cubic feet per minute (scfm)	m_n^3/s dm_n^3/s cm_n^3/s	scfm × 0.4718 = dm_n^3/s	8
volumetric flow (liquids)	Q	gallons per minute (USGPM)	$\underline{\ell/min}$ ℓ/s mℓ/s	USGPM × 3.785 = ℓ/min	7
work	W	foot-pound ft-lb(f)	$\underline{J}$	ft-lb(f) × 1.356 = J	

NOTES

1 Indicate capacity (displacement) of a rotary device as "per revolution". Express non-rotary devices as "per cycle".

2 The centipoise (practical unit "cP") is a non-SI unit, use of which is permitted by ISO 1000. The centipoise is equal to 10^{-3} Ns/m².

3 Efficiencies are normally stated as "percent", but the use of a ratio is also permissible.

4 The centistokes (practical unit "cSt") is a non-SI unit, use of which is permitted by ISO 1000. The centistokes is equal to 10^{-6} m²/s.

5 The litre is a non-SI unit, use of which is permitted by ISO 1000. The litre is a special name for a unit of liquid measure and is exactly equal to the cubic decimetre.

6 The bar is a non-SI unit, use of which is permitted by ISO 1000. The bar is a special name of a unit of liquid pressure and is assumed to be "gauge" unless otherwise specified.
   1 bar = 100 kPa   1 bar = 10^5 N/m²

7 Method for indicating surface roughness to be determined in accordance with ISO/R 1302.

8 Use the kelvin (practical unit "K") not only to express a temperature but also to express an interval of temperature in accordance with the 13th Conférence Générale des Poids et Mesures (CGPM), held in 1967. Use the term "degree Celsius" (practical unit "°C") for quoting values of the Celsius temperature. Usages resulting from earlier decisions of the CGPM remain allowable for the time being (e.g., degK and degC, which are synonymous).

9 The abbreviation "ANR" means that the result of the measurement has been referred to the Standard Reference Atmosphere (Atmosphère Normale de Référence) as defined in 2.2 of ISO/R 554. Insert this abbreviation immediately following the unit used or the expression of the quantity.

APPENDIX B
Pipe Flow Data

(Reproduced by Courtesy of Crane Co.)

Commercial Wrought Steel Pipe Data

Schedule Wall Thickness—Per ASA B36.10-1950

	Nominal Pipe Size Inches	Outside Diameter Inches	Thickness Inches	Inside Diameter d Inches	Inside Diameter D Feet	Inside Diameter Functions (In Inches) d^2	d^3	d^4	d^5	Transverse Internal Area a Sq. In.	A Sq. Ft.
Schedule 80—cont.	1½	1.900	0.200	1.500	0.1250	2.250	3.375	5.062	7.594	1.767	0.01225
	2	2.375	0.218	1.939	0.1616	3.760	7.290	14.136	27.41	2.953	0.02050
	2½	2.875	0.276	2.323	0.1936	5.396	12.536	29.117	67.64	4.238	0.02942
	3	3.5	0.300	2.900	0.2417	8.410	24.389	70.728	205.1	6.605	0.04587
	3½	4.0	0.318	3.364	0.2803	11.32	38.069	128.14	430.8	8.888	0.06170
	4	4.5	0.337	3.826	0.3188	14.64	56.006	214.33	819.8	11.497	0.07986
	5	5.563	0.375	4.813	0.4011	23.16	111.49	536.38	2583.	18.194	0.1263
	6	6.625	0.432	5.761	0.4801	33.19	191.20	1101.6	6346.	26.067	0.1810
	8	8.625	0.500	7.625	0.6354	58.14	443.32	3380.3	25775.	45.663	0.3171
	10	10.75	0.593	9.564	0.7970	91.47	874.82	8366.8	80020.	71.84	0.4989
	12	12.75	0.687	11.376	0.9480	129.41	1472.2	16747.	190523.	101.64	0.7058
	14	14.0	0.750	12.500	1.0417	156.25	1953.1	24414.	305176.	122.72	0.8522
	16	16.0	0.843	14.314	1.1928	204.89	2932.8	41980.	600904.	160.92	1.1175
	18	18.0	0.937	16.126	1.3438	260.05	4193.5	67626.	1090518.	204.24	1.4183
	20	20.0	1.031	17.938	1.4948	321.77	5771.9	103536.	1857248.	252.72	1.7550
	24	24.0	1.218	21.564	1.7970	465.01	10027.	216234.	4662798.	365.22	2.5362
Schedule 100	8	8.625	0.593	7.439	0.6199	55.34	411.66	3062.	22781.	43.46	0.3018
	10	10.75	0.718	9.314	0.7762	86.75	807.99	7526.	69357.	68.13	0.4732
	12	12.75	0.843	11.064	0.9220	122.41	1354.4	14985.	165791.	96.14	0.6677
	14	14.0	0.937	12.126	1.0105	147.04	1783.0	21621.	262173.	115.49	0.8020
	16	16.0	1.031	13.938	1.1615	194.27	2707.7	37740.	526020.	152.58	1.0596
	18	18.0	1.156	15.688	1.3057	246.11	3861.0	60572.	950250.	193.30	1.3423
	20	20.0	1.281	17.438	1.4532	304.08	5302.6	92467.	1612438.	238.83	1.6585
	24	24.0	1.531	20.938	1.7448	438.40	9179.2	192195.	4024179.	344.32	2.3911
Schedule 120	4	4.50	0.438	3.624	0.302	13.13	47.595	172.49	625.1	10.315	0.07163
	5	5.563	0.500	4.563	0.3802	20.82	95.006	433.5	1978.	16.35	0.1136
	6	6.625	0.562	5.501	0.4584	30.26	166.47	915.7	5037.	23.77	0.1650
	8	8.625	0.718	7.189	0.5991	51.68	371.54	2671.	19202.	40.59	0.2819
	10	10.75	0.843	9.064	0.7553	82.16	744.66	6750.	61179.	64.53	0.4481
	12	12.75	1.000	10.750	0.8959	115.56	1242.3	13355.	143563.	90.76	0.6303
	14	14.0	1.093	11.814	0.9845	139.57	1648.9	19480.	230137.	109.62	0.7612
	16	16.0	1.218	13.564	1.1303	183.98	2495.5	33849.	459133.	144.50	1.0035
	18	18.0	1.375	15.250	1.2708	232.56	3546.6	54086.	824804.	182.66	1.2684
	20	20.0	1.500	17.000	1.4166	289.00	4913.0	83521.	1419857.	226.98	1.5762
	24	24.0	1.812	20.376	1.6980	415.18	8459.7	172375.	3512313.	326.08	2.2645
Schedule 140	8	8.625	0.812	7.001	0.5834	49.01	343.15	2402.	16819.	38.50	0.2673
	10	10.75	1.000	8.750	0.7292	76.56	669.92	5862.	51291.	60.13	0.4176
	12	12.75	1.125	10.500	0.8750	110.25	1157.6	12155.	127628.	86.59	0.6013
	14	14.0	1.250	11.500	0.9583	132.25	1520.9	17490.	201136.	103.87	0.7213
	16	16.0	1.438	13.124	1.0937	172.24	2260.5	29666.	389340.	135.28	0.9394
	18	18.0	1.562	14.876	1.2396	221.30	3292.0	48972.	728502.	173.80	1.2070
	20	20.0	1.750	16.5	1.3750	272.25	4492.1	74120.	1222981.	213.82	1.4849
	24	24.0	2.062	19.876	1.6563	395.06	7852.1	156009.	3102022.	310.28	2.1547
Schedule 160	½	0.840	0.187	0.466	0.0388	0.2172	0.1012	0.04716	0.02197	0.1706	0.00118
	¾	1.050	0.218	0.614	0.0512	0.3770	0.2315	0.1421	0.08726	0.2961	0.00206
	1	1.315	0.250	0.815	0.0679	0.6642	0.5413	0.4412	0.3596	0.5217	0.00362
	1¼	1.660	0.250	1.160	0.0966	1.346	1.561	1.811	2.100	1.057	0.00734
	1½	1.900	0.281	1.338	0.1115	1.790	2.395	3.205	4.288	1.406	0.00976
	2	2.375	0.343	1.689	0.1407	2.853	4.818	8.138	13.74	2.241	0.01556
	2½	2.875	0.375	2.125	0.1771	4.516	9.596	20.39	43.33	3.546	0.02463
	3	3.50	0.438	2.624	0.2187	6.885	18.067	47.41	124.4	5.408	0.03755
	4	4.50	0.531	3.438	0.2865	11.82	40.637	139.7	480.3	9.283	0.06447
	5	5.563	0.625	4.313	0.3594	18.60	80.230	346.0	1492.	14.61	0.1015
	6	6.625	0.718	5.189	0.4324	26.93	139.72	725.0	3762.	21.15	0.1469
	8	8.625	0.906	6.813	0.5677	46.42	316.24	2155.	14679.	36.46	0.2532
	10	10.75	1.125	8.500	0.7083	72.25	614.12	5220.	44371.	56.75	0.3941
	12	12.75	1.312	10.126	0.8438	102.54	1038.3	10514.	106461.	80.53	0.5592
	14	14.0	1.406	11.188	0.9323	125.17	1400.4	15668.	175292.	98.31	0.6827
	16	16.0	1.593	12.814	1.0678	164.20	2104.0	26961.	345482.	128.96	0.8956
	18	18.0	1.781	14.438	1.2032	208.45	3009.7	43454.	627387.	163.72	1.1369
	20	20.0	1.968	16.064	1.3387	258.05	4145.3	66590.	1069715.	202.67	1.4074
	24	24.0	2.343	19.314	1.6095	373.03	7204.7	139152.	2687582.	292.98	2.0346

Commercial Wrought Steel Pipe Data

(Per ASA B36.10-1950)

Nominal Pipe Size Inches	Outside Diameter Inches	Thickness Inches	Inside Diameter		Inside Diameter Functions (In Inches)				Transverse Internal Area	
			d Inches	D Feet	d^2	d^3	d^4	d^5	a Sq. In.	A Sq. Ft.
Standard Wall Pipe										
1/8	0.405	0.068	0.269	0.0224	0.0724	0.0195	0.00524	0.00141	0.057	0.00040
1/4	0.540	0.088	0.364	0.0303	0.1325	0.0482	0.01756	0.00639	0.104	0.00072
3/8	0.675	0.091	0.493	0.0411	0.2430	0.1198	0.05905	0.02912	0.191	0.00133
1/2	0.840	0.109	0.622	0.0518	0.3869	0.2406	0.1497	0.0931	0.304	0.00211
3/4	1.050	0.113	0.824	0.0687	0.679	0.5595	0.4610	0.3799	0.533	0.00371
1	1.315	0.133	1.049	0.0874	1.100	1.154	1.210	1.270	0.864	0.00600
1 1/4	1.660	0.140	1.380	0.1150	1.904	2.628	3.625	5.005	1.495	0.01040
1 1/2	1.900	0.145	1.610	0.1342	2.592	4.173	6.718	10.82	2.036	0.01414
2	2.375	0.154	2.067	0.1722	4.272	8.831	18.250	37.72	3.355	0.02330
2 1/2	2.875	0.203	2.469	0.2057	6.096	15.051	37.161	91.75	4.788	0.03322
3	3.500	0.216	3.068	0.2557	9.413	28.878	88.605	271.8	7.393	0.05130
3 1/2	4.000	0.226	3.548	0.2957	12.59	44.663	158.51	562.2	9.886	0.06870
4	4.500	0.237	4.026	0.3355	16.21	65.256	262.76	1058.	12.730	0.08840
5	5.563	0.258	5.047	0.4206	25.47	128.56	648.72	3275.	20.006	0.1390
6	6.625	0.280	6.065	0.5054	36.78	223.10	1352.8	8206.	28.891	0.2006
8	8.625	0.277	8.071	0.6725	65.14	525.75	4243.0	34248.	51.161	0.3553
	8.625S	0.322	7.981	0.6651	63.70	508.36	4057.7	32380.	50.027	0.3474
10	10.75	0.279	10.192	0.8493	103.88	1058.7	10789.	109876.	81.585	0.5666
	10.75	0.307	10.136	0.8446	102.74	1041.4	10555.	106987.	80.691	0.5604
	10.75S	0.365	10.020	0.8350	100.4	1006.0	10080.	101000.	78.855	0.5475
12	12.75	0.330	12.090	1.0075	146.17	1767.2	21366.	258300.	114.80	0.7972
	12.75S	0.375	12.000	1.000	144.0	1728.0	20736.	248800.	113.10	0.7854
Extra Strong Pipe										
1/8	0.405	0.095	0.215	0.0179	0.0462	0.00994	0.002134	0.000459	0.036	0.00025
1/4	0.540	0.119	0.302	0.0252	0.0912	0.0275	0.008317	0.002513	0.072	0.00050
3/8	0.675	0.126	0.423	0.0353	0.1789	0.0757	0.03201	0.01354	0.141	0.00098
1/2	0.840	0.147	0.546	0.0455	0.2981	0.1628	0.08886	0.04852	0.234	0.00163
3/4	1.050	0.154	0.742	0.0618	0.5506	0.4085	0.3032	0.2249	0.433	0.00300
1	1.315	0.179	0.957	0.0797	0.9158	0.8765	0.8387	0.8027	0.719	0.00499
1 1/4	1.660	0.191	1.278	0.1065	1.633	2.087	2.6667	3.409	1.283	0.00891
1 1/2	1.900	0.200	1.500	0.1250	2.250	3.375	5.062	7.594	1.767	0.01225
2	2.375	0.218	1.939	0.1616	3.760	7.290	14.136	27.41	2.953	0.02050
2 1/2	2.875	0.276	2.323	0.1936	5.396	12.536	29.117	67.64	4.238	0.02942
3	3.500	0.300	2.900	0.2417	8.410	24.389	70.728	205.1	6.605	0.04587
3 1/2	4.000	0.318	3.364	0.2803	11.32	38.069	128.14	430.8	8.888	0.06170
4	4.500	0.337	3.826	0.3188	14.64	56.006	214.33	819.8	11.497	0.07986
5	5.563	0.375	4.813	0.4011	23.16	111.49	536.6	2583.	18.194	0.1263
6	6.625	0.432	5.761	0.4801	33.19	191.20	1101.6	6346.	26.067	0.1810
8	8.625	0.500	7.625	0.6354	58.14	443.32	3380.3	25775.	45.663	0.3171
10	10.75	0.500	9.750	0.8125	95.06	926.86	9036.4	88110.	74.662	0.5185
12	12.75	0.500	11.750	0.9792	138.1	1622.2	19072.	223970.	108.434	0.7528
Double Extra Strong Pipe										
1/2	0.840	0.294	0.252	0.0210	0.0635	0.0160	0.004032	0.00102	0.050	0.00035
3/4	1.050	0.308	0.434	0.0362	0.1884	0.0817	0.03549	0.01540	0.148	0.00103
1	1.315	0.358	0.599	0.0499	0.3588	0.2149	0.1287	0.07711	0.282	0.00196
1 1/4	1.660	0.382	0.896	0.0747	0.8028	0.7193	0.6445	0.5775	0.630	0.00438
1 1/2	1.900	0.400	1.100	0.0917	1.210	1.331	1.4641	1.611	0.950	0.00660
2	2.375	0.436	1.503	0.1252	2.259	3.395	5.1031	7.670	1.774	0.01232
2 1/2	2.875	0.552	1.771	0.1476	3.136	5.554	9.8345	17.42	2.464	0.01710
3	3.500	0.600	2.300	0.1917	5.290	12.167	27.984	64.36	4.155	0.02885
3 1/2	4.000	0.636	2.728	0.2273	7.442	20.302	55.383	151.1	5.845	0.04059
4	4.500	0.674	3.152	0.2627	9.935	31.315	98.704	311.1	7.803	0.05419
5	5.563	0.750	4.063	0.3386	16.51	67.072	272.58	1107.	12.966	0.09006
6	6.625	0.864	4.897	0.4081	23.98	117.43	575.04	2816.	18.835	0.1308
8	8.625	0.875	6.875	0.5729	47.27	324.95	2234.4	15360.	37.122	0.2578

PIPE FLOW DATA 273

B-15

Flow of Air Through Schedule 40 Steel Pipe

For lengths of pipe other than 100 feet, the pressure drop is proportional to the length. Thus, for 50 feet of pipe, the pressure drop is approximately one-half the value given in the table . . . for 300 feet, three times the given value, etc.

The pressure drop is also inversely proportional to the absolute pressure and directly proportional to the absolute temperature.

Therefore, to determine the pressure drop for inlet or average pressures other than 100 psi and at temperatures other than 60 F, multiply the values given in the table by the ratio:

$$\left(\frac{100+14.7}{P+14.7}\right)\left(\frac{460+t}{520}\right)$$

where:

"P" is the inlet or average gauge pressure in pounds per square inch, and,

"t" is the temperature in degrees Fahrenheit under consideration.

The cubic feet per minute of compressed air at any pressure is inversely proportional to the absolute pressure and directly proportional to the absolute temperature.

To determine the cubic feet per minute of compressed air at any temperature and pressure other than standard conditions, multiply the value of cubic feet per minute of free air by the ratio:

$$\left(\frac{14.7}{14.7+P}\right)\left(\frac{460+t}{520}\right)$$

Calculations for Pipe Other than Schedule 40

To determine the velocity of water, or the pressure drop of water or air, through pipe other than Schedule 40, use the following formulas:

$$v_a = v_{40}\left(\frac{d_{40}}{d_a}\right)^2$$

$$\Delta P_a = \Delta P_{40}\left(\frac{d_{40}}{d_a}\right)^5$$

Subscript "a" refers to the Schedule of pipe through which velocity or pressure drop is desired.

Subscript "40" refers to the velocity or pressure drop through Schedule 40 pipe, as given here and on p. 279

Free Air q'_m Cubic Feet Per Minute at 60 F and 14.7 psia	Compressed Air Cubic Feet Per Minute at 60 F and 100 psig	Pressure Drop of Air In Pounds per Square Inch Per 100 Feet of Schedule 40 Pipe For Air at 100 Pounds per Square Inch Gauge Pressure and 60 F Temperature									
		1/8"	1/4"	3/8"	1/2"						
1	0.128	0.361	0.083	0.018							
2	0.256	1.31	0.285	0.064	0.020						
3	0.384	3.06	0.605	0.133	0.042	3/4"					
4	0.513	4.83	1.04	0.226	0.071						
5	0.641	7.45	1.58	0.343	0.106	0.027	1"				
6	0.769	10.6	2.23	0.408	0.148	0.037					
8	1.025	18.6	3.89	0.848	0.255	0.062	0.019				
10	1.282	28.7	5.96	1.26	0.356	0.094	0.029	1 1/4"	1 1/2"		
15	1.922	...	13.0	2.73	0.834	0.201	0.062				
20	2.563	...	22.8	4.76	1.43	0.345	0.102	0.026			
25	3.204		35.6	7.34	2.21	0.526	0.156	0.039	0.019		
30	3.845		...	10.5	3.15	0.748	0.219	0.055	0.026		
35	4.486		...	14.2	4.24	1.00	0.293	0.073	0.035		
40	5.126		...	18.4	5.49	1.30	0.379	0.095	0.044		
45	5.767		...	23.1	6.90	1.62	0.474	0.116	0.055	2"	
50	6.408		28.5	8.49	1.99	0.578	0.149	0.067	0.019		
60	7.690	2 1/2"	40.7	12.2	2.85	0.819	0.200	0.094	0.027		
70	8.971		...	16.5	3.83	1.10	0.270	0.126	0.036		
80	10.25	0.019	...	21.4	4.96	1.43	0.350	0.162	0.046		
90	11.53	0.023		27.0	6.25	1.80	0.437	0.203	0.058		
100	12.82	0.029	3"	33.2	7.69	2.21	0.534	0.247	0.070		
125	16.02	0.044		...	11.9	3.39	0.825	0.380	0.107		
150	19.22	0.062	0.021		17.0	4.87	1.17	0.537	0.151		
175	22.43	0.083	0.028	3 1/2"	23.1	6.60	1.58	0.727	0.205		
200	25.63	0.107	0.036		30.0	8.54	2.05	0.937	0.264		
225	28.84	0.134	0.045	0.022	37.9	10.8	2.59	1.19	0.331		
250	32.04	0.164	0.055	0.027		13.3	3.18	1.45	0.404		
275	35.24	0.191	0.066	0.032		16.0	3.83	1.75	0.484		
300	38.45	0.232	0.078	0.037		19.0	4.56	2.07	0.573		
325	41.65	0.270	0.090	0.043	4"	22.3	5.32	2.42	0.673		
350	44.87	0.313	0.104	0.050		25.8	6.17	2.80	0.776		
375	48.06	0.356	0.119	0.057	0.030	29.6	7.05	3.20	0.887		
400	51.26	0.402	0.134	0.064	0.034	33.6	8.02	3.64	1.00		
425	54.47	0.452	0.151	0.072	0.038	37.9	9.01	4.09	1.13		
450	57.67	0.507	0.168	0.081	0.042	...	10.2	4.59	1.26		
475	60.88	0.562	0.187	0.089	0.047		11.3	5.09	1.40		
500	64.08	0.623	0.206	0.099	0.052		12.5	5.61	1.55		
550	70.49	0.749	0.248	0.118	0.062		15.1	6.79	1.87		
600	76.90	0.887	0.293	0.139	0.073		18.0	8.04	2.21		
650	83.30	1.04	0.342	0.163	0.086	5"	21.1	9.43	2.60		
700	89.71	1.19	0.395	0.188	0.099	0.032	24.3	10.9	3.00		
750	96.12	1.36	0.451	0.214	0.113	0.036	27.9	12.6	3.44		
800	102.5	1.55	0.513	0.244	0.127	0.041	31.8	14.2	3.90		
850	108.9	1.74	0.576	0.274	0.144	0.046	6"	35.9	16.0	4.40	
900	115.3	1.95	0.642	0.305	0.160	0.051		40.2	18.0	4.91	
950	121.8	2.18	0.715	0.340	0.178	0.057	0.023		20.0	5.47	
1 000	128.2	2.40	0.788	0.375	0.197	0.063	0.025		22.1	6.06	
1 100	141.0	2.89	0.948	0.451	0.236	0.075	0.030		26.7	7.29	
1 200	153.8	3.44	1.13	0.533	0.279	0.089	0.035		31.8	8.63	
1 300	166.6	4.01	1.32	0.626	0.327	0.103	0.041		37.3	10.1	
1 400	179.4	4.65	1.52	0.718	0.377	0.119	0.047			11.8	
1 500	192.2	5.31	1.74	0.824	0.431	0.136	0.054			13.5	
1 600	205.1	6.04	1.97	0.932	0.490	0.154	0.061	8"		15.3	
1 800	230.7	7.65	2.50	1.18	0.616	0.193	0.075			19.3	
2 000	256.3	9.44	3.06	1.45	0.757	0.237	0.094	0.023	10"	23.9	
2 500	320.4	14.7	4.76	2.25	1.17	0.366	0.143	0.035			37.3
3 000	384.5	21.1	6.82	3.20	1.67	0.524	0.204	0.051	0.016		
3 500	448.6	28.8	9.23	4.33	2.26	0.709	0.276	0.068	0.022		
4 000	512.6	37.6	12.1	5.66	2.94	0.919	0.358	0.088	0.028		
4 500	576.7	47.6	15.3	7.16	3.69	1.16	0.450	0.111	0.035	12"	
5 000	640.8	...	18.8	8.85	4.56	1.42	0.552	0.136	0.043	0.018	
6 000	769.0	...	27.1	12.7	6.57	2.03	0.794	0.195	0.061	0.025	
7 000	897.1	...	36.9	17.2	8.94	2.76	1.07	0.262	0.082	0.034	
8 000	1025		...	22.5	11.7	3.59	1.39	0.339	0.107	0.044	
9 000	1153			28.5	14.9	4.54	1.76	0.427	0.134	0.055	
10 000	1282			35.2	18.4	5.60	2.16	0.526	0.164	0.067	
11 000	1410			...	22.2	6.78	2.62	0.633	0.197	0.081	
12 000	1538			...	26.4	8.07	3.09	0.753	0.234	0.096	
13 000	1666			...	31.0	9.47	3.63	0.884	0.273	0.112	
14 000	1794			...	36.0	11.0	4.21	1.02	0.316	0.129	
15 000	1922			...	...	12.6	4.84	1.17	0.364	0.148	
16 000	2051			...	...	14.3	5.50	1.33	0.411	0.167	
18 000	2307			...	...	18.2	6.96	1.68	0.520	0.213	
20 000	2563			...	...	22.4	8.60	2.01	0.642	0.260	
22 000	2820			...	...	27.1	10.4	2.50	0.771	0.314	
24 000	3076			...	...	32.3	12.4	2.97	0.918	0.371	
26 000	3332			...	...	37.9	14.5	3.49	1.12	0.435	
28 000	3588			...	...	...	16.9	4.04	1.25	0.505	
30 000	3845			...	...	...	19.3	4.64	1.42	0.520	

Commercial Wrought Steel Pipe Data
Schedule Wall Thickness—Per ASA B36.10-1950

	Nominal Pipe Size	Outside Diameter	Thickness	Inside Diameter		Inside Diameter Functions (In Inches)				Transverse Internal Area	
				d	D	d^2	d^3	d^4	d^5	a	A
	Inches	Inches	Inches	Inches	Feet					Sq. In.	Sq. Ft.
Schedule 10	14	14	0.250	13.5	1.125	182.25	2460.4	33215.	448400.	143.14	0.994
	16	16	0.250	15.5	1.291	240.25	3723.9	57720.	894660.	188.69	1.310
	18	18	0.250	17.5	1.4583	306.25	5359.4	93789.	1641309.	240.53	1.670
	20	20	0.250	19.5	1.625	380.25	7414.9	144590.	2819500.	298.65	2.074
	24	24	0.250	23.5	1.958	552.25	12977.	304980.	7167030.	433.74	3.012
	30	30	0.312	29.376	2.448	862.95	25350.	744288.	21864218.	677.76	4.707
Schedule 20	8	8.625	0.250	8.125	0.6771	66.02	536.38	4359.3	35409.	51.85	0.3601
	10	10.75	0.250	10.25	0.8542	105.06	1076.9	11038.	113141.	82.52	0.5731
	12	12.75	0.250	12.25	1.021	150.06	1838.3	22518.	275855.	117.86	0.8185
	14	14	0.312	13.376	1.111	178.92	2393.2	32012.	428185.	140.52	0.9758
	16	16	0.312	15.376	1.281	236.42	3635.2	55894.	859442.	185.69	1.290
	18	18	0.312	17.376	1.448	301.92	5246.3	91156.	1583978.	237.13	1.647
	20	20	0.375	19.250	1.604	370.56	7133.3	137317.	2643352.	291.04	2.021
	24	24	0.375	23.25	1.937	540.56	12568.	292205.	6793832.	424.56	2.948
	30	30	0.500	29.00	2.417	841.0	24389.	707281.	20511149.	660.52	4.587
Schedule 30	8	8.625	0.277	8.071	0.6726	65.14	525.75	4243.2	34248.	51.16	0.3553
	10	10.75	0.307	10.136	0.8447	102.74	1041.4	10555.	106987.	80.69	0.5603
	12	12.75	0.330	12.09	1.0075	146.17	1767.2	21366.	258304.	114.80	0.7972
	14	14	0.375	13.25	1.1042	175.56	2326.2	30821.	408394.	137.88	0.9575
	16	16	0.375	15.25	1.2708	232.56	3546.6	54084.	824801.	182.65	1.268
	18	18	0.438	17.124	1.4270	293.23	5021.3	85984.	1472397.	230.30	1.599
	20	20	0.500	19.00	1.5833	361.00	6859.0	130321.	2476099.	283.53	1.969
	24	24	0.562	22.876	1.9063	523.31	11971.	273853.	6264703.	411.00	2.854
	30	30	0.625	28.75	2.3958	826.56	23764.	683207.	19642160.	649.18	4.508
Schedule 40	1/8	0.405	0.068	0.269	0.0224	0.0724	0.0195	0.005242	0.00141	0.057	0.00040
	1/4	0.540	0.088	0.364	0.0303	0.1325	0.0482	0.01756	0.00639	0.104	0.00072
	3/8	0.675	0.091	0.493	0.0411	0.2430	0.1198	0.05905	0.02912	0.191	0.00133
	1/2	0.840	0.109	0.622	0.0518	0.3869	0.2406	0.1497	0.09310	0.304	0.00211
	3/4	1.050	0.113	0.824	0.0687	0.679	0.5595	0.4610	0.3799	0.533	0.00371
	1	1.315	0.133	1.049	0.0874	1.100	1.154	1.210	1.270	0.864	0.00600
	1 1/4	1.660	0.140	1.380	0.1150	1.904	2.628	3.625	5.005	1.495	0.01040
	1 1/2	1.900	0.145	1.610	0.1342	2.592	4.173	6.718	10.82	2.036	0.01414
	2	2.375	0.154	2.067	0.1722	4.272	8.831	18.250	37.72	3.355	0.02330
	2 1/2	2.875	0.203	2.469	0.2057	6.096	15.051	37.161	91.75	4.788	0.03322
	3	3.500	0.216	3.068	0.2557	9.413	28.878	88.605	271.8	7.393	0.05130
	3 1/2	4.000	0.226	3.548	0.2957	12.59	44.663	158.51	562.2	9.886	0.06870
	4	4.500	0.237	4.026	0.3355	16.21	65.256	262.76	1058.	12.730	0.08840
	5	5.563	0.258	5.047	0.4206	25.47	128.56	648.72	3275.	20.006	0.1390
	6	6.625	0.280	6.065	0.5054	36.78	223.10	1352.8	8206.	28.891	0.2006
	8	8.625	0.322	7.981	0.6651	63.70	508.36	4057.7	32380.	50.027	0.3474
	10	10.75	0.365	10.02	0.8350	100.4	1006.0	10080.	101000.	78.855	0.5475
	12	12.75	0.406	11.938	0.9965	142.5	1701.3	20306.	242470.	111.93	0.7773
	14	14.0	0.438	13.124	1.0937	172.24	2260.5	29666.	389340.	135.28	0.9394
	16	16.0	0.500	15.000	1.250	225.0	3375.0	50625.	759375.	176.72	1.2272
	18	18.0	0.562	16.876	1.4063	284.8	4806.3	81111.	1368820.	223.68	1.5533
	20	20.0	0.593	18.814	1.5678	354.0	6659.5	125320.	2357244.	278.00	1.9305
	24	24.0	0.687	22.626	1.8855	511.9	11583.	262040.	5929784.	402.07	2.7921
Schedule 60	8	8.625	0.406	7.813	0.6511	61.04	476.93	3725.9	29113.	47.94	0.3329
	10	10.75	0.500	9.750	0.8125	95.06	926.86	9036.4	88110.	74.66	0.5185
	12	12.75	0.562	11.626	0.9688	135.16	1571.4	18268.	212399.	106.16	0.7372
	14	14.0	0.593	12.814	1.0678	164.20	2104.0	26962.	345480.	128.96	0.8956
	16	16.0	0.656	14.688	1.2240	215.74	3168.8	46544.	683618.	169.44	1.1766
	18	18.0	0.750	16.500	1.3750	272.25	4492.1	74120.	1222982.	213.83	1.4849
	20	20.0	0.812	18.376	1.5313	337.68	6205.2	114028.	2095342.	265.21	1.8417
	24	24.0	0.968	22.064	1.8387	486.82	10741.	236994.	5229036.	382.35	2.6552
Schedule 80	1/8	0.405	0.095	0.215	0.0179	0.0462	0.00994	0.002134	0.000459	0.036	0.00025
	1/4	0.540	0.119	0.302	0.0252	0.0912	0.0275	0.008317	0.002513	0.072	0.00050
	3/8	0.675	0.126	0.423	0.0353	0.1789	0.0757	0.03200	0.01354	0.141	0.00098
	1/2	0.840	0.147	0.546	0.0455	0.2981	0.1628	0.08886	0.04852	0.234	0.00163
	3/4	1.050	0.154	0.742	0.0618	0.5506	0.4085	0.3032	0.2249	0.433	0.00300
	1	1.315	0.179	0.957	0.0797	0.9158	0.8765	0.8387	0.8027	0.719	0.00499
	1 1/4	1.660	0.191	1.278	0.1065	1.633	2.087	2.6667	3.409	1.283	0.00891

Types of Valves

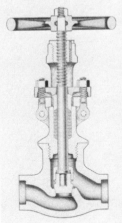
Conventional Globe Valve

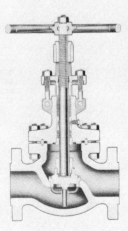
Conventional Globe Valve With Disc Guide

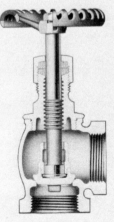
Conventional Angle Valve

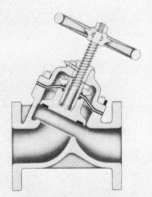
Y-Pattern Globe Valve With Stem 60 degrees from Run

Y-Pattern Globe Valve With Stem 45 degrees from Run

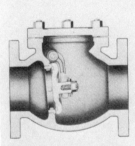
Conventional Swing Check Valve

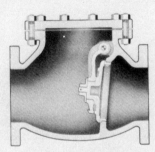
Clearway Swing Check Valve

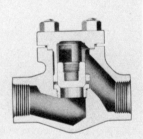
Globe Type Lift Check Valve

276 VALVE ENGINEERING AND DESIGN DATA

PHYSICAL PROPERTIES OF FLUIDS AND FLOW CHARACTERISTICS OF VALVES, FITTINGS, AND PIPE — A-29

Types of Valves

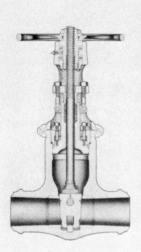
Conventional Gate Valve

Plug Gate Valve

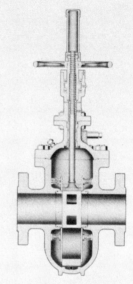
Pipe Line Gate Valve

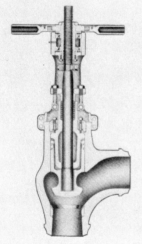
Angle Stop-Check Valve

Foot Valve With Strainer Poppet Type

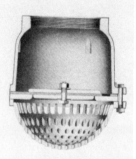
Foot Valve With Strainer Hinged Type

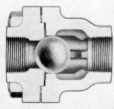
In-Line Ball Check Valve

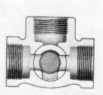
Three-Way Cock Sectional and Outside Views

PIPE FLOW DATA 277

A - 30   PHYSICAL PROPERTIES OF FLUIDS AND FLOW CHARACTERISTICS OF VALVES, FITTINGS, AND PIPE

Schedule (Thickness) of Steel Pipe Used in Obtaining Resistance Of Valves and Fittings of Various Pressure Classes by Test*

Valve or Fitting ANSI Pressure Classification		Schedule No. of Pipe Thickness
Steam Rating	Cold Rating	
250-Pound and Lower	500 psig	Schedule 40
300-Pound to 600-Pound	1440 psig	Schedule 80
900-Pound	2160 psig	Schedule 120
1500-Pound	3600 psig	Schedule 160
2500-Pound ½ to 6"	6000 psig	xx (Double Extra Strong)
2500-Pound 8" and larger	3600 psig	Schedule 160

*These schedule numbers have been arbitrarily selected only for the purpose of identifying the various pressure classes of valves and fittings with specific pipe dimensions for the interpretation of flow test data; they should not be construed as a recommendation for installation purposes.

Representative Equivalent Length‡ in Pipe Diameters (L/D) Of Various Valves and Fittings

		Description of Product		Equivalent Length In Pipe Diameters (L/D)
Globe Valves	Stem Perpendicular to Run	With no obstruction in flat, bevel, or plug type seat	Fully open	340
		With wing or pin guided disc	Fully open	450
	Y-Pattern	(No obstruction in flat, bevel, or plug type seat)		
		− With stem 60 degrees from run of pipe line	Fully open	175
		− With stem 45 degrees from run of pipe line	Fully open	145
Angle Valves		With no obstruction in flat, bevel, or plug type seat	Fully open	145
		With wing or pin guided disc	Fully open	200
Gate Valves	Wedge, Disc, Double Disc, or Plug Disc		Fully open	13
			Three-quarters open	35
			One-half open	160
			One-quarter open	900
	Pulp Stock		Fully open	17
			Three-quarters open	50
			One-half open	260
			One-quarter open	1200
Conduit Pipe Line Gate, Ball, and Plug Valves			Fully open	3**
Check Valves	Conventional Swing		0.5†...Fully open	135
	Clearway Swing		0.5†...Fully open	50
	Globe Lift or Stop; Stem Perpendicular to Run or Y-Pattern		2.0†...Fully open	Same as Globe
	Angle Lift or Stop		2.0†...Fully open	Same as Angle
	In-Line Ball		2.5 vertical and 0.25 horizontal†...Fully open	150
Foot Valves with Strainer		With poppet lift-type disc	0.3†...Fully open	420
		With leather-hinged disc	0.4†...Fully open	75
Butterfly Valves (8-inch and larger)			Fully open	40
Cocks	Straight-Through	Rectangular plug port area equal to 100% of pipe area	Fully open	18
	Three-Way	Rectangular plug port area equal to 80% of pipe area (fully open)	Flow straight through	44
			Flow through branch	140
Fittings	90 Degree Standard Elbow			30
	45 Degree Standard Elbow			16
	90 Degree Long Radius Elbow			20
	90 Degree Street Elbow			50
	45 Degree Street Elbow			26
	Square Corner Elbow			57
	Standard Tee	With flow through run		20
		With flow through branch		60
	Close Pattern Return Bend			50
Pipe	90 Degree Pipe Bends			See Page A-27
	Miter Bends			See Page A-27
	Sudden Enlargements and Contractions			See Page A-26
	Entrance and Exit Losses			See Page A-26

**Exact equivalent length is equal to the length between flange faces or welding ends.

†Minimum calculated pressure drop (psi) across valve to provide sufficient flow to lift disc fully.

‡For limitations, see page 2-11. For effect of end connections, see page 2-10 of Crane paper 410.

For resistance factor "K", equivalent length in feet of pipe, and equivalent flow coefficient "C_v", see pages A-31 and A-32.

*Equivalent Lengths *L* and *L/D* and Resistance Coefficient *K*

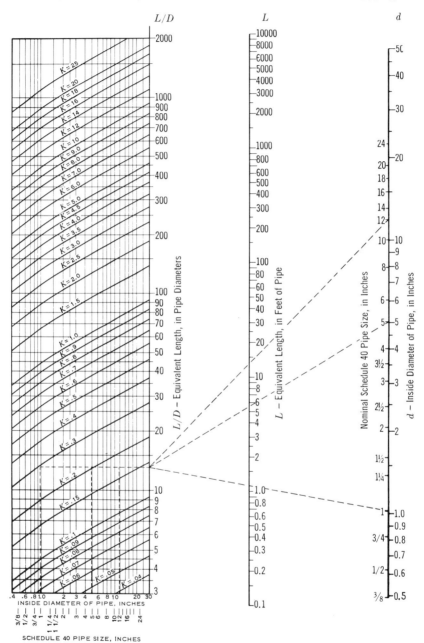

Problem: Find the equivalent length in pipe diameters and feet of Schedule 40 pipe, and the resistance factor K for 1, 5, and 12-inch fully-opened gate valves.

*For limitations, see page 2-11 of Crane paper 410.

Solution

Valve Size	1″	5″	12″	Refer to
Equivalent length, pipe diameters	13	13	13	Page A-30
Equivalent length, feet of Sched. 40 pipe	1.1	5.5	13	Dotted lines
Resist. factor K, based on Sched. 40 pipe	0.30	0.20	0.17	on chart.

PIPE FLOW DATA 279

A-32 PHYSICAL PROPERTIES OF FLUIDS AND FLOW CHARACTERISTICS OF VALVES, FITTINGS, AND PIPE

*Equivalents of Resistance Coefficient K And Flow Coefficient C_v

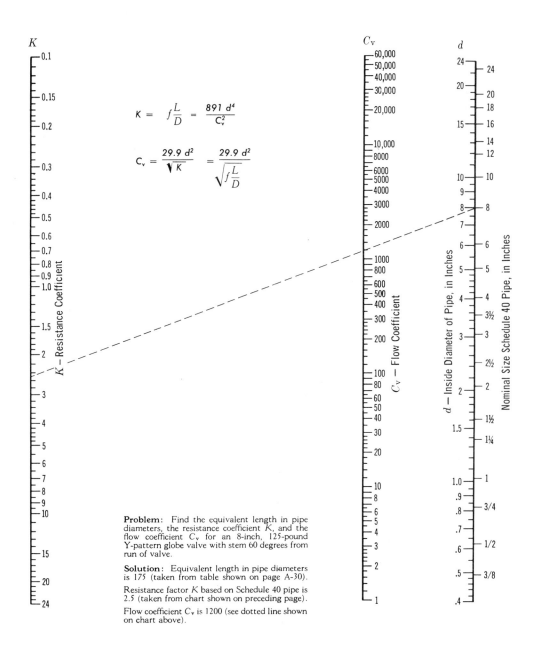

$$K = f\frac{L}{D} = \frac{891\, d^4}{C_v^2}$$

$$C_v = \frac{29.9\, d^2}{\sqrt{K}} = \frac{29.9\, d^2}{\sqrt{f\frac{L}{D}}}$$

Problem: Find the equivalent length in pipe diameters, the resistance coefficient K, and the flow coefficient C_v for an 8-inch, 125-pound Y-pattern globe valve with stem 60 degrees from run of valve.

Solution: Equivalent length in pipe diameters is 175 (taken from table shown on page A-30).

Resistance factor K based on Schedule 40 pipe is 2.5 (taken from chart shown on preceding page).

Flow coefficient C_v is 1200 (see dotted line shown on chart above).

*For limitations, see page 2-11 of Crane paper 410.

Flow of Water Through Schedule 40 Steel Pipe

Pressure Drop per 100 feet and Velocity in Schedule 40 Pipe for Water at 60 F.

Discharge		1/8"		1/4"		3/8"		1/2"		3/4"		1"		1 1/4"		1 1/2"	
Gallons per Minute	Cubic Ft. per Second	Velocity Feet per Second	Press. Drop Lbs. per Sq. In.	Velocity Feet per Second	Press. Drop Lbs. per Sq. In.	Velocity Feet per Second	Press. Drop Lbs. per Sq. In.	Velocity Feet per Second	Press. Drop Lbs. per Sq. In.	Velocity Feet per Second	Press. Drop Lbs. per Sq. In.	Velocity Feet per Second	Press. Drop Lbs. per Sq. In.	Velocity Feet per Second	Press. Drop Lbs. per Sq. In.	Velocity Feet per Second	Press. Drop Lbs. per Sq. In.
.2	0.000446	1.13	1.86	0.616	0.359												
.3	0.000668	1.69	4.22	0.924	0.903	0.504	0.159	0.317	0.061								
.4	0.000891	2.26	6.98	1.23	1.61	0.672	0.345	0.422	0.086								
.5	0.00111	2.82	10.5	1.54	2.39	0.840	0.539	0.528	0.167	0.301	0.033						
.6	0.00134	3.39	14.7	1.85	3.29	1.01	0.751	0.633	0.240	0.361	0.041	1"					
.8	0.00178	4.52	25.0	2.46	5.44	1.34	1.25	0.844	0.408	0.481	0.102			1 1/4"			
1	0.00223	5.65	37.2	3.08	8.28	1.68	1.85	1.06	0.600	0.602	0.155	0.371	0.048			1 1/2"	
2	0.00446	11.29	134.4	6.16	30.1	3.36	6.58	2.11	2.10	1.20	0.526	0.743	0.164	0.429	0.044		
3	0.00668			9.25	64.1	5.04	13.9	3.17	4.33	1.81	1.09	1.114	0.336	0.644	0.090	0.473	0.043
4	0.00891			12.33	111.2	6.72	23.9	4.22	7.42	2.41	1.83	1.49	0.565	0.858	0.150	0.630	0.071
5	0.01114		2"			8.40	36.7	5.28	11.2	3.01	2.75	1.86	0.835	1.073	0.223	0.788	0.104
6	0.01337	0.574	0.044		2 1/2"	10.08	51.9	6.33	15.8	3.61	3.84	2.23	1.17	1.29	0.309	0.946	0.145
8	0.01782	0.765	0.073			13.44	91.1	8.45	27.7	4.81	6.60	2.97	1.99	1.72	0.518	1.26	0.241
10	0.02228	0.956	0.108	0.670	0.046			10.56	42.4	6.02	9.99	3.71	2.99	2.15	0.774	1.58	0.361
15	0.03342	1.43	0.224	1.01	0.094		3"			9.03	21.6	5.57	6.36	3.22	1.63	2.37	0.755
20	0.04456	1.91	0.375	1.34	0.158	0.868	0.056		3 1/2"	12.03	37.8	7.43	10.9	4.29	2.78	3.16	1.28
25	0.05570	2.39	0.561	1.68	0.234	1.09	0.083	0.812	0.041		4"	9.28	16.7	5.37	4.22	3.94	1.93
30	0.06684	2.87	0.786	2.01	0.327	1.30	0.114	0.974	0.056			11.14	23.8	6.44	5.92	4.73	2.72
35	0.07798	3.35	1.05	2.35	0.436	1.52	0.151	1.14	0.704	0.882	0.041	12.99	32.2	7.51	7.90	5.52	3.64
40	0.08912	3.83	1.35	2.68	0.556	1.74	0.192	1.30	0.095	1.01	0.052	14.85	41.5	8.59	10.24	6.30	4.65
45	0.1003	4.30	1.67	3.02	0.668	1.95	0.239	1.46	0.117	1.13	0.064			9.67	12.80	7.09	5.85
50	0.1114	4.78	2.03	3.35	0.839	2.17	0.288	1.62	0.142	1.26	0.076		5"	10.74	15.66	7.88	7.15
60	0.1337	5.74	2.87	4.02	1.18	2.60	0.406	1.95	0.204	1.51	0.107			12.89	22.2	9.47	10.21
70	0.1560	6.70	3.84	4.69	1.59	3.04	0.540	2.27	0.261	1.76	0.143	1.12	0.047			11.05	13.71
80	0.1782	7.65	4.97	5.36	2.03	3.47	0.687	2.60	0.334	2.02	0.180	1.28	0.060		6"	12.62	17.59
90	0.2005	8.60	6.20	6.03	2.53	3.91	0.861	2.92	0.416	2.27	0.224	1.44	0.074			14.20	22.0
100	0.2228	9.56	7.59	6.70	3.09	4.34	1.05	3.25	0.509	2.52	0.272	1.60	0.090	1.11	0.036	15.78	26.9
125	0.2785	11.97	11.76	8.38	4.71	5.43	1.61	4.06	0.769	3.15	0.415	2.01	0.135	1.39	0.055	19.72	41.4
150	0.3342	14.36	16.70	10.05	6.69	6.51	2.24	4.87	1.08	3.78	0.580	2.41	0.190	1.67	0.077		
175	0.3899	16.75	22.3	11.73	8.97	7.60	3.00	5.68	1.44	4.41	0.774	2.81	0.253	1.94	0.102		
200	0.4456	19.14	28.8	13.42	11.68	8.68	3.87	6.49	1.85	5.04	0.985	3.21	0.323	2.22	0.130		8"
225	0.5013			15.09	14.63	9.77	4.83	7.30	2.32	5.67	1.23	3.61	0.401	2.50	0.162	1.44	0.043
250	0.557					10.85	5.93	8.12	2.84	6.30	1.46	4.01	0.495	2.78	0.195	1.60	0.051
275	0.6127					11.94	7.14	8.93	3.40	6.93	1.79	4.41	0.583	3.05	0.234	1.76	0.061
300	0.6684					13.00	8.36	9.74	4.02	7.56	2.11	4.81	0.683	3.33	0.275	1.92	0.072
325	0.7241					14.12	9.89	10.53	4.09	8.19	2.47	5.21	0.797	3.61	0.320	2.08	0.083
350	0.7798							11.36	5.41	8.82	2.84	5.62	0.919	3.89	0.367	2.24	0.095
375	0.8355							12.17	6.18	9.45	3.25	6.02	1.05	4.16	0.416	2.40	0.108
400	0.8912							12.98	7.03	10.08	3.68	6.42	1.19	4.44	0.471	2.56	0.121
425	0.9469							13.80	7.89	10.71	4.12	6.82	1.33	4.72	0.529	2.73	0.136
450	1.003		10"					14.61	8.80	11.34	4.60	7.22	1.48	5.00	0.590	2.89	0.151
475	1.059	1.93	0.054							11.97	5.12	7.62	1.64	5.27	0.653	3.04	0.166
500	1.114	2.03	0.059							12.60	5.65	8.02	1.81	5.55	0.720	3.21	0.182
550	1.225	2.24	0.071							13.85	6.79	8.82	2.17	6.11	0.861	3.53	0.219
600	1.337	2.44	0.083							15.12	8.04	9.63	2.55	6.66	1.02	3.85	0.258
650	1.448	2.64	0.097		12"							10.43	2.98	7.22	1.18	4.17	0.301
700	1.560	2.85	0.112	2.01	0.047		14"					11.23	3.43	7.78	1.35	4.49	0.343
750	1.671	3.05	0.127	2.15	0.054							12.03	3.92	8.33	1.55	4.81	0.392
800	1.782	3.25	0.143	2.29	0.061							12.83	4.43	8.88	1.75	5.13	0.443
850	1.894	3.46	0.160	2.44	0.068	2.02	0.042					13.64	5.00	9.44	1.96	5.45	0.497
900	2.005	3.66	0.179	2.58	0.075	2.13	0.047					14.44	5.58	9.99	2.18	5.77	0.554
950	2.117	3.86	0.198	2.72	0.083	2.25	0.052					15.24	6.21	10.55	2.42	6.09	0.613
1 000	2.228	4.07	0.218	2.87	0.091	2.37	0.057		16"			16.04	6.84	11.10	2.68	6.41	0.675
1 100	2.451	4.48	0.260	3.15	0.110	2.61	0.068					17.65	8.23	12.22	3.22	7.05	0.807
1 200	2.674	4.88	0.306	3.44	0.128	2.85	0.080	2.18	0.042					13.33	3.81	7.70	0.948
1 300	2.896	5.29	0.355	3.73	0.150	3.08	0.093	2.36	0.048					14.43	4.45	8.33	1.11
1 400	3.119	5.70	0.409	4.01	0.171	3.32	0.107	2.54	0.055					15.55	5.13	8.98	1.28
1 500	3.342	6.10	0.466	4.30	0.195	3.56	0.122	2.72	0.063		18"			16.66	5.85	9.62	1.46
1 600	3.565	6.51	0.527	4.59	0.219	3.79	0.138	2.90	0.071					17.77	6.61	10.26	1.65
1 800	4.010	7.32	0.663	5.16	0.276	4.27	0.172	3.27	0.088	2.58	0.050			19.99	8.37	11.54	2.08
2 000	4.456	8.14	0.808	5.73	0.339	4.74	0.209	3.63	0.107	2.87	0.060		20"			22.21	10.3	12.82	2.55
2 500	5.570	10.17	1.24	7.17	0.515	5.93	0.321	4.54	0.163	3.59	0.091				24"			16.03	3.94
3 000	6.684	12.20	1.76	8.60	0.731	7.11	0.451	5.45	0.232	4.30	0.129	3.46	0.075			19.24	5.59
3 500	7.798	14.24	2.38	10.03	0.982	8.30	0.607	6.35	0.312	5.02	0.173	4.04	0.101			22.44	7.56
4 000	8.912	16.27	3.08	11.47	1.27	9.48	0.787	7.26	0.401	5.74	0.222	4.62	0.129	3.19	0.052	25.65	9.80
4 500	10.03	18.31	3.87	12.90	1.60	10.67	0.99	8.17	0.503	6.46	0.280	5.20	0.162	3.59	0.065	28.87	12.2
5 000	11.14	20.35	4.71	14.33	1.95	11.85	1.21	9.08	0.617	7.17	0.340	5.77	0.199	3.99	0.079		
6 000	13.37	24.41	6.74	17.20	2.77	14.23	1.71	10.89	0.877	8.61	0.483	6.93	0.280	4.79	0.111		
7 000	15.60	28.49	9.11	20.07	3.74	16.60	2.31	12.71	1.18	10.04	0.652	8.08	0.376	5.59	0.150		
8 000	17.82			22.93	4.84	18.96	2.99	14.52	1.51	11.47	0.839	9.23	0.488	6.38	0.192		
9 000	20.05			25.79	6.09	21.34	3.76	16.34	1.90	12.91	1.05	10.39	0.608	7.18	0.242		
10 000	22.28			28.66	7.46	23.71	4.61	18.15	2.34	14.34	1.28	11.54	0.739	7.98	0.294		
12 000	26.74			34.40	10.7	28.45	6.59	21.79	3.33	17.21	1.83	13.85	1.06	9.58	0.416		
14 000	31.19					33.19	8.89	25.42	4.49	20.08	2.45	16.16	1.43	11.17	0.562		
16 000	35.65							29.05	5.83	22.95	3.18	18.47	1.85	12.77	0.723		
18 000	40.10							32.68	7.31	25.82	4.03	20.77	2.32	14.36	0.907		
20 000	44.56							36.31	9.03	28.69	4.93	23.08	2.86	15.96	1.12		

For pipe lengths other than 100 feet, the pressure drop is proportional to the length. Thus, for 50 feet of pipe, the pressure drop is approximately one-half the value given in the table ... for 300 feet, three times the given value, etc.

Velocity is a function of the cross sectional flow area; thus, it is constant for a given flow rate and is independent of pipe length.

For calculations for pipe other than Schedule 40, see explanation on page 273.

APPENDIX C
Liquid and Gas Flow Charts

Steam flow is in pounds per hour (lb/hr)

Flow coefficient — Cv

Inlet Pressure (psig)	Outlet Pressure (psig & inches mercury vacuum)	0.84	1.6	2.5	4.4	5.0	6.4	9.5	15	25	30	35	50	55	85	115	130	200	395
5 (227° F)	3	15	30	46	81	91	115	171	272	452	542	639	898	985	1530	2040	2310	3480	6980
	1	21	39	60	106	118	151	224	350	584	695	809	1142	1258	1917	2555	2840	4350	8420
	10"-28" Hg Vac.	25	49	72	125	140	177	262	402	652	779	912	1264	1395	2106	2772	3030	4591	8969
10 (240° F)	8	17	33	51	90	103	132	193	316	509	608	732	1015	1120	1710	2300	2588	3970	7860
	5	26	49	75	131	146	188	278	434	722	863	1001	1420	1555	2376	3165	3529	5391	10450
	4"-28" Hg Vac.	32	61	90	156	176	221	325	502	811	967	1131	1571	1734	2615	3449	3770	5704	11150
15 (250° F)	12	23	44	69	120	136	173	255	408	673	806	948	1335	1477	2283	3020	3427	5196	10400
	8	32	61	94	165	185	236	349	548	904	1080	1261	1780	1952	2972	3978	4380	6740	13020
	2-0	38	72	106	184	205	265	389	602	978	1163	1366	1897	2000	3158	4139	4552	6885	13460
25 (266° F)	20	34	64	99	174	194	249	367	579	965	1158	1348	1924	2098	3256	4320	4391	7400	14560
	15	43	83	127	222	248	316	469	731	1206	1447	1688	2383	2605	3940	5274	5810	8860	17260
	7-0	50	96	142	247	275	348	514	791	1286	1526	1795	2490	2735	4155	5440	5890	9040	17690
50 (298° F)	40	59	109	173	303	338	433	643	1004	1678	2063	2333	3334	3633	5573	7530	8383	12750	24810
	30	75	143	216	377	420	541	800	1246	2037	2442	2817	4027	4360	6580	8703	9580	14450	28480
	20-0	80	156	229	400	446	566	835	1280	2078	2481	2901	4027	4420	6710	8780	9660	14500	28570
75 (320° F)	60	84	157	245	431	481	619	915	1431	2380	2844	3310	4690	5133	7906	10620	11800	17990	34810
	50	99	189	289	501	561	722	1062	1663	2719	3260	3800	5340	5894	8909	11860	12980	19760	38530
	30-0	109	212	313	543	604	767	1132	1745	2832	3363	3953	5487	6018	9145	11980	13150	19880	38940
100 (338° F)	80	108	203	315	555	617	789	1175	1844	3064	3665	4258	6045	6609	10120	13660	15090	23130	44870
	70	124	235	360	628	703	900	1327	2088	3419	4142	4769	6740	7363	11160	14900	16450	24970	48830
	50-0	139	270	398	691	770	976	1442	2215	3605	4281	4972	6984	7660	11640	15240	16820	25310	49560
125 (353° F)	100	132	250	388	683	759	975	1439	2250	3756	4509	5211	7379	8089	12380	16670	18400	28240	54660
	90	149	282	430	756	838	1075	1595	2510	4100	4920	5740	8110	8840	13450	17950	19850	30250	58800
	60-0	169	328	483	838	934	1184	1749	2687	4372	5193	6104	8472	9292	14030	18420	20220	30610	60130
150 (366° F)	125	146	275	428	749	840	1070	1584	2480	4120	4980	5780	8239	8968	13910	18560	20740	31530	61420
	110	173	330	502	878	994	1258	1868	2910	4810	5768	6704	9443	10370	15720	20940	23220	35300	68590
	75-0	199	385	567	984	1095	1391	2008	3155	5111	6130	7169	9951	10910	16250	21450	23600	35630	70620
175 (377° F)	150	156	299	465	812	910	1167	1734	2590	4489	5420	6297	8979	9840	15250	20290	22750	34680	68020
	125	204	387	591	1034	1155	1481	2177	3431	5620	6754	7859	11060	12070	18320	24470	26920	40970	79990
	90-0	229	443	652	1132	1261	1599	2369	3628	5904	7072	8241	11440	12540	19060	24970	27550	41450	81180
200 (388° F)	170	182	347	541	946	1057	1350	2002	3102	5237	6271	7301	10370	11370	17520	23370	26000	40880	78180
	150	221	419	645	1128	1260	1612	2391	3933	6157	7381	8618	12120	13270	20150	26970	29610	45590	88260
	100-0	259	500	737	1279	1425	1807	2669	4101	6672	7923	9201	12900	14180	21550	28220	31140	46840	91740
250 (406° F)	200	251	476	737	1291	1444	1855	2751	4250	7135	8517	9850	14010	15290	23500	31450	34850	53200	103600
	180	282	536	816	1430	1598	2050	3010	4745	7770	9340	10860	15230	16700	25320	33840	37220	56600	110500
	130-0	314	595	883	1538	1734	2210	3245	4995	8160	9680	11390	15800	17330	26340	34500	37900	57410	112100
300 (422° F)	255	271	514	800	1392	1555	1985	2980	4590	7753	9240	10760	15630	16710	26110	34700	38550	60200	115100
	225	326	614	948	1662	1855	2388	3508	5508	9057	10850	12650	17850	19580	29780	39780	43860	67320	130100
	150-0	379	734	1081	1876	2091	2652	3876	6018	9792	11730	13660	18970	20810	31620	41410	45690	68750	134600
350 (436° F)	300	304	573	901	1556	1767	2270	3346	5265	8775	10530	12170	17320	18950	29480	39310	43750	67150	131300
	250	393	749	1135	1989	2223	2854	4212	6552	10760	12910	15020	21080	23160	35100	46800	51480	78390	152500
	175-0	435	842	1240	2153	2399	3042	4493	6903	11230	13330	15680	21760	23868	36270	47500	52180	72390	152500
400 (446° F)	325	389	728	1126	1988	2212	2835	4190	6520	10920	13110	15150	21500	23520	36150	48250	53500	82100	159000
	300	424	811	1240	2170	2432	3110	4572	7156	11860	14240	16560	23350	25660	38410	51400	56900	87500	169400
	200-0	493	954	1405	2439	2780	3442	5000	7820	12720	15150	17620	24650	27020	41190	53500	59110	89000	175100

Sizing table for saturated steam.

(Reproduced by Courtesy of Jordan Valve, Division of Richards Industries.)

VALVE ENGINEERING AND DESIGN DATA

Air flow is standard cubic feet per minute (SCFM)

Inlet Pressure (psig)	Outlet Pressure (psig & inches mercury vacuum)	Flow coefficient — Cv																
		0.84	1.6	2.5	4.4	5.0	6.4	9.5	15	25	30	35	50	55	85	115	130	200	395
5	3	5	9	14	25	29	36	54	86	142	176	200	282	311	481	636	722	1092	2193
	1	6	12	19	33	37	48	71	111	184	220	255	361	396	605	805	898	1372	2660
	10"-28" Hg Vac.	8	15	23	39	44	56	83	127	212	245	288	400	441	666	877	959	1459	2838
10	8	5	10	16	28	32	41	54	97	161	194	230	323	355	544	731	819	1253	2479
	5	8	15	23	41	46	59	88	138	230	275	319	452	495	757	1009	1125	1719	3333
	4"-28" Hg Vac.	10	19	28	49	55	70	104	159	259	309	361	501	554	836	1095	1202	1820	3555
15	12	7	14	22	38	43	55	81	129	213	256	301	425	469	729	961	1088	1650	3304
	8	10	21	29	52	58	75	111	174	285	343	400	563	617	941	1257	1391	2132	4115
	2-0	12	22	34	59	65	84	123	191	311	369	434	602	661	1009	1323	1455	2193	4276
25	20	10	20	32	56	63	80	119	188	312	374	435	621	678	1052	1400	1580	2407	4737
	15	14	26	41	71	80	102	152	237	389	467	545	759	841	1271	1704	1886	2875	5577
	7-0	16	32	45	79	89	112	166	255	415	496	580	805	876	1348	1761	1939	2929	5670
50	40	19	35	56	98	109	141	209	326	546	655	759	1085	1182	1814	2439	2729	4152	8179
	30	24	46	70	122	136	176	260	405	663	795	917	1288	1421	2154	2834	2980	4751	9306
	20-0	26	51	74	130	145	184	271	416	676	807	944	1311	1436	2260	2873	3154	4770	9306
75	60	27	53	82	143	160	205	303	475	790	944	1099	1566	1705	2632	3528	3920	5978	11622
	50	33	62	96	166	186	240	352	552	903	1083	1266	1767	1950	2940	3920	4312	6566	12790
	30-0	36	70	104	181	197	256	378	579	941	1123	1313	1822	2054	3141	3994	4384	6570	12790
100	80	36	67	105	184	205	262	391	613	1020	1220	1417	2002	2200	3370	4550	5025	7700	14940
	70	41	78	120	209	235	300	442	695	1137	1370	1587	2250	2452	3717	4982	5475	8315	16150
	50-0	46	88	131	227	257	326	476	740	1210	1432	1670	2325	2567	3870	5072	5600	8432	16500
125	100	44	83	129	228	254	322	481	751	1256	1509	1747	2470	2708	4145	5581	6161	9455	18300
	90	50	95	144	253	283	360	532	838	1372	1646	1921	2717	2961	4500	6025	6650	10140	19700
	60-0	56	109	161	281	314	396	588	901	1464	1738	2013	2810	3093	4691	6161	6766	10280	20130
150	125	49	92	144	252	283	361	532	837	1404	1688	1944	2772	3024	4680	6246	6976	10610	20660
	110	58	111	170	295	331	423	628	978	1616	1940	2260	3185	3480	5284	7050	7800	11870	23050
	75-0	69	129	190	331	370	469	693	1063	1723	2063	2412	3348	3704	5570	7337	8003	12180	23760
175	150	52	100	157	273	307	393	583	872	1525	1827	2150	3022	3312	5141	6831	7659	11710	23060
	125	68	131	199	348	388	499	733	1156	1893	2278	2646	3726	4071	6152	8240	9062	13800	26900
	90-0	77	148	219	380	424	538	795	1223	1987	2359	2773	3850	4205	6436	8437	9261	14000	27320
200	170	61	117	182	317	356	455	674	1044	1771	2112	2459	3493	3830	5890	7860	8760	13760	26300
	150	74	141	217	380	424	542	804	1324	2073	2485	2901	4081	4488	6805	9112	9701	14730	29810
	100-0	87	168	248	431	482	608	902	1382	2246	2686	3135	4352	4797	7254	9453	10390	15820	30890
250	200	85	162	249	440	491	628	929	1448	2427	2895	3358	4762	5202	7970	10690	11890	18180	35260
	180	96	182	277	485	543	696	1023	1611	2640	3179	3691	5202	5675	8620	11500	12650	19240	37550
	130-0	107	206	306	531	601	753	1113	1708	2774	3294	3869	5375	6011	8950	11780	12920	19550	38150
300	255	91	173	269	468	523	668	1003	1545	2610	3110	3625	5260	5620	8790	11670	12960	20300	38720
	225	110	213	321	562	625	802	1189	1860	3057	3674	4272	6045	6615	10050	13390	14840	22670	43970
	150-0	127	246	364	634	707	893	1324	2030	3297	3916	4602	6389	7010	10600	13900	15370	23240	45340
350	300	103	197	306	535	600	766	1132	1788	2975	3582	4147	5906	6451	10010	13290	14840	22830	44650
	250	133	254	386	675	756	979	1428	2232	3664	4390	5110	7164	7880	11940	15920	17510	26640	51840
	175-0	148	286	422	734	819	1035	1534	2352	3820	4561	5333	7403	8150	12360	16220	17830	26930	52530
400	325	132	249	386	670	759	963	1430	2244	3728	4480	5185	7348	8054	12300	16740	18280	28050	54390
	300	145	277	424	741	831	1062	1561	2441	4050	4859	5657	7970	8759	13120	17550	19460	29670	57820
	200-0	168	325	479	835	932	1176	1744	2674	4344	5160	6020	8416	9220	14050	18240	20270	30410	59730

Sizing table for air with inlet temperature of 60°F

LIQUID AND GAS FLOW CHARTS

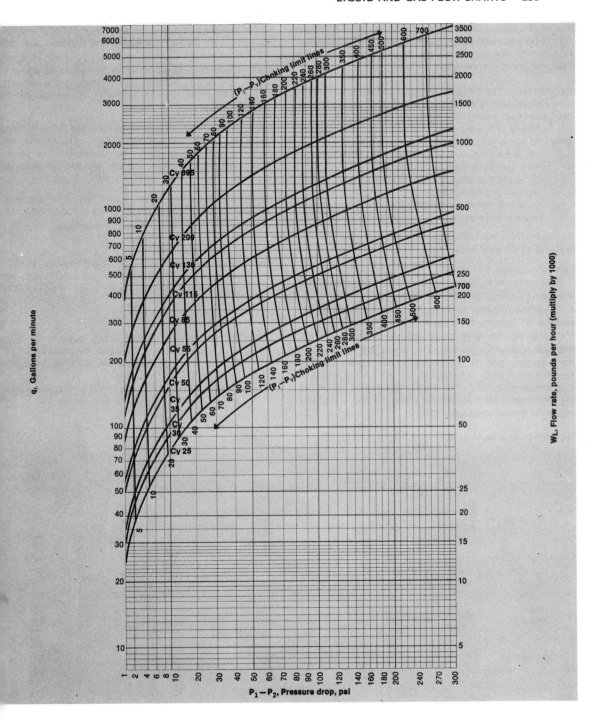

Basic liquid flow rate, for large valves.

284 VALVE ENGINEERING AND DESIGN DATA

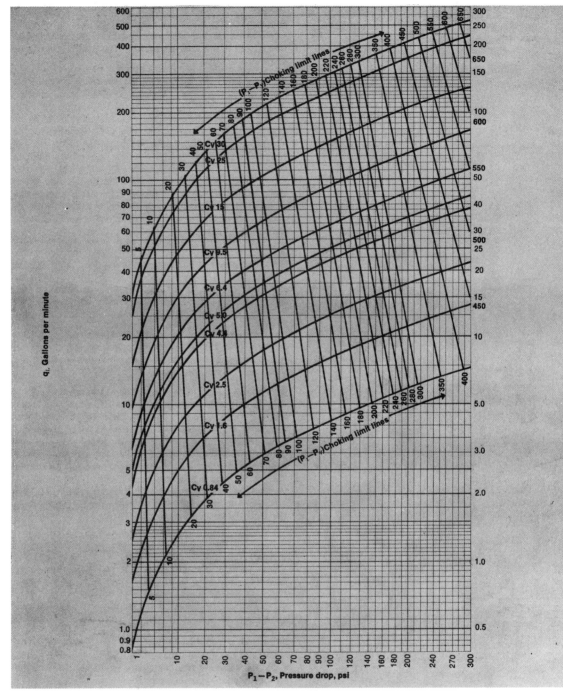

Basic liquid flow rate, for small valves.

LIQUID AND GAS FLOW CHARTS

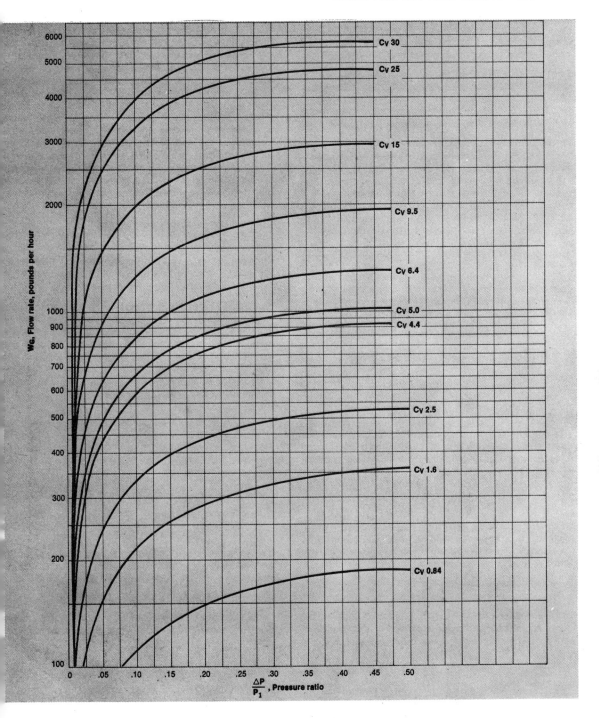

Basic flow rate, gases and vapors for small valves.

Index*

acceleration, conversion, 252
accessories, *136–137*
actuators, 88, *132–134*
aerodynamic noise, 241, 242–243
Aerospace Recommended Practice 24B, 215–225
Aerospace Recommended Practice 868, 212–213
air, sizing table, 282
air conditioning, *125*
air core solenoid valve, parts identification, 14
alloys
   corrosion resistant, 98
   nonferrous, 98
American National Standard Interfaces for Four-Way General Purpose Industrial Pneumatic Directional Control Valves, 153–154
   identification statement, 154
   introduction, 153
   port identification, 154
   purpose, 154
   scope, 153–154
   terms and definitions, 154
   use of optional holes, 154
American National Standard Symbols for Marking Electrical Leads and Ports on Fluid Power Valves, 149–151
   examples, *151*
   introduction, 149
   port symbols, 149–151
   purpose, 149
   rules, 149
   scope, 149
   terms and abbreviations, 149
American Standard Graphical Symbols for Pipe Fittings, Valves, and Piping, 119–126
amplifier device, 195
angle stop check valve, 276
angle valve, *122*, 275
angular momentum, calculation, 16
area, conversion, 250–251
attenuation, defined, 240

Babcock formula, calculation from, 104
ball valve
   description, usage, advantages, disadvantages, 3–4
   parts identification of representative types, 4
basic flow test systems, 227, 233
bending moment, conversion, 262
Bernoulli equation, 104
Bernoulli's law, calculation from, 18
Blasius equation, 104
bulk modulus, calculation, 19–20
bushing, *119*
butterfly valve
   description, usage, advantages, disadvantages, 4–5, 7
   parts identification, 5

cavitation, 241
cap, *119*
capacity tests, 226
check valves, *123*
   diagram, 159–160
chemicals, use and nonuse with valve materials, 109–118
closure members, 3, 89
cock, *123*
coefficient of linear expansion, calculation, 23
conductor, fluid, *129*
construction, factors related to, 88–89
contamination, 87
continuity equation, calculation from, 23
control, directional, 87
controls, *132–134*
control valve capacity test procedure for compressible fluids, 226–232
   flow measurement, 232
   manifold piping requirements, 228–231
   pressure taps, 232
   scope, 227
   size of pipe, 232
   test system, 227–232
   throttling, valves, 232
control valve capacity test procedure for incompressible fluids, 233–238
   flow measurement, 238
   manifold piping requirements, 234–237
   pressure taps, 238

*Page numbers in italic type indicate those containing graphical symbols.

control valve capacity test procedure (*Continued*)
  scope, 233
  size of pipe, 238
  throttling valves, 238
corrosion resistance, 109–118
critical point equations, 104
cross, *119*
crossover, *119*
C_v method, 90–91
  values, 91

Darcy's equation, 104
  calculations from, 27
dbA, defined, 240
db, defined, 240
delay
  timing in, 194
  timing out, 194
density, conversion, 263
design factors, 87–89
devices, miscellaneous, 207–210
diagramming, attached method, 191–197
  diagram, 209
diagramming, detached method, 197–207
  diagram, 210
diagram rules, attached method, 195
diaphragm actuated valves, 247, 248
diaphragm valve, *123*
directional control valves, 171, 172
  diagram, 164–166
double ported valve, schematic, 30
dynamic force, formulas for calculation, 104

elbow, *119–120*
electricity, conversion, 260
end fittings, 89
energy, conversion, 255
energy/area time, conversion, 262
equivalent lengths, 277, 278
equivalent orifice method, 92–95
example problems, 100, 102
  no. 1, 90–91
  no. 2, 91
  no. 3, 91
  no. 4, 93
  no. 5, 93–94
  no. 6, 95
expansion valve, schematic, 32

flip-flop device, 193
flow, conversion, 264
flow coefficient (C_v), 278
flow data, 270–279
flow of gases, formulas for calculating, 104–105
flow of liquids, formulas for calculating, 105–106
flow rate, 244, 246
  gases and vapors for small valves, 285
  liquid, for large valves, 283
  liquid, for small valves, 284
fluid conditioners, *131–132*

fluid memory device, 193
fluid power usage, SI units for, 268
foot valves, 276
force, conversion, 256
force/length, conversion, 262
formulas, useful, 103–108
frequency, conversion, 254
friction factor equations, 106
Froude number, calculation, 37
fuels, pressure drop, 211–214

gases, 242–243
gate valves, *123–124*, 276
  description, usage, advantages, disadvantages, 7
  parts identification, 6
Gibbs function, calculation, 38
globe valves, *124*, 275
  description, usage, advantages, disadvantages, 7
glossary, 12–79

Hagen-Poiseville laws, 39, 106
  calculations from, 39
Hazen-Williams formula, 106
head loss equation
  laminar flow, 106
  turbulent flow, 106
heat, conversion, 263
heating, *125–126*
hose valve, *124*
hydraulic fluids pressure drop, 214–225
  chart method of solution for straight smooth pipes, 217
  determination, 215–225
  equipment, 215
  friction factor f, 222–223
  general, 215
  nomenclature, 216
  piezometer tube, 225
  pressure tap fitting, 224
  purpose, 215
  systems of units, 218
  tare schematic, 220
  theory, 216–217
  viscosity conversion factors, 219
hydraulic horsepower, calculation, 40
hydrodynamic noise, 241–242
hydrostatic force, formula for calculation, 106
Hz, defined, 240

ideal gas equation, 106
inlet pressure, 245, 246
in-line ball check valves, 276
Instrument Society of America S39.4, 226–232
Instrument Society of America, S39.2, 233–238
instruments, *136–137*
interfaces, 188–191
  types, 155–157
internal and operating valve, schematic, 81
isothermal equation, 107

joint, *120*

INDEX

K-factor, calculation, 42

lateral, *120*
leakage, 3
  low, 87
length, conversion, 250, 253
lifetimes, 3
lift check valves, 275
light, conversion, 261
limit values, typical, *185*
linear devices, *132*
liquids, 241
lock shield valve, *124*
logic devices, 191–196, 197–207
logic diagrams, attached method, 196–197

magnetism, conversion, 260
manifold piping requirements, 228–231, 234–237
mass, conversion, 255
mass/area, conversion, 264
mass capacity, 263
mass/time, conversion, 264
mass/volume, conversion, 263
materials, 89, 98–99
  properties, 109–118
measurement, miscellaneous units, 262–264
mechanical vibration, 241
method of diagramming for moving parts fluid controls, 172–210
  check valves, *208*
  electric-to-relay valves, *206–207*
  flow sensors, *187*
  fluid conditioners, *208*
  graphic symbols, attached method, *191–195*
  identification statement, 208
  introduction, 173
  line identification, *177*
  line techniques, *176–177*
  manual controls, *177–183*
  off-return fluid memory relay valves, *201–202*
  one-shot relay valves, *205–206*
  positioning sensors, *183–186*
  power control valves, *207–208*
  pressure control valve, *207–208*
  pressure gage, *207*
  pressure indicator, *207*
  pressure sensors, *186–187*
  purpose, 173–174
  relay valves, *197–201*
  relay valves in a timing circuit, *204–205*
  resistance devices, *204*
  shuttle valve, *208*
  symbol rules, 175–176
  temperature sensors, *187*
  terms and definitions, 174–175
  test point, *207*
  time delay relay valves, *202–204*
  visual indicators, *207*
metric-English (U.S.) conversion factors, 249–269
mounting surfaces for subplate type hydraulic directional control valves—for 315 bar hydraulic service, 170–172
  general, 171
  identification statement, 172
  introduction, 170
  purpose, 171
  scope, 170–171
  subplate identification coding, 171–172
  terms and definitions, 171
  tolerances-breaks-radii, 172
  units, 171

noise calculation, 239–246
  aerodynamic, 242–243
  definition of terms, 240
  methods, 239–243
noises, common, sound pressure level, 241

Occupational Safety and Health Act (OSHA), 239, 240
one-shot device, 194
orifice flange, *120*

packing, 89
physical properties and flow characteristics, 275–278
piezometer, double, 214
pilot-operated test check valves, diagram, 160–161
pinch valves, description, usage, advantages, disadvantages, 7
pipes and piping
  commercial wrought steel, 271, 274
  flow data, 270–279
  schedule 40, flow of air, 273
  schedule 40, flow of water, 279
  size, 232, 238
plugs, *121*
plumbing, *126*
pneumatic tubes, *126*
poppet valves
  description, usage, advantages, disadvantages, 7, 8
  parts identification, 8
power, conversion, 256
power control valves, 188–191
practice, attached, *195*
Prandtl number, calculation, 51
pressure, 256, 265
pressure, high, 87
pressure compensated control valve, principal parts, 71
pressure control valves
  poppet type, diagram, 163–164
  spool type, diagram, 162–163
pressure drop, formulas for calculating, 107
pressure drop test for fuel system components, 211–225
pressure drop test setup, 214
Ps, defined, 240
pushbuttons, typical, *179*

quick opening valve, *124*

reducer, *121*
reducing flange, *121*

reducing valve, schematic, 56
relay valve, schematic, 74
relief, 87
relief valve, representative parts, 57
resistance coefficient (K), 277, 278
Reynolds number, calculation, 58, 107
rotary devices, *134-136*

safety, 87
safety valve, *124*
    representative parts, 59
seals, 89
seat, 89
shutoff, 87
single-seated valve, 248
sizing, 90-95
    C_v method, 90-91
    equivalent orifice method, 92-95
sleeve, *121*
sonic flow equation, 107
sound, defined, 240
specific gravity, formulas for calculating, 108
speed, conversion, 252, 267
Spitz glass formula, 108
SPL, defined, 240
spring rate, formulas for calculating, 108
springs
    design, 96-102
    materials, 98-99
    nested, 100-101
    stresses, formulas for calculation, 108
    torsion, 101-102
sprinklers, *126*
steam, saturated, sizing table, 281
steam service, 87
steel
    high alloy, 99
    low alloy, 98-99
stop valve, *124*
storage, energy and fluid, *130*
stress, conversion, 256-257
subplate code suffixes, 171
swing check valves, 275
    representative parts, 66
swing valve
    description, usage, advantages, disadvantages, 11
    parts identification, 10
symbols, definition of, 90, 96, 103-104
    representative composite, *141-148*
symbols-input devices, 177-187

taper plug valve
    description, usage, advantages, disadvantages, 9, 11
    parts identification, 10
tee, *121-122*
temperature, conversion, 258-259
temperatures, high, 87

terminology, 12-79
terms, defined, 240
test methods, 211-238
three-port flow control valves, 168-170
three-way cock valve, 276
throttling, 87
time, conversion, 252, 254
Toricelli's theorem, calculations from, 68
torque, conversion, 262
torque/length, conversion, 262
truth tables, 192-195
turbulence, 241
two-port flow control valves, 166-168

union, *122*
usage, factors related to, 87
USA Standard Dimensions for Mounting Surfaces of Sub-plate Type Hydraulic Fluid Power Valves, 154-170
    general, 158
    introduction, 154
    mounting surface drawing identification coding, 158
    purpose, 158
    scope, 154, 158
    tolerances-breaks-radii, 158-170
USA Standard Graphic Symbols for Fluid Power Diagrams, 127-151
    introduction, 127
    symbol rules, 127-128

valve noise, 240-241
    aerodynamic, 241, 242-243
    causes, 241
    hydrodynamic, 241-242
    see also noise, calculation
valves, *137-141*
    major types, 3-11
    miscellaneous, 79-84
    number, 3
    ratings, 88-89
    springs, 96-102
    types, 88-89, 275
velocity, sonic, forms for calculation, 108
velocity of flow, forms for calculation, 108
velocity head, calculation, 77
venturi throat valve, schematic, 76
viscosity
    conversion, 264
    formulas for calculation, 108
volume, conversion, 251, 266
volume/time, conversion, 264

Walsh-Healy Public Contracts Act, 240
Weber number, calculation, 77-78
weight, conversion, 255, 257
Weymouth formula, 108
work, conversion, 255-256